Universitext

Jan Hendrik Bruinier
Gerard van der Geer
Günter Harder
Don Zagier

The 1-2-3
of Modular Forms

Lectures at a Summer School
in Nordfjordeid, Norway

Springer

Jan Hendrik Bruinier

Fachbereich Mathematik
Technische Universität Darmstadt
Schloßgartenstraße 7
64289 Darmstadt, Germany
bruinier@mathematik.tu-darmstadt.de

Gerard van der Geer

Korteweg-de Vries Instituut
Universiteit van Amsterdam
Plantage Muidergracht 24
1018 TV Amsterdam, The Netherlands
geer@science.uva.nl

Editor:

Kristian Ranestad

Department of Mathematics
University of Oslo
P.O. Box 1053 Blindern
0316 Oslo, Norway
ranestad@math.uio.no

Günter Harder

Mathematisches Institut
Universität Bonn
Beringstraße 1
53115 Bonn, Germany
harder@math.uni-bonn.de
and
Max-Planck-Institut für Mathematik
Vivatsgasse 7
53111 Bonn, Germany
harder@mpim-bonn.mpg.de

Don Zagier

Max-Planck-Institut für Mathematik
Vivatsgasse 7
53111 Bonn, Germany
zagier@mpim-bonn.mpg.de
and
Collège de France
3, rue d'Ulm
75231 Paris Cedex 05, France
don.zagier@college-de-france.fr

ISBN 978-3-540-74117-6   ISBN 978-3-540-74119-0 (eBook)

DOI 10.1007/978-3-540-74119-0

Library of Congress Control Number: 2007939406

Mathematics Subject Classification (2000): 14-01, 11Gxx, 14Gxx

© 2008 Springer-Verlag Berlin Heidelberg

Typesetting and Production: LE-TeX Jelonek, Schmidt & Vöckler GbR, Leipzig, Germany
Cover design: WMX Design GmbH, Heidelberg, Germany

Printed on acid-free paper

9 8 7 6 5 4 3 2 1

springer.com

Preface

This book grew out of lectures given at the summer school on "Modular Forms and their Applications" at the Sophus Lie Conference center in Nordfjordeid in June 2004. This center, set beautifully in the fjords of the west coast of Norway, has been the site of annual summer schools in algebra and algebraic geometry since 1996. The schools are a joint effort between the universities in Bergen, Oslo, Tromsø and Trondheim. They are primarily aimed at graduate students in Norway, but also attract a large number of students from other parts of the world. The theme varies among central topics in contemporary mathematics, but the format is the same: three leading experts give independent but connected series of lectures, and give exercises that the students work on in evening sessions.

In 2004 the organizing committee consisted of Stein Arild Strømme (Bergen), Geir Ellingsrud and Kristian Ranestad (Oslo) and Alexei Rudakov (Trondheim). We wanted to have a summer school that introduced the students both to the beauty of modular forms and to their varied applications in other areas of mathematics, and were very fortunate to have Don Zagier, Jan Bruinier and Gerard van der Geer give the lectures.

The lectures were organized in three series that are reflected in the title of this book both by their numbering and their content. The first series treats the classical one-variable theory and some of its many applications in number theory, algebraic geometry and mathematical physics.

The second series, which has a more geometric flavor, gives an introduction to the theory of Hilbert modular forms in two variables and to Hilbert modular surfaces. In particular, it discusses Borcherds products and some geometric and arithmetic applications.

The third gives an introduction to Siegel modular forms, both scalar- and vector-valued, especially Siegel modular forms of degree 2, which are functions of three complex variables. It presents a beautiful application of the theory of curves over finite fields to Siegel modular forms by providing evidence for a conjecture of Harder on congruences between elliptic and Siegel modular forms.

Günter Harder came forward with this conjecture in a colloquium lecture in Bonn in 2003. He kindly allowed us to include his notes for this colloquium talk in Bonn on the subject. Even though the three lecture series are strongly connected, each of them is self contained and can be read independently of the others.

There is quote ascribed (perhaps apocryphally) to Martin Eichler, saying that there are five fundamental operations in mathematics: addition, subtraction, multiplication, division and modular forms. We hope this book will help convince newcomers and oldtimers alike that this is only partially an exaggeration.

Oslo, July 2007   *Kristian Ranestad*

Contents

A Congruence Between a Siegel and an Elliptic Modular Form

Elliptic Modular Forms and Their Applications

Don Zagier

Max-Planck-Institut für Mathematik, Vivatsgasse 7, 53111 Bonn, Germany
E-mail: zagier@mpim-bonn.mpg.de

Foreword

These notes give a brief introduction to a number of topics in the classical theory of modular forms. Some of theses topics are (planned) to be treated in much more detail in a book, currently in preparation, based on various courses held at the Collège de France in the years 2000–2004. Here each topic is treated with the minimum of detail needed to convey the main idea, and longer proofs are omitted.

Classical (or "elliptic") modular forms are functions in the complex upper half-plane which transform in a certain way under the action of a discrete subgroup Γ of $\mathrm{SL}(2, \mathbb{R})$ such as $\mathrm{SL}(2, \mathbb{Z})$. From the point of view taken here, there are two cardinal points about them which explain why we are interested. First of all, the space of modular forms of a given weight on Γ is finite dimensional and algorithmically computable, so that it is a mechanical procedure to prove any given identity among modular forms. Secondly, modular forms occur naturally in connection with problems arising in many other areas of mathematics. Together, these two facts imply that modular forms have a huge number of applications in other fields. The principal aim of these notes – as also of the notes on Hilbert modular forms by Bruinier and on Siegel modular forms by van der Geer – is to give a feel for some of these applications, rather than emphasizing only the theory. For this reason, we have tried to give as many and as varied examples of interesting applications as possible. These applications are placed in separate mini-subsections following the relevant sections of the main text, and identified both in the text and in the table of contents by the symbol ♠. (The end of such a mini-subsection is correspondingly indicated by the symbol ♡: these are *major* applications.) The subjects they cover range from questions of pure number theory and combinatorics to differential equations, geometry, and mathematical physics.

The notes are organized as follows. Section 1 gives a basic introduction to the theory of modular forms, concentrating on the full modular group

$\Gamma_1 = \mathrm{SL}(2,\mathbb{Z})$. Much of what is presented there can be found in standard textbooks and will be familiar to most readers, but we wanted to make the exposition self-contained. The next two sections describe two of the most important constructions of modular forms, Eisenstein series and theta series. Here too most of the material is quite standard, but we also include a number of concrete examples and applications which may be less well known. Section 4 gives a brief account of Hecke theory and of the modular forms arising from algebraic number theory or algebraic geometry whose L-series have Euler products. In the last two sections we turn to topics which, although also classical, are somewhat more specialized; here there is less emphasis on proofs and more on applications. Section 5 treats the aspects of the theory connected with differentiation of modular forms, and in particular the differential equations which these functions satisfy. This is perhaps the most important single source of applications of the theory of modular forms, ranging from irrationality and transcendence proofs to the power series arising in mirror symmetry. Section 6 treats the theory of complex multiplication. This too is a classical theory, going back to the turn of the (previous) century, but we try to emphasize aspects that are more recent and less familiar: formulas for the norms and traces of the values of modular functions at CM points, Borcherds products, and explicit Taylor expansions of modular forms. (The last topic is particularly pretty and has applications to quite varied problems of number theory.) A planned seventh section would have treated the integrals, or "periods," of modular forms, which have a rich combinatorial structure and many applications, but had to be abandoned for reasons of space and time. Apart from the first two, the sections are largely independent of one another and can be read in any order. The text contains 29 numbered "Propositions" whose proofs are given or sketched and 20 unnumbered "Theorems" which are results quoted from the literature whose proofs are too difficult (in many cases, *much* too difficult) to be given here, though in many cases we have tried to indicate what the main ingredients are. To avoid breaking the flow of the exposition, references and suggestions for further reading have not been given within the main text but collected into a single section at the end. Notations are standard (e.g., $\mathbb{Z}$, $\mathbb{Q}$, $\mathbb{R}$ and $\mathbb{C}$ for the integers, rationals, reals and complex numbers, respectively, and $\mathbb{N}$ for the strictly positive integers). Multiplication precedes division hierarchically, so that, for instance, $1/4\pi$ means $1/(4\pi)$ and not $(1/4)\pi$.

The presentation in Sections 1–5 is based partly on notes taken by Christian Grundh, Magnus Dehli Vigeland and my wife, Silke Wimmer-Zagier, of the lectures which I gave at Nordfjordeid, while that of Section 6 is partly based on the notes taken by John Voight of an earlier course on complex multiplication which I gave in Berekeley in 1992. I would like to thank all of them here, but especially Silke, who read each section of the notes as it was written and made innumerable useful suggestions concerning the exposition. And of course special thanks to Kristian Ranestad for the wonderful week in Nordfjordeid which he organized.

1 Basic Definitions

In this section we introduce the basic objects of study – the group $\mathrm{SL}(2, \mathbb{R})$ and its action on the upper half plane, the modular group, and holomorphic modular forms – and show that the space of modular forms of any weight and level is finite-dimensional. This is the key to the later applications.

1.1 Modular Groups, Modular Functions and Modular Forms

The *upper half plane*, denoted $\mathfrak{H}$, is the set of all complex numbers with positive imaginary part:

$$\mathfrak{H} = \left\{ z \in \mathbb{C} \mid \Im(z) > 0 \right\}.$$

The special linear group $\mathrm{SL}(2, \mathbb{R})$ acts on $\mathfrak{H}$ in the standard way by *Möbius transformations* (or *fractional linear transformations*):

$$\gamma = \begin{pmatrix} a & b \\ c & d \end{pmatrix} : \mathfrak{H} \to \mathfrak{H}, \qquad z \mapsto \gamma z = \gamma(z) = \frac{az + b}{cz + d}.$$

To see that this action is well-defined, we note that the denominator is non-zero and that $\mathfrak{H}$ is mapped to $\mathfrak{H}$ because, as a sinple calculation shows,

$$\Im(\gamma z) = \frac{\Im(z)}{|cz + d|^2}. \tag{1}$$

The transitivity of the action also follows by direct calculations, or alternatively we can view $\mathfrak{H}$ as the set of classes of $\left\{ \begin{pmatrix} \omega_1 \\ \omega_2 \end{pmatrix} \in \mathbb{C}^2 \mid \omega_2 \neq 0, \Im(\omega_1/\omega_2) > 0 \right\}$ under the equivalence relation of multiplication by a non-zero scalar, in which case the action is given by ordinary matrix multiplication from the left. Notice that the matrices $\pm\gamma$ act in the same way on $\mathfrak{H}$, so we can, and often will, work instead with the group $\mathrm{PSL}(2, \mathbb{R}) = \mathrm{SL}(2, \mathbb{R})/\{\pm 1\}$.

Elliptic modular functions and modular forms are functions in $\mathfrak{H}$ which are either invariant or transform in a specific way under the action of a discrete subgroup Γ of $\mathrm{SL}(2, \mathbb{R})$. In these introductory notes we will consider only the group $\Gamma_1 = \mathrm{SL}(2, \mathbb{Z})$ (the "full modular group") and its congruence subgroups (subgroups of finite index of Γ_1 which are described by congruence conditions on the entries of the matrices). We should mention, however, that there are other interesting discrete subgroups of $\mathrm{SL}(2, \mathbb{R})$, most notably the non-congruence subgroups of $\mathrm{SL}(2, \mathbb{Z})$, whose corresponding modular forms have rather different arithmetic properties from those on the congruence subgroups, and subgroups related to quaternion algebras over $\mathbb{Q}$, which have a compact fundamental domain. The latter are important in the study of both Hilbert and Siegel modular forms, treated in the other contributions in this volume.

The modular group takes its name from the fact that the points of the quotient space $\Gamma_1\backslash\mathfrak{H}$ are *moduli* (= parameters) for the isomorphism classes of elliptic curves over $\mathbb{C}$. To each point $z \in \mathfrak{H}$ one can associate the lattice $\Lambda_z = \mathbb{Z}.z + \mathbb{Z}.1 \subset \mathbb{C}$ and the quotient space $E_z = \mathbb{C}/\Lambda_z$, which is an elliptic curve, i.e., it is at the same time a complex curve and an abelian group. Conversely, every elliptic curve over $\mathbb{C}$ can be obtained in this way, but not uniquely: if E is such a curve, then E can be written as the quotient $\mathbb{C}/\Lambda$ for some lattice (discrete rank 2 subgroup) $\Lambda \subset \mathbb{C}$ which is unique up to "homotheties" $\Lambda \mapsto \lambda\Lambda$ with $\lambda \in \mathbb{C}^*$, and if we choose an oriented basis (ω_1, ω_2) of Λ (one with $\Im(\omega_1/\omega_2) > 0$) and use $\lambda = \omega_2^{-1}$ for the homothety, then we see that $E \cong E_z$ for some $z \in \mathfrak{H}$, but choosing a different oriented basis replaces z by γz for some $\gamma \in \Gamma_1$. The quotient space $\Gamma_1\backslash\mathfrak{H}$ is the simplest example of what is called a *moduli space*, i.e., an algebraic variety whose points classify isomorphism classes of other algebraic varieties of some fixed type. A complex-valued function on this space is called a *modular function* and, by virtue of the above discussion, can be seen as any one of four equivalent objects: a function from $\Gamma_1\backslash\mathfrak{H}$ to $\mathbb{C}$, a function $f : \mathfrak{H} \to \mathbb{C}$ satisfying the transformation equation $f(\gamma z) = f(z)$ for every $z \in \mathfrak{H}$ and every $\gamma \in \Gamma_1$, a function assigning to every elliptic curve E over $\mathbb{C}$ a complex number depending only on the isomorphism type of E, or a function on lattices in $\mathbb{C}$ satisfying $F(\lambda\Lambda) = F(\Lambda)$ for all lattices Λ and all $\lambda \in \mathbb{C}^\times$, the equivalence between f and F being given in one direction by $f(z) = F(\Lambda_z)$ and in the other by $F(\Lambda) = f(\omega_1/\omega_2)$ where (ω_1, ω_2) is any oriented basis of Λ. Generally the term "modular function", on Γ_1 or some other discrete subgroup $\Gamma \subset \mathrm{SL}(2, \mathbb{R})$, is used only for meromorphic modular functions, i.e., Γ-invariant meromorphic functions in $\mathfrak{H}$ which are of exponential growth at infinity (i.e., $f(x + iy) = \mathrm{O}(e^{Cy})$ as $y \to \infty$ and $f(x + iy) = \mathrm{O}(e^{C/y})$ as $y \to 0$ for some $C > 0$), this latter condition being equivalent to the requirement that f extends to a meromorphic function on the compactified space $\overline{\Gamma\backslash\mathfrak{H}}$ obtained by adding finitely many "cusps" to $\Gamma\backslash\mathfrak{H}$ (see below).

It turns out, however, that for the purposes of doing interesting arithmetic the modular functions are not enough and that one needs a more general class of functions called *modular forms*. The reason is that modular functions have to be allowed to be meromorphic, because there are no global holomorphic functions on a compact Riemann surface, whereas modular forms, which have a more flexible transformation behavior, are holomorphic functions (on $\mathfrak{H}$ and, in a suitable sense, also at the cusps). Every modular function can be represented as a quotient of two modular forms, and one can think of the modular functions and modular forms as in some sense the analogues of rational numbers and integers, respectively. From the point of view of functions on lattices, modular forms are simply functions $\Lambda \mapsto F(\Lambda)$ which transform under homotheties by $F(\lambda\Lambda) = \lambda^{-k}F(\Lambda)$ rather than simply by $F(\lambda\Lambda) = F(\Lambda)$ as before, where k is a fixed integer called the *weight* of the modular form. If we translate this back into the language of functions on $\mathfrak{H}$ via $f(z) = F(\Lambda_z)$ as before, then we see that f is now required to satisfy the *modular transformation property*

$$f\left(\frac{az+b}{cz+d}\right) = (cz+d)^k f(z) \qquad (2)$$

for all $z \in \mathfrak{H}$ and all $\left(\begin{smallmatrix} a & b \\ c & d \end{smallmatrix}\right) \in \Gamma_1$; conversely, given a function $f : \mathfrak{H} \to \mathbb{C}$ satisfying (2), we can define a funcion on lattices, homogeneous of degree $-k$ with respect to homotheties, by $F(\mathbb{Z}.\omega_1 + \mathbb{Z}.\omega_2) = \omega_2^{-k} f(\omega_1/\omega_2)$. As with modular functions, there is a standard convention: when the word "modular form" (on some discrete subgroup Γ of $\mathrm{SL}(2,\mathbb{R})$) is used with no further adjectives, one generally means "holomorphic modular form", i.e., a function f on $\mathfrak{H}$ satisfying (2) for all $\left(\begin{smallmatrix} a & b \\ c & d \end{smallmatrix}\right) \in \Gamma$ which is holomorphic in $\mathfrak{H}$ and of subexponential growth at infinity (i.e., f satisfies the same estimate as above, but now for *all* rather than *some* $C > 0$). This growth condition, which corresponds to holomorphy at the cusps in a sense which we do not explain now, implies that the growth at infinity is in fact polynomial; more precisely, f automatically satisfies $f(z) = O(1)$ as $y \to \infty$ and $f(x+iy) = O(y^{-k})$ as $y \to 0$. We denote by $M_k(\Gamma)$ the space of holomorphic modular forms of weight k on Γ. As we will see in detail for $\Gamma = \Gamma_1$, this space is finite-dimensional, effectively computable for all k, and zero for $k < 0$, and the algebra $M_*(\Gamma) := \bigoplus_k M_k(\Gamma)$ of all modular forms on Γ is finitely generated over $\mathbb{C}$.

If we specialize (2) to the matrix $\left(\begin{smallmatrix} 1 & 1 \\ 0 & 1 \end{smallmatrix}\right)$, which belongs to Γ_1, then we see that any modular form on Γ_1 satisfies $f(z+1) = f(z)$ for all $z \in \mathfrak{H}$, i.e., it is a periodic function of period 1. It is therefore a function of the quantity $e^{2\pi i z}$, traditionally denoted q; more precisely, we have the *Fourier development*

$$f(z) = \sum_{n=0}^{\infty} a_n e^{2\pi i n z} = \sum_{n=0}^{\infty} a_n q^n \qquad \left(z \in \mathfrak{H},\ q = e^{2\pi i z}\right), \qquad (3)$$

where the fact that only terms q^n with $n \geq 0$ occur is a consequence of (and in the case of Γ_1, in fact equivalent to) the growth conditions on f just given. It is this Fourier development which is responsible for the great importance of modular forms, because it turns out that there are many examples of modular forms f for which the Fourier coefficients a_n in (3) are numbers that are of interest in other domains of mathematics.

1.2 The Fundamental Domain of the Full Modular Group

In the rest of §1 we look in more detail at the modular group. Because Γ_1 contains the element $-1 = \left(\begin{smallmatrix} -1 & 0 \\ 0 & -1 \end{smallmatrix}\right)$ which fixes every point of $\mathfrak{H}$, we can also consider the action of the quotient group $\overline{\Gamma}_1 = \Gamma_1/\{\pm 1\} = \mathrm{PSL}(2,\mathbb{Z}) \subset \mathrm{PSL}(2,\mathbb{R})$ on $\mathfrak{H}$. It is clear from (2) that a modular form of odd weight on Γ_1 (or on any subgroup of $\mathrm{SL}(2,\mathbb{R})$ containing -1) must vanish, so we can restrict our attention to even k. But then the "automorphy factor" $(cz+d)^k$ in (2) is unchanged when we replace $\gamma \in \Gamma_1$ by $-\gamma$, so that we can consider equation (2) for k even and $\left(\begin{smallmatrix} a & b \\ c & d \end{smallmatrix}\right) \in \overline{\Gamma}_1$. By a slight abuse of notation, we will use the same notation for an element γ of Γ_1 and its image $\pm\gamma$ in $\overline{\Gamma}_1$, and, for k even, will not distinguish between the isomorphic spaces $M_k(\Gamma_1)$ and $M_k(\overline{\Gamma}_1)$.

The group $\overline{\Gamma}_1$ is generated by the two elements $T = \left(\begin{smallmatrix} 1 & 1 \\ 0 & 1 \end{smallmatrix}\right)$ and $S = \left(\begin{smallmatrix} 0 & 1 \\ -1 & 0 \end{smallmatrix}\right)$, with the relations $S^2 = (ST)^3 = 1$. The actions of S and T on $\mathfrak{H}$ are given by

$$S : z \mapsto -1/z, \qquad T : z \mapsto z+1.$$

Therefore f is a modular form of weight k on Γ_1 precisely when f is periodic with period 1 and satisfies the single further functional equation

$$f(-1/z) = z^k f(z) \qquad (z \in \mathfrak{H}). \tag{4}$$

If we know the value of a modular form f on some group Γ at one point $z \in \mathfrak{H}$, then equation (2) tells us the value at all points in the same Γ_1-orbit as z. So to be able to completely determine f it is enough to know the value at one point from each orbit. This leads to the concept of a *fundamental domain* for Γ, namely an open subset $\mathcal{F} \subset \mathfrak{H}$ such that no two distinct points of $\mathcal{F}$ are equivalent under the action of Γ and every point $z \in \mathfrak{H}$ is Γ-equivalent to some point in the closure $\overline{\mathcal{F}}$ of $\mathcal{F}$.

Proposition 1. *The set*

$$\mathcal{F}_1 = \left\{ z \in \mathfrak{H} \mid |z| > 1, \ |\Re(z)| < \tfrac{1}{2} \right\}$$

is a fundamental domain for the full modular group Γ_1. (See Fig. 1A.)

Proof. Take a point $z \in \mathfrak{H}$. Then $\{mz + n \mid m, n \in \mathbb{Z}\}$ is a lattice in $\mathbb{C}$. Every lattice has a point different from the origin of minimal modulus. Let $cz + d$ be such a point. The integers c, d must be relatively prime (otherwise we could divide $cz + d$ by an integer to get a new point in the lattice of even smaller modulus). So there are integers a and b such that $\gamma_1 = \left(\begin{smallmatrix} a & b \\ c & d \end{smallmatrix}\right) \in \Gamma_1$. By the transformation property (1) for the imaginary part $y = \Im(z)$ we get that $\Im(\gamma_1 z)$ is a maximal member of $\{\Im(\gamma z) \mid \gamma \in \Gamma_1\}$. Set $z^* = T^n \gamma_1 z = \gamma_1 z + n$,

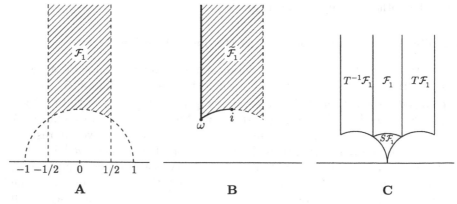

Fig. 1. The standard fundamental domain for $\overline{\Gamma}_1$ and its neighbors

where n is such that $|\Re(z^*)| \leq \frac{1}{2}$. We cannot have $|z^*| < 1$, because then we would have $\Im(-1/z^*) = \Im(z^*)/|z^*|^2 > \Im(z^*)$ by (1), contradicting the maximality of $\Im(z^*)$. So $z^* \in \mathcal{F}_1$, and z is equivalent under Γ_1 to z^*.

Now suppose that we had two Γ_1-equivalent points z_1 and $z_2 = \gamma z_1$ in $\mathcal{F}_1$, with $\gamma \neq \pm 1$. This γ cannot be of the form T^n since this would contradict the condition $|\Re(z_1)|$, $|\Re(z_2)| < \frac{1}{2}$, so $\gamma = \left(\begin{smallmatrix} a & b \\ c & d \end{smallmatrix}\right)$ with $c \neq 0$. Note that $\Im(z) > \sqrt{3}/2$ for all $z \in \mathcal{F}_1$. Hence from (1) we get

$$\frac{\sqrt{3}}{2} < \Im(z_2) = \frac{\Im(z_1)}{|cz_1 + d|^2} \leq \frac{\Im(z_1)}{c^2 \Im(z_1)^2} < \frac{2}{c^2 \sqrt{3}},$$

which can only be satisfied if $c = \pm 1$. Without loss of generality we may assume that $\Im z_1 \leq \Im z_2$. But $|\pm z_1 + d| \geq |z_1| > 1$, and this gives a contradiction with the transformation property (1).

Remarks. 1. The points on the borders of the fundamental region are Γ_1-equivalent as follows: First, the points on the two lines $\Re(z) = \pm\frac{1}{2}$ are equivalent by the action of $T : z \mapsto z + 1$. Secondly, the points on the left and right halves of the arc $|z| = 1$ are equivalent under the action of $S : z \mapsto -1/z$. In fact, these are the only equivalences for the points on the boundary. For this reason we define $\widetilde{\mathcal{F}_1}$ to be the semi-closure of $\mathcal{F}_1$ where we have added only the boundary points with non-positive real part (see Fig. 1B). Then every point of $\mathfrak{H}$ is Γ_1-equivalent to a *unique* point of $\widetilde{\mathcal{F}_1}$, i.e., $\widetilde{\mathcal{F}_1}$ is a *strict fundamental domain* for the action of Γ_1. (But terminology varies, and many people use the words "fundamental domain" for the strict fundamental domain or for its closure, rather than for the interior.)

2. The description of the fundamental domain $\mathcal{F}_1$ also implies the above-mentioned fact that Γ_1 (or $\overline{\Gamma}_1$) is generated by S and T. Indeed, by the very definition of a fundamental domain we know that $\overline{\mathcal{F}_1}$ and its translates $\gamma\overline{\mathcal{F}_1}$ by elements γ of Γ_1 cover $\mathfrak{H}$, disjointly except for their overlapping boundaries (a so-called "tesselation" of the upper half-plane). The neighbors of $\mathcal{F}_1$ are $T^{-1}\mathcal{F}_1$, $S\mathcal{F}_1$ and $T\mathcal{F}_1$ (see Fig. 1C), so one passes from any translate $\gamma\mathcal{F}_1$ of $\mathcal{F}_1$ to one of its three neighbors by applying $\gamma S\gamma^{-1}$ or $\gamma T^{\pm 1}\gamma^{-1}$. In particular, if the element γ describing the passage from $\mathcal{F}_1$ to a given translated fundamental domain $\mathcal{F}_1' = \gamma\mathcal{F}_1$ can be written as a word in S and T, then so can the element of Γ_1 which describes the motion from $\mathcal{F}_1$ to any of the neighbors of $\mathcal{F}_1'$. Therefore by moving from neighbor to neighbor across the whole upper half-plane we see inductively that this property holds for every $\gamma \in \Gamma_1$, as asserted. More generally, one sees that if one has given a fundamental domain $\mathcal{F}$ for any discrete group Γ, then the elements of Γ which identify in pairs the sides of $\overline{\mathcal{F}}$ always generate Γ.

♠ **Finiteness of Class Numbers**

Let D be a negative discriminant, i.e., a negative integer which is congruent to 0 or 1 modulo 4. We consider binary quadratic forms of the form $Q(x, y) =$

$Ax^2 + Bxy + Cy^2$ with $A, B, C \in \mathbb{Z}$ and $B^2 - 4AC = D$. Such a form is definite (i.e., $Q(x,y) \neq 0$ for non-zero $(x,y) \in \mathbb{R}^2$) and hence has a fixed sign, which we take to be positive. (This is equivalent to $A > 0$.) We also assume that Q is primitive, i.e., that $\gcd(A, B, C) = 1$. Denote by $\mathfrak{Q}_D$ the set of these forms. The group Γ_1 (or indeed $\overline{\Gamma}_1$) acts on $\mathfrak{Q}_D$ by $Q \mapsto Q \circ \gamma$, where $(Q \circ \gamma)(x,y) = Q(ax + by, cx + dy)$ for $\gamma = \pm \left(\begin{smallmatrix} a & b \\ c & d \end{smallmatrix} \right) \in \overline{\Gamma}_1$. We claim that the number of equivalence classes under this action is finite. This number, called the *class number* of D and denoted $h(D)$, also has an interpretation as the number of ideal classes (either for the ring of integers or, if D is a non-trivial square multiple of some other discriminant, for a non-maximal order) in the imaginary quadratic field $\mathbb{Q}(\sqrt{D})$, so this claim is a special case – historically the first one, treated in detail in Gauss's *Disquisitiones Arithmeticae* – of the general theorem that the number of ideal classes in any number field is finite. To prove it, we observe that we can associate to any $Q \in \mathfrak{Q}_D$ the unique root $\mathfrak{z}_Q = (-B + \sqrt{D})/2A$ of $Q(\mathfrak{z}, 1) = 0$ in the upper half-plane (here $\sqrt{D} = +i\sqrt{|D|}$ by definition and $A > 0$ by assumption). One checks easily that $\mathfrak{z}_{Q \circ \gamma} = \gamma^{-1}(\mathfrak{z}_Q)$ for any $\gamma \in \overline{\Gamma}_1$, so each $\overline{\Gamma}_1$-equivalence class of forms $Q \in \mathfrak{Q}_D$ has a unique representative belonging to the set

$$\mathfrak{Q}_D^{\mathrm{red}} = \{ [A, B, C] \in \mathfrak{Q}_D \mid -A < B \leq A < C \quad \text{or} \quad 0 \leq B \leq A = C \} \tag{5}$$

of $Q \in \mathfrak{Q}_D$ for which $\mathfrak{z}_Q \in \widetilde{\mathcal{F}}_1$ (the so-called *reduced* quadratic forms of discriminant D), and this set is finite because $C \geq A \geq |B|$ implies $|D| = 4AC - B^2 \geq 3A^2$, so that both A and B are bounded in absolute value by $\sqrt{|D|/3}$, after which C is fixed by $C = (B^2 - D)/4A$. This even gives us a way to compute $h(D)$ effectively, e.g., $\mathfrak{Q}_{-47}^{\mathrm{red}} = \{[1, 1, 12], [2, \pm 1, 6], [3, \pm 1, 4]\}$ and hence $h(-47) = 5$. We remark that the class numbers $h(D)$, or a small modification of them, are themselves the coefficients of a modular form (of weight $3/2$), but this will not be discussed further in these notes. ♡

1.3 The Finite Dimensionality of $M_k(\Gamma)$

We end this section by applying the description of the fundamental domain to show that $M_k(\Gamma_1)$ is finite-dimensional for every k and to get an upper bound for its dimension. In §2 we will see that this upper bound is in fact the correct value.

If f is a modular form of weight k on Γ_1 or any other discrete group Γ, then f is not a well-defined function on the quotient space $\Gamma \backslash \mathfrak{H}$, but the transformation formula (2) implies that the order of vanishing $\mathrm{ord}_z(f)$ at a point $z \in \mathfrak{H}$ depends only on the orbit Γz. We can therefore define a local order of vanishing, $\mathrm{ord}_P(f)$, for each $P \in \Gamma \backslash \mathfrak{H}$. The key assertion is that the total number of zeros of f, i.e., the sum of all of these local orders, depends only on Γ and k. But to make this true, we have to look more carefully at the geometry of the quotient space $\Gamma \backslash \mathfrak{H}$, taking into account the fact that some points (the so-called *elliptic fixed points*, corresponding to the points

$z \in \mathfrak{H}$ which have a non-trivial stabilizer for the image of Γ in $\mathrm{PSL}(2,\mathbb{R})$) are singular and also that $\Gamma\backslash\mathfrak{H}$ is not compact, but has to be compactified by the addition of one or more further points called *cusps*. We explain this for the case $\Gamma = \Gamma_1$.

In §1.2 we identified the quotient space $\Gamma_1\backslash\mathfrak{H}$ as a set with the semi-closure $\widetilde{\mathcal{F}_1}$ of $\mathcal{F}_1$ and as a topological space with the quotient of $\overline{\mathcal{F}_1}$ obtained by identifying the opposite sides (lines $\Re(z) = \pm\frac{1}{2}$ or halves of the arc $|z| = 1$) of the boundary $\partial\overline{\mathcal{F}_1}$. For a generic point of $\widetilde{\mathcal{F}_1}$ the stabilizer subgroup of $\overline{\Gamma}_1$ is trivial. But the two points $\omega = \frac{1}{2}(-1 + i\sqrt{3}) = e^{2\pi i/3}$ and i are stabilized by the cyclic subgroups of order 3 and 2 generated by ST and S respectively. This means that in the quotient manifold $\Gamma_1\backslash\mathfrak{H}$, ω and i are singular. (From a metric point of view, they have neighborhoods which are not discs, but quotients of a disc by these cyclic subgroups, with total angle 120° or 180° instead of 360°.) If we define an integer n_P for every $P \in \Gamma_1\backslash\mathfrak{H}$ as the order of the stabilizer in $\overline{\Gamma}_1$ of any point in $\mathfrak{H}$ representing P, then n_P equals 2 or 3 if P is Γ_1-equivalent to i or ω and $n_P = 1$ otherwise. We also have to consider the compactified quotient $\overline{\Gamma_1\backslash\mathfrak{H}}$ obtained by adding a point at infinity ("cusp") to $\Gamma_1\backslash\mathfrak{H}$. More precisely, for $Y > 1$ the image in $\Gamma_1\backslash\mathfrak{H}$ of the part of $\mathfrak{H}$ above the line $\Im(z) = Y$ can be identified via $q = e^{2\pi i z}$ with the punctured disc $0 < q < e^{-2\pi Y}$. Equation (3) tells us that a holomorphic modular form of any weight k on Γ_1 is not only a well-defined function on this punctured disc, but extends holomorphically to the point $q = 0$. We therefore define $\overline{\Gamma_1\backslash\mathfrak{H}} = \Gamma_1\backslash\mathfrak{H} \cup \{\infty\}$, where the point "$\infty$" corresponds to $q = 0$, with q as a local parameter. One can also think of $\overline{\Gamma_1\backslash\mathfrak{H}}$ as the quotient of $\overline{\mathfrak{H}}$ by Γ_1, where $\overline{\mathfrak{H}} = \mathfrak{H}\cup\mathbb{Q}\cup\{\infty\}$ is the space obtained by adding the full Γ_1-orbit $\mathbb{Q}\cup\{\infty\}$ of ∞ to $\mathfrak{H}$. We define the *order of vanishing at infinity* of f, denoted $\mathrm{ord}_\infty(f)$, as the smallest integer n such that $a_n \neq 0$ in the Fourier expansion (3).

Proposition 2. *Let f be a non-zero modular form of weight k on Γ_1. Then*

$$\sum_{P\in\Gamma_1\backslash\mathfrak{H}} \frac{1}{n_P}\,\mathrm{ord}_P(f) + \mathrm{ord}_\infty(f) = \frac{k}{12}. \tag{6}$$

Proof. Let D be the closed set obtained from $\overline{\mathcal{F}_1}$ by deleting ε-neighborhoods of all zeros of f and also the "neighborhood of infinity" $\Im(z) > Y = \varepsilon^{-1}$, where ε is chosen sufficiently small that all of these neighborhoods are disjoint (see Fig. 2.) Since f has no zeros in D, Cauchy's theorem implies that the integral of $d(\log f(z)) = \dfrac{f'(z)}{f(z)}\,dz$ over the boundary of D is 0. This boundary consists of several parts: the horizontal line from $-\frac{1}{2} + iY$ to $\frac{1}{2} + iY$, the two vertical lines from ω to $-\frac{1}{2}+iY$ and from $\omega+1$ to $\frac{1}{2}+iY$ (with some ε-neighborhoods removed), the arc of the circle $|z| = 1$ from ω to $\omega + 1$ (again with some ε-neighborhoods deleted), and the boundaries of the ε-neighborhoods of the zeros P of f. These latter have total angle 2π if P is not an elliptic fixed point

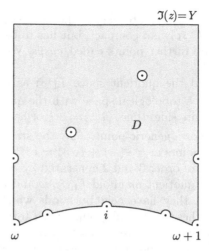

Fig. 2. The zeros of a modular form

(they consist of a full circle if P is an interior point of $\overline{\mathcal{F}_1}$ and of two half-circles if P corresponds to a boundary point of $\overline{\mathcal{F}_1}$ different from ω, $\omega + 1$ or i), and total angle π or $2\pi/3$ if $P \sim i$ or ω. The corresponding contributions to the integral are as follows. The two vertical lines together give 0, because f takes on the same value on both and the orientations are opposite. The horizontal line from $-\frac{1}{2}+iY$ to $\frac{1}{2}+iY$ gives a contribution $2\pi i \operatorname{ord}_\infty(f)$, because $d(\log f)$ is the sum of $\operatorname{ord}_\infty(f)\, dq/q$ and a function of q which is holomorphic at 0, and this integral corresponds to an integral around a small circle $|q| = e^{-2\pi Y}$ around $q = 0$. The integral on the boundary of the deleted ε-neighborhood of a zero P of f contributes $2\pi i \operatorname{ord}_P(f)$ if $n_P = 1$ by Cauchy's theorem, because $\operatorname{ord}_P(f)$ is the residue of $d(\log f(z))$ at $z = P$, while for $n_P > 1$ we must divide by n_P because we are only integrating over one-half or one-third of the full circle around P. Finally, the integral along the bottom arc contributes $\pi i k/6$, as we see by breaking up this arc into its left and right halves and applying the formula $d \log f(Sz) = d \log f(z) + k\,dz/z$, which is a consequence of the transformation equation (4). Combining all of these terms with the appropriate signs dictated by the orientation, we obtain (6). The details are left to the reader.

Corollary 1. *The dimension of $M_k(\Gamma_1)$ is 0 for $k < 0$ or k odd, while for even $k \geq 0$ we have*

$$\dim M_k(\Gamma_1) \leq \begin{cases} [k/12] + 1 & \text{if } k \not\equiv 2 \pmod{12} \\ [k/12] & \text{if } k \equiv 2 \pmod{12}. \end{cases} \tag{7}$$

Proof. Let $m = [k/12] + 1$ and choose m distinct non-elliptic points $P_i \in \Gamma_1 \backslash \mathfrak{H}$. Given any modular forms $f_1, \ldots, f_{m+1} \in M_k(\Gamma_1)$, we can find a linear combination f of them which vanishes in all P_i, by linear algebra. But then $f \equiv 0$ by the proposition, since $m > k/12$, so the f_i are linearly dependent. Hence $\dim M_k(\Gamma_1) \leq m$. If $k \equiv 2 \pmod{12}$ we can improve the estimate by 1 by noticing that the only way to satisfy (6) is to have (at least) a simple zero at i and a double zero at ω (contributing a total of $1/2 + 2/3 = 7/6$ to $\sum \operatorname{ord}_P(f)/n_P$ together with $k/12 - 7/6 = m - 1$ further zeros, so that the same argument now gives $\dim M_k(\Gamma_1) \leq m - 1$.

Corollary 2. *The space $M_{12}(\Gamma_1)$ has dimension ≤ 2, and if $f, g \in M_{12}(\Gamma_1)$ are linearly independent, then the map $z \mapsto f(z)/g(z)$ gives an isomorphism from $\Gamma_1 \backslash \mathfrak{H} \cup \{\infty\}$ to $\mathbb{P}^1(\mathbb{C})$.*

Proof. The first statement is a special case of Corollary 1. Suppose that f and g are linearly independent elements of $M_{12}(\Gamma_1)$. For any $(0,0) \neq (\lambda, \mu) \in \mathbb{C}^2$ the modular form $\lambda f - \mu g$ of weight 12 has exactly one zero in $\Gamma_1 \backslash \mathfrak{H} \cup \{\infty\}$ by Proposition 2, so the modular function $\psi = f/g$ takes on every value $(\mu : \lambda) \in \mathbb{P}^1(\mathbb{C})$ exactly once, as claimed.

We will make an explicit choice of f, g and ψ in §2.4, after we have introduced the "discriminant function" $\Delta(z) \in M_{12}(\Gamma_1)$.

The true interpretation of the factor $1/12$ multiplying k in equation (6) is as $1/4\pi$ times the volume of $\Gamma_1 \backslash \mathfrak{H}$, taken with respect to the hyperbolic metric. We say only a few words about this, since these ideas will not be used again. To give a metric on a manifold is to specify the distance between any two sufficiently near points. The *hyperbolic metric* in $\mathfrak{H}$ is defined by saying that the hyperbolic distance between two points in a small neighborhood of a point $z = x + iy \in \mathfrak{H}$ is very nearly $1/y$ times the Euclidean distance between them, so the volume element, which in Euclidean geometry is given by the 2-form $dx\,dy$, is given in hyperbolic geometry by $d\mu = y^{-2}dx\,dy$. Thus

$$\operatorname{Vol}(\Gamma_1 \backslash \mathfrak{H}) = \int_{\mathcal{F}_1} d\mu = \int_{-1/2}^{1/2} \left(\int_{\sqrt{1-x^2}}^{\infty} \frac{dy}{y^2} \right) dx$$

$$= \int_{-1/2}^{1/2} \frac{dx}{\sqrt{1-x^2}} = \arcsin(x) \Big|_{-1/2}^{1/2} = \frac{\pi}{3}.$$

Now we can consider other discrete subgroups of $\mathrm{SL}(2, \mathbb{R})$ which have a fundamental domain of finite volume. (Such groups are usually called *Fuchsian groups of the first kind*, and sometimes "lattices", but we will reserve this latter term for discrete cocompact subgroups of Euclidean spaces.) Examples are the subgroups $\Gamma \subset \Gamma_1$ of finite index, for which the volume of $\Gamma \backslash \mathfrak{H}$ is $\pi/3$ times the index of Γ in Γ_1 (or more precisely, of the image of Γ in $\mathrm{PSL}(2, \mathbb{R})$ in $\overline{\Gamma}_1$). If Γ is any such group, then essentially the same proof as for Proposition 2 shows that the number of Γ-inequivalent zeros of any non-zero modular form $f \in M_k(\Gamma)$ equals $k \operatorname{Vol}(\Gamma \backslash \mathfrak{H})/4\pi$, where just as in the case of Γ_1 we must count the zeros at elliptic fixed points or cusps of Γ with appropriate multiplicities. The same argument as for Corollary 1 of Proposition 2 then tells us $M_k(\Gamma)$ is finite dimensional and gives an explicit upper bound:

Proposition 3. *Let Γ be a discrete subgroup of $SL(2, \mathbb{R})$ for which $\Gamma \backslash \mathfrak{H}$ has finite volume V. Then* $\dim M_k(\Gamma) \leq \dfrac{kV}{4\pi} + 1$ *for all $k \in \mathbb{Z}$.*

In particular, we have $M_k(\Gamma) = \{0\}$ for $k < 0$ and $M_0(\Gamma) = \mathbb{C}$, i.e., there are no holomorphic modular forms of negative weight on any group Γ, and the only modular forms of weight 0 are the constants. A further consequence is that any three modular forms on Γ are algebraically dependent. (If f, g, h were algebraically independent modular forms of positive weights, then for large k the dimension of $M_k(\Gamma)$ would be at least the number of monomials in f, g, h of total weight k, which is bigger than some positive multiple of k^2, contradicting the dimension estimate given in the proposition.) Equivalently, any two modular functions on Γ are algebraically dependent, since every modular function is a quotient of two modular forms. This is a special case of the general fact that there cannot be more than n algebraically independent algebraic functions on an algebraic variety of dimension n. But the most important consequence of Proposition 3 from our point of view is that it is the origin of the (unreasonable?) effectiveness of modular forms in number theory: if we have two interesting arithmetic sequences $\{a_n\}_{n \geq 0}$ and $\{b_n\}_{n \geq 0}$ and conjecture that they are identical (and clearly many results of number theory can be formulated in this way), then if we can show that both $\sum a_n q^n$ and $\sum b_n q^n$ are modular forms of the same weight and group, we need only verify the equality $a_n = b_n$ for a finite number of n in order to know that it is true in general. There will be many applications of this principle in these notes.

2 First Examples: Eisenstein Series and the Discriminant Function

In this section we construct our first examples of modular forms: the Eisenstein series $E_k(z)$ of weight $k > 2$ and the discriminant function $\Delta(z)$ of weight 12, whose definition is closely connected to the non-modular Eisenstein series $E_2(z)$.

2.1 Eisenstein Series and the Ring Structure of $M_*(\Gamma_1)$

There are two natural ways to introduce the Eisenstein series. For the first, we observe that the characteristic transformation equation (2) of a modular form can be written in the form $f|_k\gamma = f$ for $\gamma \in \Gamma$, where $f|_k\gamma : \mathfrak{H} \to \mathbb{C}$ is defined by

$$(f|_kg)(z) = (cz+d)^{-k} f\left(\frac{az+b}{cz+d}\right) \qquad \left(z \in \mathbb{C}, \quad g = \begin{pmatrix} a & b \\ c & d \end{pmatrix} \in \mathrm{SL}(2,\mathbb{R})\right). \tag{8}$$

One checks easily that for fixed $k \in \mathbb{Z}$, the map $f \mapsto f|_kg$ defines an operation of the group $\mathrm{SL}(2,\mathbb{R})$ (i.e., $f|_k(g_1g_2) = (f|_kg_1)|_kg_2$ for all $g_1, g_2 \in \mathrm{SL}(2,\mathbb{R})$) on the vector space of holomorphic functions in $\mathfrak{H}$ having subexponential or polynomial growth. The space $M_k(\Gamma)$ of holomorphic modular forms of weight k on a group $\Gamma \subset \mathrm{SL}(2,\mathbb{R})$ is then simply the subspace of this vector space fixed by Γ.

If we have a linear action $v \mapsto v|g$ of a *finite* group G on a vector space V, then an obvious way to construct a G-invariant vector in V is to start with an arbitrary vector $v_0 \in V$ and form the sum $v = \sum_{g \in G} v_0|g$ (and to hope that the result is non-zero). If the vector v_0 is invariant under some subgroup $G_0 \subset G$, then the vector $v_0|g$ depends only on the coset $G_0g \in G_0\backslash G$ and we can form instead the smaller sum $v = \sum_{g \in G_0\backslash G} v_0|g$, which again is G-invariant. If G is infinite, the same method sometimes applies, but we now have to be careful about convergence. If the vector v_0 is fixed by an infinite subgroup G_0 of G, then this improves our chances because the sum over $G_0\backslash G$ is much smaller than a sum over all of G (and in any case $\sum_{g \in G} v|g$ has no chance of converging since every term occurs infinitely often). In the context when $G = \Gamma \subset \mathrm{SL}(2,\mathbb{R})$ is a Fuchsian group (acting by $|_k$) and v_0 a rational function, the modular forms obtained in this way are called *Poincaré series*. An especially easy case is that when v_0 is the constant function "1" and $\Gamma_0 = \Gamma_\infty$, the stabilizer of the cusp at infinity. In this case the series $\sum_{\Gamma_\infty\backslash\Gamma} 1|_k\gamma$ is called an *Eisenstein series*.

Let us look at this series more carefully when $\Gamma = \Gamma_1$. A matrix $\begin{pmatrix} a & b \\ c & d \end{pmatrix} \in \mathrm{SL}(2,\mathbb{R})$ sends ∞ to a/c, and hence belongs to the stabilizer of ∞ if and only if $c = 0$. In Γ_1 these are the matrices $\pm\begin{pmatrix} 1 & n \\ 0 & 1 \end{pmatrix}$ with $n \in \mathbb{Z}$, i.e., up to sign the matrices T^n. We can assume that k is even (since there are no modular forms of odd weight on Γ_1) and hence work with $\overline{\Gamma_1} = \mathrm{PSL}(2,\mathbb{Z})$, in which case the stabilizer $\overline{\Gamma}_\infty$ is the infinite cyclic group generated by T. If we multiply an arbitrary matrix $\gamma = \begin{pmatrix} a & b \\ c & d \end{pmatrix}$ on the left by $\begin{pmatrix} 1 & n \\ 0 & 1 \end{pmatrix}$, then the resulting matrix $\gamma' = \begin{pmatrix} a+nc & b+nd \\ c & d \end{pmatrix}$ has the same bottom row as γ. Conversely, if $\gamma' = \begin{pmatrix} a' & b' \\ c & d \end{pmatrix} \in \Gamma_1$ has the same bottom row as γ, then from $(a'-a)d-(b'-b)c = \det(\gamma)-\det(\gamma') = 0$ and $(c,d) = 1$ (the elements of any row or column of a matrix in $\mathrm{SL}(2,\mathbb{Z})$ are coprime!) we see that $a'-a = nc$, $b'-b = nd$ for some $n \in \mathbb{Z}$, i.e., $\gamma' = T^n\gamma$. Since every coprime pair of integers occurs as the bottom row of a matrix in $\mathrm{SL}(2,\mathbb{Z})$, these considerations give the formula

$$E_k(z) = \sum_{\gamma \in \Gamma_\infty\backslash\Gamma_1} 1\big|_k\gamma = \sum_{\gamma \in \overline{\Gamma}_\infty\backslash\overline{\Gamma}_1} 1\big|_k\gamma = \frac{1}{2} \sum_{\substack{c,d \in \mathbb{Z} \\ (c,d)=1}} \frac{1}{(cz+d)^k} \tag{9}$$

for the Eisenstein series (the factor $\frac{1}{2}$ arises because $(c\ d)$ and $(-c\ -d)$ give the same element of $\Gamma_1 \backslash \overline{\Gamma}_1$). It is easy to see that this sum is absolutely convergent for $k > 2$ (the number of pairs (c, d) with $N \leq |cz + d| < N + 1$ is the number of lattice points in an annulus of area $\pi(N + 1)^2 - \pi N^2$ and hence is $O(N)$, so the series is majorized by $\sum_{N=1}^{\infty} N^{1-k}$), and this absolute convergence guarantees the modularity (and, since it is locally uniform in z, also the holomorphy) of the sum. The function $E_k(z)$ is therefore a modular form of weight k for all even $k \geq 4$. It is also clear that it is non-zero, since for $\Im(z) \to \infty$ all the terms in (9) except $(c\ d) = (\pm 1\ 0)$ tend to 0, the convergence of the series being sufficiently uniform that their sum also goes to 0 (left to the reader), so $E_k(z) = 1 + o(1) \neq 0$.

The second natural way of introducing the Eisenstein series comes from the interpretation of modular forms given in the beginning of §1.1, where we identified solutions of the transformation equation (2) with functions on lattices $\Lambda \subset \mathbb{C}$ satisfying the homogeneity condition $F(\lambda \Lambda) = \lambda^{-k} F(\Lambda)$ under homotheties $\Lambda \mapsto \lambda\Lambda$. An obvious way to produce such a homogeneous function – if the series converges – is to form the sum $G_k(\Lambda) = \frac{1}{2} \sum_{\lambda \in \Lambda \backslash 0} \lambda^{-k}$ of the $(-k)$th powers of the non-zero elements of Λ. (The factor "$\frac{1}{2}$" has again been introduce to avoid counting the vectors λ and $-\lambda$ doubly when k is even; if k is odd then the series vanishes anyway.) In terms of $z \in \mathfrak{H}$ and its associated lattice $\Lambda_z = \mathbb{Z}.z + \mathbb{Z}.1$, this becomes

$$G_k(z) = \frac{1}{2} \sum_{\substack{m, n \in \mathbb{Z} \\ (m,n) \neq (0,0)}} \frac{1}{(mz + n)^k} \qquad (k > 2,\ z \in \mathfrak{H}), \qquad (10)$$

where the sum is again absolutely and locally uniformly convergent for $k > 2$, guaranteeing that $G_k \in M_k(\Gamma_1)$. The modularity can also be seen directly by noting that $(G_k|_k\gamma)(z) = \sum_{m,n}(m'z + n')^{-k}$ where $(m', n') = (m, n)\gamma$ runs over the non-zero vectors of $\mathbb{Z}^2 \backslash \{(0,0)\}$ as (m, n) does.

In fact, the two functions (9) and (10) are proportional, as is easily seen: any non-zero vector $(m, n) \in \mathbb{Z}^2$ can be written uniquely as $r(c, d)$ with r (the greatest common divisor of m and n) a positive integer and c and d coprime integers, so

$$G_k(z) = \zeta(k)\, E_k(z), \qquad (11)$$

where $\zeta(k) = \sum_{r \geq 1} 1/r^k$ is the value at k of the Riemann zeta function. It may therefore seem pointless to have introduced both definitions. But in fact, this is not the case. First of all, each definition gives a distinct point of view and has advantages in certain settings which are encountered at later points in the theory: the E_k definition is better in contexts like the famous Rankin-Selberg method where one integrates the product of the Eisenstein series with another modular form over a fundamental domain, while the G_k definition is better for analytic calculations and for the Fourier development given in §2.2.

Moreover, if one passes to other groups, then there are σ Eisenstein series of each type, where σ is the number of cusps, and, although they span the same vector space, they are not individually proportional. In fact, we will actually want to introduce a *third* normalization

$$\mathbb{G}_k(z) = \frac{(k-1)!}{(2\pi i)^k} G_k(z) \tag{12}$$

because, as we will see below, it has Fourier coefficients which are rational numbers (and even, with one exception, integers) and because it is a normalized eigenfunction for the Hecke operators discussed in §4.

As a first application, we can now determine the ring structure of $M_*(\Gamma_1)$

Proposition 4. *The ring $M_*(\Gamma_1)$ is freely generated by the modular forms E_4 and E_6.*

Corollary. *The inequality (7) for the dimension of $M_k(\Gamma_1)$ is an equality for all even $k \geq 0$.*

Proof. The essential point is to show that the modular forms $E_4(z)$ and $E_6(z)$ are algebraically independent. To see this, we first note that the forms $E_4(z)^3$ and $E_6(z)^2$ of weight 12 cannot be proportional. Indeed, if we had $E_6(z)^2 = \lambda E_4(z)^3$ for some (necessarily non-zero) constant λ, then the meromorphic modular form $f(z) = E_6(z)/E_4(z)$ of weight 2 would satisfy $f^2 = \lambda E_4$ (and also $f^3 = \lambda^{-1} E_6$) and would hence be holomorphic (a function whose square is holomorphic cannot have poles), contradicting the inequality $\dim M_2(\Gamma_1) \leq 0$ of Corollary 1 of Proposition 2. But *any* two modular forms f_1 and f_2 of the same weight which are not proportional are necessarily algebraically independent. Indeed, if $P(X, Y)$ is any polynomial in $\mathbb{C}[X, Y]$ such that $P(f_1(z), f_2(z)) \equiv 0$, then by considering the weights we see that $P_d(f_1, f_2)$ has to vanish identically for each homogeneous component P_d of P. But $P_d(f_1, f_2)/f_2^d = p(f_1/f_2)$ for some polynomial $p(t)$ in one variable, and since p has only finitely many roots we can only have $P_d(f_1, f_2) \equiv 0$ if f_1/f_2 is a constant. It follows that E_4^3 and E_6^2, and hence also E_4 and E_6, are algebraically independent. But then an easy calculation shows that the dimension of the weight k part of the subring of $M_*(\Gamma_1)$ which they generate equals the right-hand side of the inequality (7), so that the proposition and corollary follow from this inequality.

2.2 Fourier Expansions of Eisenstein Series

Recall from (3) that any modular form on Γ_1 has a Fourier expansion of the form $\sum_{n=0}^{\infty} a_n q^n$, where $q = e^{2\pi i z}$. The coefficients a_n often contain interesting arithmetic information, and it is this that makes modular forms important for classical number theory. For the Eisenstein series, normalized by (12), the coefficients are given by:

Proposition 5. *The Fourier expansion of the Eisenstein series* $\mathbb{G}_k(z)$ (k *even,* $k > 2$) *is*

$$\mathbb{G}_k(z) = -\frac{B_k}{2k} + \sum_{n=1}^{\infty} \sigma_{k-1}(n)\, q^n, \tag{13}$$

where B_k *is the* kth *Bernoulli number and where* $\sigma_{k-1}(n)$ *for* $n \in \mathbb{N}$ *denotes the sum of the* $(k-1)$st *powers of the positive divisors of* n.

We recall that the Bernoulli numbers are defined by the generating function $\sum_{k=0}^{\infty} B_k x^k/k! = x/(e^x - 1)$ and that the first values of B_k ($k > 0$ even) are given by $B_2 = \frac{1}{6}$, $B_4 = -\frac{1}{30}$, $B_6 = \frac{1}{42}$, $B_8 = -\frac{1}{30}$, $B_{10} = \frac{5}{66}$, $B_{12} = -\frac{691}{2730}$, and $B_{14} = \frac{7}{6}$.

Proof. A well known and easily proved identity of Euler states that

$$\sum_{n \in \mathbb{Z}} \frac{1}{z+n} = \frac{\pi}{\tan \pi z} \qquad (z \in \mathbb{C} \setminus \mathbb{Z}), \tag{14}$$

where the sum on the left, which is not absolutely convergent, is to be interpreted as a Cauchy principal value ($= \lim \sum_{-M}^{N}$ where M, N tend to infinity with $M - N$ bounded). The function on the right is periodic of period 1 and its Fourier expansion for $z \in \mathfrak{H}$ is given by

$$\frac{\pi}{\tan \pi z} = \pi \frac{\cos \pi z}{\sin \pi z} = \pi i \frac{e^{\pi i z} + e^{-\pi i z}}{e^{\pi i z} - e^{-\pi i z}} = -\pi i \frac{1+q}{1-q} = -2\pi i \left(\frac{1}{2} + \sum_{r=1}^{\infty} q^r \right),$$

where $q = e^{2\pi i z}$. Substitute this into (14), differentiate $k-1$ times and divide by $(-1)^{k-1}(k-1)!$ to get

$$\sum_{n \in \mathbb{Z}} \frac{1}{(z+n)^k} = \frac{(-1)^{k-1}}{(k-1)!} \frac{d^{k-1}}{dz^{k-1}} \left(\frac{\pi}{\tan \pi z} \right) = \frac{(-2\pi i)^k}{(k-1)!} \sum_{r=1}^{\infty} r^{k-1} q^r$$

$$(k \geq 2,\ z \in \mathfrak{H}),$$

an identity known as Lipschitz's formula. Now the Fourier expansion of G_k ($k > 2$ even) is obtained immediately by splitting up the sum in (10) into the terms with $m = 0$ and those with $m \neq 0$:

$$G_k(z) = \frac{1}{2} \sum_{\substack{n \in \mathbb{Z} \\ n \neq 0}} \frac{1}{n^k} + \frac{1}{2} \sum_{\substack{m,\, n \in \mathbb{Z} \\ m \neq 0}} \frac{1}{(mz+n)^k} = \sum_{n=1}^{\infty} \frac{1}{n^k} + \sum_{m=1}^{\infty} \sum_{n=-\infty}^{\infty} \frac{1}{(mz+n)^k}$$

$$= \zeta(k) + \frac{(2\pi i)^k}{(k-1)!} \sum_{m=1}^{\infty} \sum_{r=1}^{\infty} r^{k-1} q^{mr}$$

$$= \frac{(2\pi i)^k}{(k-1)!} \left(-\frac{B_k}{2k} + \sum_{n=1}^{\infty} \sigma_{k-1}(n)\, q^n \right),$$

where in the last line we have used Euler's evaluation of $\zeta(k)$ ($k > 0$ even) in terms of Bernoulli numbers. The result follows.

The first three examples of Proposition 5 are the expansions

$$G_4(z) = \frac{1}{240} + q + 9q^2 + 28q^3 + 73q^4 + 126q^5 + 252q^6 + \cdots,$$

$$G_6(z) = -\frac{1}{504} + q + 33q^2 + 244q^3 + 1057q^4 + \cdots,$$

$$G_8(z) = \frac{1}{480} + q + 129q^2 + 2188q^3 + \cdots.$$

The other two normalizations of these functions are given by

$$G_4(z) = \frac{16\,\pi^4}{3!}\, \mathbb{G}_4(z) = \frac{\pi^4}{90}\, E_4(z), \qquad E_4(z) = 1 + 240q + 2160q^2 + \cdots,$$

$$G_6(z) = -\frac{64\,\pi^6}{5!}\, \mathbb{G}_6(z) = \frac{\pi^6}{945}\, E_6(z), \qquad E_6(z) = 1 - 504q - 16632q^2 - \cdots,$$

$$G_8(z) = \frac{256\,\pi^8}{7!}\, \mathbb{G}_8(z) = \frac{\pi^8}{9450}\, E_8(z), \qquad E_8(z) = 1 + 480q + 61920q^2 + \cdots.$$

Remark. We have discussed only Eisenstein series on the full modular group in detail, but there are also various kinds of Eisenstein series for subgroups $\Gamma \subset \Gamma_1$. We give one example. Recall that a *Dirichlet character* modulo $N \in \mathbb{N}$ is a homomorphism $\chi : (\mathbb{Z}/N\mathbb{Z})^* \to \mathbb{C}^*$, extended to a map $\chi : \mathbb{Z} \to \mathbb{C}$ (traditionally denoted by the same letter) by setting $\chi(n)$ equal to $\chi(n \bmod N)$ if $(n, N) = 1$ and to 0 otherwise. If χ is a non-trivial Dirichlet character and k a positive integer with $\chi(-1) = (-1)^k$, then there is an Eisenstein series having the Fourier expansion

$$\mathbb{G}_{k,\chi}(z) = c_k(\chi) + \sum_{n=1}^{\infty}\left(\sum_{d|n}\chi(d)\,d^{k-1}\right)q^n$$

which is a "modular form of weight k and character χ on $\Gamma_0(N)$." (This means that $\mathbb{G}_{k,\chi}(\frac{az+b}{cz+d}) = \chi(a)(cz+d)^k\mathbb{G}_{k,\chi}(z)$ for any $z \in \mathfrak{H}$ and any $\left(\begin{smallmatrix} a & b \\ c & d \end{smallmatrix}\right) \in$ SL$(2,\mathbb{Z})$ with $c \equiv 0 \pmod{N}$.) Here $c_k(\chi) \in \overline{\mathbb{Q}}$ is a suitable constant, given explicitly by $c_k(\chi) = \frac{1}{2}L(1 - k, \chi)$, where $L(s, \chi)$ is the analytic continuation of the Dirichlet series $\sum_{n=1}^{\infty}\chi(n)n^{-s}$.

The simplest example, for $N = 4$ and $\chi = \chi_{-4}$ the Dirichlet character modulo 4 given by

$$\chi_{-4}(n) = \begin{cases} +1 & \text{if } n \equiv 1 \pmod 4, \\ -1 & \text{if } n \equiv 3 \pmod 4, \\ 0 & \text{if } n \text{ is even} \end{cases} \tag{15}$$

and $k = 1$, is the series

$$\mathbb{G}_{1,\chi_{-4}}(z) = c_1(\chi_{-4}) + \sum_{n=1}^{\infty}\left(\sum_{d|n}\chi_{-4}(d)\right)q^n = \frac{1}{4} + q + q^2 + q^4 + 2q^5 + q^8 + \cdots.$$

$$\tag{16}$$

(The fact that $L(0, \chi_{-4}) = 2c_1(\chi_{-4}) = \dfrac{1}{2}$ is equivalent via the functional equation of $L(s, \chi_{-4})$ to Leibnitz's famous formula $L(1, \chi_{-4}) = 1 - \dfrac{1}{3} + \dfrac{1}{5} - \cdots = \dfrac{\pi}{4}$.) We will see this function again in §3.1.

♠ **Identities Involving Sums of Powers of Divisors**

We now have our first explicit examples of modular forms and their Fourier expansions and can immediately deduce non-trivial number-theoretic identities. For instance, each of the spaces $M_4(\Gamma_1)$, $M_6(\Gamma_1)$, $M_8(\Gamma_1)$, $M_{10}(\Gamma_1)$ and $M_{14}(\Gamma_1)$ has dimension exactly 1 by the corollary to Proposition 2, and is therefore spanned by the Eisenstein series $E_k(z)$ with leading coefficient 1, so we immediately get the identities

$$E_4(z)^2 = E_8(z), \quad E_4(z)E_6(z) = E_{10}(z),$$
$$E_6(z)E_8(z) = E_4(z)E_{10}(z) = E_{14}(z).$$

Each of these can be combined with the Fourier expansion given in Proposition 5 to give an identity involving the sums-of-powers-of-divisors functions $\sigma_{k-1}(n)$, the first and the last of these being

$$\sum_{m=1}^{n-1} \sigma_3(m)\sigma_3(n-m) = \frac{\sigma_7(n) - \sigma_3(n)}{120},$$

$$\sum_{m=1}^{n-1} \sigma_3(m)\sigma_9(n-m) = \frac{\sigma_{13}(n) - 11\sigma_9(n) + 10\sigma_3(n)}{2640}.$$

Of course similar identities can be obtained from modular forms in higher weights, even though the dimension of $M_k(\Gamma_1)$ is no longer equal to 1. For instance, the fact that $M_{12}(\Gamma_1)$ is 2-dimensional and contains the three modular forms E_4E_8, E_6^2 and E_{12} implies that the three functions are linearly dependent, and by looking at the first two terms of the Fourier expansions we find that the relation between them is given by $441E_4E_8 + 250E_6^2 = 691E_{12}$, a formula which the reader can write out explicitly as an identity among sums-of-powers-of-divisors functions if he or she is so inclined. It is not easy to obtain any of these identities by direct number-theoretical reasoning (although in fact it can be done). ♡

2.3 The Eisenstein Series of Weight 2

In §2.1 and §2.2 we restricted ourselves to the case when $k > 2$, since then the series (9) and (10) are absolutely convergent and therefore define modular forms of weight k. But the final formula (13) for the Fourier expansion of $\mathbb{G}_k(z)$ converges rapidly and defines a holomorphic function of z also for $k = 2$, so in this weight we can simply *define* the Eisenstein series $\mathbb{G}_2$, G_2 and E_2 by equations (13), (12), and (11), respectively, i.e.,

$$\mathbb{G}_2(z) = -\frac{1}{24} + \sum_{n=1}^{\infty} \sigma_1(n)\, q^n = -\frac{1}{24} + q + 3q^2 + 4q^3 + 7q^4 + 6q^5 + \cdots ,$$

$$G_2(z) = -4\pi^2\, \mathbb{G}_2(z), \quad E_2(z) = \frac{6}{\pi^2}\, G_2(z) = 1 - 24q - 72q^2 - \cdots .$$

$$(17)$$

Moreover, the same proof as for Proposition 5 still shows that $G_2(z)$ is given by the expression (10), if we agree to carry out the summation over n first and then over m:

$$G_2(z) = \frac{1}{2} \sum_{n\neq 0} \frac{1}{n^2} + \frac{1}{2} \sum_{m\neq 0} \sum_{n\in\mathbb{Z}} \frac{1}{(mz+n)^2} . \tag{18}$$

The only difference is that, because of the non-absolute convergence of the double series, we can no longer interchange the order of summation to get the modular transformation equation $G_2(-1/z) = z^2 G_2(z)$. (The equation $G_2(z+1) = G_2(z)$, of course, still holds just as for higher weights.) Nevertheless, the function $G_2(z)$ and its multiples $E_2(z)$ and $\mathbb{G}_2(z)$ do have some modular properties and, as we will see later, these are important for many applications.

Proposition 6. *For $z \in \mathfrak{H}$ and $\left(\begin{smallmatrix} a & b \\ c & d \end{smallmatrix}\right) \in SL(2,\mathbb{Z})$ we have*

$$G_2\left(\frac{az+b}{cz+d}\right) = (cz+d)^2\, G_2(z) - \pi i c(cz+d). \tag{19}$$

Proof. There are many ways to prove this. We sketch one, due to Hecke, since the method is useful in many other situations. The series (10) for $k = 2$ does not converge absolutely, but it is just at the edge of convergence, since $\sum_{m,n} |mz+n|^{-\lambda}$ converges for any real number $\lambda > 2$. We therefore modify the sum slightly by introducing

$$G_{2,\varepsilon}(z) = \frac{1}{2} {\sum_{m,n}}' \frac{1}{(mz+n)^2\, |mz+n|^{2\varepsilon}} \qquad (z \in \mathfrak{H},\ \varepsilon > 0). \tag{20}$$

(Here $\sum'$ means that the value $(m,n) = (0,0)$ is to be omitted from the summation.) The new series converges absolutely and transforms by $G_{2,\varepsilon}\left(\frac{az+b}{cz+d}\right) = (cz+d)^2 |cz+d|^{2\varepsilon} G_{2,\varepsilon}(z)$. We claim that $\lim_{\varepsilon\to 0} G_{2,\varepsilon}(z)$ exists and equals $G_2(z) - \pi/2y$, where $y = \Im(z)$. It follows that each of the three non-holomorphic functions

$$G_2^*(z) = G_2(z) - \frac{\pi}{2y}, \quad E_2^*(z) = E_2(z) - \frac{3}{\pi y}, \quad \mathbb{G}_2^*(z) = \mathbb{G}_2(z) + \frac{1}{8\pi y}$$

$$(21)$$

transforms like a modular form of weight 2, and from this one easily deduces the transformation equation (19) and its analogues for E_2 and $\mathbb{G}_2$. To prove the claim, we define a function I_ε by

$$I_\varepsilon(z) = \int_{-\infty}^{\infty} \frac{dt}{(z+t)^2 \, |z+t|^{2\varepsilon}} \qquad (z \in \mathfrak{H}, \ \varepsilon > -\tfrac{1}{2}).$$

Then for $\varepsilon > 0$ we can write

$$G_{2,\varepsilon} - \sum_{m=1}^{\infty} I_\varepsilon(mz) = \sum_{n=1}^{\infty} \frac{1}{n^{2+2\varepsilon}}$$

$$+ \sum_{m=1}^{\infty} \sum_{n=-\infty}^{\infty} \left[\frac{1}{(mz+n)^2 \, |mz+n|^{2\varepsilon}} - \int_n^{n+1} \frac{dt}{(mz+t)^2 |mz+t|^{2\varepsilon}} \right].$$

Both sums on the right converge absolutely and locally uniformly for $\varepsilon > -\tfrac{1}{2}$ (the second one because the expression in square brackets is $O\left(|mz+n|^{-3-2\varepsilon}\right)$ by the mean-value theorem, which tells us that $f(t) - f(n)$ for any differentiable function f is bounded in $n \le t \le n+1$ by $\max_{n \le u \le n+1} |f'(u)|$), so the limit of the expression on the right as $\varepsilon \to 0$ exists and can be obtained simply by putting $\varepsilon = 0$ in each term, where it reduces to $G_2(z)$ by (18). On the other hand, for $\varepsilon > -\tfrac{1}{2}$ we have

$$I_\varepsilon(x+iy) = \int_{-\infty}^{\infty} \frac{dt}{(x+t+iy)^2 \, ((x+t)^2+y^2)^\varepsilon}$$

$$= \int_{-\infty}^{\infty} \frac{dt}{(t+iy)^2 \, (t^2+y^2)^\varepsilon} = \frac{I(\varepsilon)}{y^{1+2\varepsilon}},$$

where $I(\varepsilon) = \int_{-\infty}^{\infty} (t+i)^{-2}(t^2+1)^{-\varepsilon} dt$, so $\sum_{m=1}^{\infty} I_\varepsilon(mz) = I(\varepsilon)\zeta(1+2\varepsilon)/y^{1+2\varepsilon}$ for $\varepsilon > 0$. Finally, we have $I(0) = 0$ (obvious),

$$I'(0) = -\int_{-\infty}^{\infty} \frac{\log(t^2+1)}{(t+i)^2} \, dt = \left(\frac{1+\log(t^2+1)}{t+i} - \tan^{-1} t \right) \Bigg|_{-\infty}^{\infty} = -\pi,$$

and $\zeta(1+2\varepsilon) = \dfrac{1}{2\varepsilon} + O(1)$, so the product $I(\varepsilon)\zeta(1+2\varepsilon)/y^{1+2\varepsilon}$ tends to $-\pi/2y$ as $\varepsilon \to 0$. The claim follows.

Remark. The transformation equation (18) says that G_2 is an example of what is called a *quasimodular* form, while the functions G_2^*, E_2^* and $\mathbb{G}_2^*$ defined in (21) are so-called *almost holomorphic modular forms* of weight 2. We will return to this topic in Section 5.

2.4 The Discriminant Function and Cusp Forms

For $z \in \mathfrak{H}$ we define the *discriminant function* $\Delta(z)$ by the formula

$$\Delta(z) = e^{2\pi i z} \prod_{n=1}^{\infty} \left(1 - e^{2\pi i n z}\right)^{24}. \tag{22}$$

(The name comes from the connection with the discriminant of the elliptic curve $E_z = \mathbb{C}/(\mathbb{Z}.z + \mathbb{Z}.1)$, but we will not discuss this here.) Since $|e^{2\pi i z}| < 1$ for $z \in \mathfrak{H}$, the terms of the infinite product are all non-zero and tend exponentially rapidly to 1, so the product converges everywhere and defines a holomorphic and everywhere non-zero function in the upper half-plane. This function turns out to be a modular form and plays a special role in the entire theory.

Proposition 7. *The function $\Delta(z)$ is a modular form of weight 12 on $SL(2, \mathbb{Z})$.*

Proof. Since $\Delta(z) \neq 0$, we can consider its logarithmic derivative. We find

$$\frac{1}{2\pi i}\frac{d}{dz}\log\Delta(z) = 1 - 24\sum_{n=1}^{\infty}\frac{n\,e^{2\pi i n z}}{1 - e^{2\pi i n z}} = 1 - 24\sum_{m=1}^{\infty}\sigma_1(m)\,e^{2\pi i m z} = E_2(z),$$

where the second equality follows by expanding $\dfrac{e^{2\pi i n z}}{1 - e^{2\pi i n z}}$ as a geometric series $\sum_{r=1}^{\infty} e^{2\pi i r n z}$ and interchanging the order of summation, and the third equality from the definition of $E_2(z)$ in (17). Now from the transformation equation for E_2 (obtained by comparing (19) and (11)) we find

$$\frac{1}{2\pi i}\frac{d}{dz}\log\left(\frac{\Delta\left(\frac{az+b}{cz+d}\right)}{(cz+d)^{12}\Delta(z)}\right) = \frac{1}{(cz+d)^2}E_2\left(\frac{az+b}{cz+d}\right) - \frac{12}{2\pi i}\frac{c}{cz+d} - E_2(z)$$
$$= 0.$$

In other words, $(\Delta|_{12}\gamma)(z) = C(\gamma)\,\Delta(z)$ for all $z \in \mathfrak{H}$ and all $\gamma \in \Gamma_1$, where $C(\gamma)$ is a non-zero complex number depending only on γ, and where $\Delta|_{12}\gamma$ is defined as in (8). It remains to show that $C(\gamma) = 1$ for all γ. But $C : \Gamma_1 \to \mathbb{C}^*$ is a homomorphism because $\Delta \mapsto \Delta|_{12}\gamma$ is a group action, so it suffices to check this for the generators $T = \left(\begin{smallmatrix}1&1\\0&1\end{smallmatrix}\right)$ and $S = \left(\begin{smallmatrix}0&-1\\1&0\end{smallmatrix}\right)$ of Γ_1. The first is obvious since $\Delta(z)$ is a power series in $e^{2\pi i z}$ and hence periodic of period 1, while the second follows by substituting $z = i$ into the equation $\Delta(-1/z) = C(S)\,z^{12}\Delta(z)$ and noting that $\Delta(i) \neq 0$.

Let us look at this function $\Delta(z)$ more carefully. We know from Corollary 1 to Proposition 2 that the space $M_{12}(\Gamma_1)$ has dimension at most 2, so $\Delta(z)$ must be a linear combination of the two functions $E_4(z)^3$ and $E_6(z)^2$. From the Fourier expansions $E_4^3 = 1 + 720q + \cdots$, $E_6(z)^2 = 1 - 1008q + \cdots$ and $\Delta(z) = q + \cdots$ we see that this relation is given by

$$\Delta(z) = \frac{1}{1728}\left(E_4(z)^3 - E_6(z)^2\right). \tag{23}$$

This identity permits us to give another, more explicit, version of the fact that every modular form on Γ_1 is a polynomial in E_4 and E_6 (Proposition 4). Indeed, let $f(z)$ be a modular form of arbitrary even weight $k \geq 4$, with Fourier expansion as in (3). Choose integers $a, b \geq 0$ with $4a + 6b = k$ (this is always possible) and set $h(z) = \left(f(z) - a_0 E_4(z)^a E_6(z)^b\right)/\Delta(z)$. This function is holomorphic in $\mathfrak{H}$ (because $\Delta(z) \neq 0$) and also at infinity (because $f - a_0 E_4^a E_6^b$ has a Fourier expansion with no constant term and the Fourier expansion of Δ begins with q), so it is a modular form of weight $k - 12$. By induction on the weight, h is a polynomial in E_4 and E_6, and then from $f = a_0 E_4^a E_6^b + \Delta h$ and (23) we see that f also is.

In the above argument we used that $\Delta(z)$ has a Fourier expansion beginning $q + O(q^2)$ and that $\Delta(z)$ is never zero in the upper half-plane. We deduced both facts from the product expansion (22), but it is perhaps worth noting that this is not necessary: if we were simply to *define* $\Delta(z)$ by equation (23), then the fact that its Fourier expansion begins with q would follow from the knowledge of the first two Fourier coefficients of E_4 and E_6, and the fact that it never vanishes in $\mathfrak{H}$ would then follow from Proposition 2 because the total number $k/12 = 1$ of Γ_1-inequivalent zeros of Δ is completely accounted for by the first-order zero at infinity.

We can now make the concrete normalization of the isomorphism between $\overline{\Gamma_1 \backslash \mathfrak{H}}$ and $\mathbb{P}^1(\mathbb{C})$ mentioned after Corollary 2 of Proposition 2. In the notation of that proposition, choose $f(z) = E_4(z)^3$ and $g(z) = \Delta(z)$. Their quotient is then the modular function

$$j(z) = \frac{E_4(z)^3}{\Delta(z)} = q^{-1} + 744 + 196884\,q + 21493760\,q^2 + \cdots,$$

called the *modular invariant*. Since $\Delta(z) \neq 0$ for $z \in \mathfrak{H}$, this function is finite in $\mathfrak{H}$ and defines an isomorphism from $\Gamma_1 \backslash \mathfrak{H}$ to $\mathbb{C}$ as well as from $\overline{\Gamma_1 \backslash \mathfrak{H}}$ to $\mathbb{P}^1(\mathbb{C})$.

The next (and most interesting) remarks about $\Delta(z)$ concern its Fourier expansion. By multiplying out the product in (22) we obtain the expansion

$$\Delta(z) = q \prod_{n=1}^{\infty} (1 - q^n)^{24} = \sum_{n=1}^{\infty} \tau(n)\, q^n \tag{24}$$

where $q = e^{2\pi i z}$ as usual (this is the last time we will repeat this!) and the coefficients $\tau(n)$ are certain integers, the first values being given by the table

n	1	2	3	4	5	6	7	8	9	10
$\tau(n)$	1	-24	252	-1472	4830	-6048	-16744	84480	-113643	-115920

Ramanujan calculated the first 30 values of $\tau(n)$ in 1915 and observed several remarkable properties, notably the multiplicativity property that $\tau(pq) = \tau(p)\tau(q)$ if p and q are distinct primes (e.g., $-6048 = -24 \cdot 252$ for $p = 2$, $q = 3$) and $\tau(p^2) = \tau(p)^2 - p^{11}$ if p is prime (e.g., $-1472 = (-24)^2 - 2048$ for $p = 2$). This was proved by Mordell the next year and later generalized by Hecke to the theory of Hecke operators, which we will discuss in §4.

Ramanujan also observed that $|\tau(p)|$ was bounded by $2p^5\sqrt{p}$ for primes $p < 30$, and conjectured that this holds for all p. This property turned out to be immeasurably deeper than the assertion about multiplicativity and was only proved in 1974 by Deligne as a consequence of his proof of the famous Weil conjectures (and of his previous, also very deep, proof that these conjectures implied Ramanujan's). However, the weaker inequality $|\tau(p)| \leq Cp^6$ with some effective constant $C > 0$ is much easier and was proved in the 1930's by Hecke. We reproduce Hecke's proof, since it is simple. In fact, the proof applies to a much more general class of modular forms. Let us call a modular form on Γ_1 a *cusp form* if the constant term a_0 in the Fourier expansion (3) is zero. Since the constant term of the Eisenstein series $\mathbb{G}_k(z)$ is non-zero, any modular form can be written uniquely as a linear combination of an Eisenstein series and a cusp form of the same weight. For the former the Fourier coefficients are given by (13) and grow like n^{k-1} (since $n^{k-1} \leq \sigma_{k-1}(n) < \zeta(k-1)n^{k-1}$). For the latter, we have:

Proposition 8. *Let $f(z)$ be a cusp form of weight k on Γ_1 with Fourier expansion $\sum_{n=1}^{\infty} a_n q^n$. Then $|a_n| \leq Cn^{k/2}$ for all n, for some constant C depending only on f.*

Proof. From equations (1) and (2) we see that the function $z \mapsto y^{k/2}|f(z)|$ on $\mathfrak{H}$ is Γ_1-invariant. This function tends rapidly to 0 as $y = \Im(z) \to \infty$ (because $f(z) = O(q)$ by assumption and $|q| = e^{-2\pi y}$), so from the form of the fundamental domain of Γ_1 as given in Proposition 1 it is clearly bounded. Thus we have the estimate

$$|f(z)| \leq c\,y^{-k/2} \qquad (z = x + iy \in \mathfrak{H}) \tag{25}$$

for some $c > 0$ depending only on f. Now the integral representation

$$a_n = e^{2\pi n y} \int_0^1 f(x + iy)\,e^{-2\pi i n x}\,dx$$

for a_n, valid for any $y > 0$, show that $|a_n| \leq cy^{-k/2}e^{2\pi n y}$. Taking $y = 1/n$ (or, optimally, $y = k/4\pi n$) gives the estimate of the proposition with $C = c\,e^{2\pi}$ (or, optimally, $C = c\,(4\pi e/k)^{k/2}$).

Remark. The definition of cusp forms given above is actually valid only for the full modular group Γ_1 or for other groups having only one cusp. In general one must require the vanishing of the constant term of the Fourier expansion of f, suitably defined, at every cusp of the group Γ, in which case it again follows that f can be estimated as in (25). Actually, it is easier to simply *define* cusp forms of weight k as modular forms for which $y^{k/2}f(x + iy)$ is bounded, a definition which is equivalent but does not require the explicit knowledge of the Fourier expansion of the form at every cusp.

♠ Congruences for $\tau(n)$

As a mini-application of the calculations of this and the preceding sections we prove two simple congruences for the Ramanujan tau-function defined by equation (24). First of all, let us check directly that the coefficient $\tau(n)$ of q^n of the function defined by (23) is integral for all n. (This fact is, of course, obvious from equation (22).) We have

$$\Delta = \frac{(1+240A)^3 - (1-504B)^2}{1728} = 5\frac{A-B}{12} + B + 100A^2 - 147B^2 + 8000A^3 \tag{26}$$

with $A = \sum_{n=1}^{\infty} \sigma_3(n)q^n$ and $B = \sum_{n=1}^{\infty} \sigma_5(n)q^n$. But $\sigma_5(n) - \sigma_3(n)$ is divisible by 12 for every n (because 12 divides $d^5 - d^3$ for every d), so $(A-B)/12$ has integral coefficients. This gives the integrality of $\tau(n)$, and even a congruence modulo 2. Indeed, we actually have $\sigma_5(n) \equiv \sigma_3(n) \pmod{24}$, because $d^3(d^2-1)$ is divisible by 24 for every d, so $(A-B)/12$ has even coefficients and (26) gives $\Delta \equiv B + B^2 \pmod 2$ or, recalling that $(\sum a_n q^n)^2 \equiv \sum a_n q^{2n} \pmod 2$ for every power series $\sum a_n q^n$ with integral coefficients, $\tau(n) \equiv \sigma_5(n) + \sigma_5(n/2) \pmod 2$, where $\sigma_5(n/2)$ is defined as 0 if $2 \nmid n$. But $\sigma_5(n)$, for any integer n, is congruent modulo 2 to the sum of the odd divisors of n, and this is odd if and only if n is a square or twice a square, as one sees by writing $n = 2^s n_0$ with n_0 odd and pairing the complementary divisors of n_0. It follows that $\sigma_5(n) + \sigma_5(n/2)$ is odd if and only if n is an odd square, so we get the congruence:

$$\tau(n) \equiv \begin{cases} 1 \pmod 2 & \text{if } n \text{ is an odd square}, \\ 0 \pmod 2 & \text{otherwise}. \end{cases} \tag{27}$$

In a different direction, from $\dim M_{12}(\Gamma_1) = 2$ we immediately deduce the linear relation

$$\mathbb{G}_{12}(z) = \Delta(z) + \frac{691}{156}\left(\frac{E_4(z)^3}{720} + \frac{E_6(z)^3}{1008}\right)$$

and from this a famous congruence of Ramanujan,

$$\tau(n) \equiv \sigma_{11}(n) \pmod{691} \qquad (\forall n \ge 1), \tag{28}$$

where the "691" comes from the numerator of the constant term $-B_{12}/24$ of $\mathbb{G}_{12}$. ♡

3 Theta Series

If Q is a positive definite integer-valued quadratic form in m variables, then there is an associated modular form of weight $m/2$, called the *theta series* of Q, whose nth Fourier coefficient for every integer $n \ge 0$ is the number of representations of n by Q. This provides at the same time one of the main constructions of modular forms and one of the most important sources of applications of the theory. In 3.1 we consider unary theta series ($m = 1$), while the general case is discussed in 3.2. The unary case is the most classical, going back to Jacobi, and already has many applications. It is also the basis of the general theory, because any quadratic form can be diagonalized over $\mathbb{Q}$ (i.e., by passing to a suitable sublattice it becomes the direct sum of m quadratic forms in one variable).

3.1 Jacobi's Theta Series

The simplest theta series, corresponding to the unary (one-variable) quadratic form $x \mapsto x^2$, is Jacobi's theta function

$$\theta(z) = \sum_{n \in \mathbb{Z}} q^{n^2} = 1 + 2q + 2q^4 + 2q^9 + \cdots, \qquad (29)$$

where $z \in \mathfrak{H}$ and $q = e^{2\pi i z}$ as usual. Its modular transformation properties are given as follows.

Proposition 9. *The function* $\theta(z)$ *satisfies the two functional equations*

$$\theta(z+1) = \theta(z), \qquad \theta\left(\frac{-1}{4z}\right) = \sqrt{\frac{2z}{i}}\, \theta(z) \qquad (z \in \mathfrak{H}). \qquad (30)$$

Proof. The first equation in (30) is obvious since $\theta(z)$ depends only on q. For the second, we use the Poisson transformation formula. Recall that this formula says that for any function $f : \mathbb{R} \to \mathbb{C}$ which is smooth and small at infinity, we have $\sum_{n \in \mathbb{Z}} f(n) = \sum_{n \in \mathbb{Z}} \widetilde{f}(n)$, where $\widetilde{f}(y) = \int_{-\infty}^{\infty} e^{2\pi i x y} f(x)\, dx$ is the Fourier transform of f. (*Proof*: the sum $\sum_{n \in \mathbb{Z}} f(n+x)$ is convergent and defines a function $g(x)$ which is periodic of period 1 and hence has a Fourier expansion $g(x) = \sum_{n \in \mathbb{Z}} c_n e^{2\pi i n x}$ with $c_n = \int_0^1 g(x) e^{-2\pi i n x}\, dx = \widetilde{f}(-n)$, so $\sum_n f(n) = g(0) = \sum_n c_n = \sum_n \widetilde{f}(-n) = \sum_n \widetilde{f}(n)$.) Applying this to the function $f(x) = e^{-\pi t x^2}$, where t is a positive real number, and noting that

$$\widetilde{f}(y) = \int_{-\infty}^{\infty} e^{-\pi t x^2 + 2\pi i x y}\, dx = \frac{e^{-\pi y^2/t}}{\sqrt{t}} \int_{-\infty}^{\infty} e^{-\pi u^2}\, du = \frac{e^{-\pi y^2/t}}{\sqrt{t}}$$

(substitution $u = \sqrt{t}\,(x - iy/t)$ followed by a shift of the path of integration), we obtain

$$\sum_{n=-\infty}^{\infty} e^{-\pi n^2 t} = \frac{1}{\sqrt{t}} \sum_{n=-\infty}^{\infty} e^{-\pi n^2/t} \qquad (t > 0).$$

This proves the second equation in (30) for $z = it/2$ lying on the positive imaginary axis, and the general case then follows by analytic continuation.

The point is now that the two transformations $z \mapsto z+1$ and $z \mapsto -1/4z$ generate a subgroup of $\mathrm{SL}(2,\mathbb{R})$ which is commensurable with $\mathrm{SL}(2,\mathbb{Z})$, so (30) implies that the function $\theta(z)$ is a modular form of weight $1/2$. (We have not defined modular forms of half-integral weight and will not discuss their theory in these notes, but the reader can simply interpret this statement as saying that $\theta(z)^2$ is a modular form of weight 1.) More specifically, for every $N \in \mathbb{N}$ we have the "congruence subgroup" $\Gamma_0(N) \subseteq \Gamma_1 = \mathrm{SL}(2,\mathbb{Z})$, consisting of matrices $\left(\begin{smallmatrix} a & b \\ c & d \end{smallmatrix}\right) \in \Gamma_1$ with c divisible by N, and the larger group $\Gamma_0^+(N) = \langle \Gamma_0(N), W_N \rangle = \Gamma_0(N) \cup \Gamma_0(N)W_N$, where $W_N = \frac{1}{\sqrt{N}}\left(\begin{smallmatrix} 0 & -1 \\ N & 0 \end{smallmatrix}\right)$ ("Fricke involution") is an element of $\mathrm{SL}(2,\mathbb{R})$ of order 2 which normalizes $\Gamma_0(N)$. The group $\Gamma_0^+(N)$ contains the elements $T = \left(\begin{smallmatrix} 1 & 1 \\ 0 & 1 \end{smallmatrix}\right)$ and W_N for any N. In general they generate a subgroup of infinite index, so that to check the modularity of a given function it does not suffice to verify its behavior just for $z \mapsto z+1$ and $z \mapsto -1/Nz$, but for $N=4$ (like for $N=1$!) they generate the full group and this is sufficient. The proof is simple. Since $W_N^2 = -1$, it is sufficient to show that the two matrices T and $\widetilde{T} = W_4 T W_4^{-1} = \left(\begin{smallmatrix} 1 & 0 \\ 4 & 1 \end{smallmatrix}\right)$ generate the image of $\Gamma_0(4)$ in $\mathrm{PSL}(2,\mathbb{R})$, i.e., that any element $\gamma = \left(\begin{smallmatrix} a & b \\ c & d \end{smallmatrix}\right) \in \Gamma_0(4)$ is, up to sign, a word in T and $\widetilde{T}$. Now a is odd, so $|a| \neq 2|b|$. If $|a| < 2|b|$, then either $b+a$ or $b-a$ is smaller than b in absolute value, so replacing γ by $\gamma \cdot T^{\pm 1}$ decreases $a^2 + b^2$. If $|a| > 2|b| \neq 0$, then either $a + 4b$ or $a - 4b$ is smaller than a in absolute value, so replacing γ by $\gamma \cdot \widetilde{T}^{\pm 1}$ decreases $a^2 + b^2$. Thus we can keep multiplying γ on the right by powers of T and $\widetilde{T}$ until $b=0$, at which point $\pm\gamma$ is a power of $\widetilde{T}$.

Now, by the principle "a finite number of q-coefficients suffice" formulated at the end of Section 1, the mere fact that $\theta(z)$ is a modular form is already enough to let one prove non-trivial identities. (We enunciated the principle only in the case of forms of integral weight, but even without knowing the details of the theory it is clear that it then also applies to half-integral weight, since a space of modular forms of half-integral weight can be mapped injectively into a space of modular forms of the next higher integral weight by multiplying by $\theta(z)$.) And indeed, with almost no effort we obtain proofs of two of the most famous results of number theory of the 17th and 18th centuries, the theorems of Fermat and Lagrange about sums of squares.

♠ Sums of Two and Four Squares

Let $r_2(n) = \#\{(a,b) \in \mathbb{Z}^2 \mid a^2 + b^2 = n\}$ be the number of representations of an integer $n \geq 0$ as a sum of two squares. Since $\theta(z)^2 = \left(\sum_{a \in \mathbb{Z}} q^{a^2}\right)\left(\sum_{b \in \mathbb{Z}} q^{b^2}\right)$, we see that $r_2(n)$ is simply the coefficient of q^n in $\theta(z)^2$. From Proposition 9 and the just-proved fact that $\Gamma_0(4)$ is generated by $-\mathrm{Id}_2$, T and $\widetilde{T}$, we find that the function $\theta(z)^2$ is a "modular form of weight 1 and character χ_{-4} on $\Gamma_0(4)$" in the sense explained in the paragraph preceding equation (15), where χ_{-4} is the Dirichlet character modulo 4 defined by (15).

Since the Eisenstein series $\mathbb{G}_{1,\chi_{-4}}$ in (16) is also such a modular form, and since the space of all such forms has dimension at most 1 by Proposition 3 (because $\Gamma_0(4)$ has index 6 in $SL(2,\mathbb{Z})$ and hence volume 2π), these two functions must be proportional. The proportionality factor is obviously 4, and we obtain:

Proposition 10. *Let n be a positive integer. Then the number of representations of n as a sum of two squares is 4 times the sum of $(-1)^{(d-1)/2}$, where d runs over the positive odd divisors of n.*

Corollary (Theorem of Fermat). *Every prime number $p \equiv 1 \pmod 4$ is a sum of two squares.*

Proof of Corollary. We have $r_2(p) = 4\left(1 + (-1)^{(p-1)/4}\right) = 8 \neq 0$.

The same reasoning applies to other powers of θ. In particular, the number $r_4(n)$ of representations of an integer n as a sum of four squares is the coefficient of q^n in the modular form $\theta(z)^4$ of weight 2 on $\Gamma_0(4)$, and the space of all such modular forms is at most two-dimensional by Proposition 3. To find a basis for it, we use the functions $\mathbb{G}_2(z)$ and $\mathbb{G}_2^*(z)$ defined in equations (17) and (21). We showed in §2.3 that the latter function transforms with respect to $SL(2,\mathbb{Z})$ like a modular form of weight 2, and it follows easily that the three functions $\mathbb{G}_2^*(z)$, $\mathbb{G}_2^*(2z)$ and $\mathbb{G}_2^*(4z)$ transform like modular forms of weight 2 on $\Gamma_0(4)$ (exercise!). Of course these three functions are not holomorphic, but since $\mathbb{G}_2^*(z)$ differs from the holomorphic function $\mathbb{G}_2(z)$ by $1/8\pi y$, we see that the linear combinations $\mathbb{G}_2^*(z) - 2\mathbb{G}_2^*(2z) = \mathbb{G}_2(z) - 2\mathbb{G}_2(2z)$ and $\mathbb{G}_2^*(2z) - 2\mathbb{G}_2^*(4z) = \mathbb{G}_2(2z) - 2\mathbb{G}_2(4z)$ are holomorphic, and since they are also linearly independent, they provide the desired basis for $M_2(\Gamma_0(4))$. Looking at the first two Fourier coefficients of $\theta(z)^4 = 1 + 8q + \cdots$, we find that $\theta(z)^4$ equals $8\left(\mathbb{G}_2(z) - 2\mathbb{G}_2(2z)\right) + 16\left(\mathbb{G}_2(2z) - 2\mathbb{G}_2(4z)\right)$. Now comparing coefficients of q^n gives:

Proposition 11. *Let n be a positive integer. Then the number of representations of n as a sum of four squares is 8 times the sum of the positive divisors of n which are not multiples of 4.*

Corollary (Theorem of Lagrange). *Every positive integer is a sum of four squares.* ♡

For another simple application of the q-expansion principle, we introduce two variants $\theta_M(z)$ and $\theta_F(z)$ ("M" and "F" for "male" and "female" or "minus sign" and "fermionic") of the function $\theta(z)$ by inserting signs or by shifting the indices by $1/2$ in its definition:

$$\theta_M(z) = \sum_{n \in \mathbb{Z}} (-1)^n q^{n^2} = 1 - 2q + 2q^4 - 2q^9 + \cdots ,$$

$$\theta_F(z) = \sum_{n \in \mathbb{Z}+1/2} q^{n^2} = 2q^{1/4} + 2q^{9/4} + 2q^{25/4} + \cdots .$$

These are again modular forms of weight $1/2$ on $\Gamma_0(4)$. With a little experimentation, we discover the identity

$$\theta(z)^4 = \theta_M(z)^4 + \theta_F(z)^4 \tag{31}$$

due to Jacobi, and by the q-expansion principle all we have to do to prove it is to verify the equality of a finite number of coefficients (here just one). In this particular example, though, there is also an easy combinatorial proof, left as an exercise to the reader.

The three theta series θ, θ_M and θ_F, in a slightly different guise and slightly different notation, play a role in many contexts, so we say a little more about them. As well as the subgroup $\Gamma_0(N)$ of Γ_1, one also has the *principal congruence subgroup* $\Gamma(N) = \{\gamma \in \Gamma_1 \mid \gamma \equiv \mathrm{Id}_2 \pmod{N}\}$ for every integer $N \in \mathbb{N}$, which is more basic than $\Gamma_0(N)$ because it is a normal subgroup (the kernel of the map $\Gamma_1 \to \mathrm{SL}(2, \mathbb{Z}/N\mathbb{Z})$ given by reduction modulo N). Exceptionally, the group $\Gamma_0(4)$ is isomorphic to $\Gamma(2)$, simply by conjugation by $\left(\begin{smallmatrix} 2 & 0 \\ 0 & 1 \end{smallmatrix}\right) \in \mathrm{GL}(2, \mathbb{R})$, so that there is a bijection between modular forms on $\Gamma_0(4)$ of any weight and modular forms on $\Gamma(2)$ of the same weight given by $f(z) \to f(z/2)$. In particular, our three theta functions correspond to three new theta functions

$$\theta_3(z) = \theta(z/2), \qquad \theta_4(z) = \theta_M(z/2), \qquad \theta_2(z) = \theta_F(z/2) \tag{32}$$

on $\Gamma(2)$, and the relation (31) becomes $\theta_2^4 + \theta_4^4 = \theta_3^4$. (Here the index "1" is missing because the fourth member of the quartet, $\theta_1(z) = \sum (-1)^n q^{(n+1/2)^2/2}$ is identically zero, as one sees by sending n to $-n - 1$. It may look odd that one keeps a whole notation for the zero function. But in fact the functions $\theta_i(z)$ for $1 \le i \le 4$ are just the "Thetanullwerte" or "theta zero-values" of the two-variable series $\theta_i(z, u) = \sum \varepsilon_n q^{n^2/2} e^{2\pi i n u}$, where the sum is over $\mathbb{Z}$ or $\mathbb{Z} + \frac{1}{2}$ and ε_n is either 1 or $(-1)^n$, none of which vanishes identically. The functions $\theta_i(z, u)$ play a basic role in the theory of elliptic functions and are also the simplest example of *Jacobi forms*, a theory which is related to many of the themes treated in these notes and is also important in connection with Siegel modular forms of degree 2 as discussed in Part III of this book.) The quotient group $\Gamma_1/\Gamma(2) \cong \mathrm{SL}(2, \mathbb{Z}/2\mathbb{Z})$, which has order 6 and is isomorphic to the symmetric group $\mathfrak{S}_3$ on three symbols, acts as the latter on the modular forms $\theta_i(z)^8$, while the fourth powers transform by

$$\Theta(z) := \begin{pmatrix} \theta_2(z)^4 \\ -\theta_3(z)^4 \\ \theta_4(z)^4 \end{pmatrix} \quad \Rightarrow \quad \Theta(z+1) = -\begin{pmatrix} 1 & 0 & 0 \\ 0 & 0 & 1 \\ 0 & 1 & 0 \end{pmatrix} \Theta(z),$$

$$z^{-2}\Theta\left(-\frac{1}{z}\right) = -\begin{pmatrix} 0 & 0 & 1 \\ 0 & 1 & 0 \\ 1 & 0 & 0 \end{pmatrix} \Theta(z).$$

This illustrates the general principle that a modular form on a subgroup of finite index of the modular group Γ_1 can also be seen as a component of a vector-valued modular form on Γ_1 itself. The full ring $M_*(\Gamma(2))$ of modular forms on $\Gamma(2)$ is generated by the three components of $\Theta(z)$ (or by any two of them, since their sum is zero), while the subring $M_*(\Gamma(2))^{\mathfrak{S}_3} = M_*(\Gamma_1)$ is generated by the modular forms $\theta_2(z)^8 + \theta_3(z)^8 + \theta_4(z)^8$ and $\left(\theta_2(z)^4 + \theta_3(z)^4\right)\left(\theta_3(z)^4 + \theta_4(z)^4\right)\left(\theta_4(z)^4 - \theta_2(z)^4\right)$ of weights 4 and 6 (which are then equal to $2E_4(z)$ and $2E_6(z)$, respectively). Finally, we see that $\frac{1}{256}\theta_2(z)^8\theta_3(z)^8\theta_4(z)^8$ is a cusp form of weight 12 on Γ_1 and is, in fact, equal to $\Delta(z)$.

This last identity has an interesting consequence. Since $\Delta(z)$ is non-zero for all $z \in \mathfrak{H}$, it follows that each of the three theta-functions $\theta_i(z)$ has the same property. (One can also see this by noting that the "visible" zero of $\theta_2(z)$ at infinity accounts for all the zeros allowed by the formula discussed in §1.3, so that this function has no zeros at finite points, and then the same holds for $\theta_3(z)$ and $\theta_4(z)$ because they are related to θ_2 by a modular transformation.) This suggests that these three functions, or equivalently their $\Gamma_0(4)$-versions θ, θ_M and θ_F, might have a product expansion similar to that of the function $\Delta(z)$, and indeed this is the case: we have the three identities

$$\theta(z) = \frac{\eta(2z)^5}{\eta(z)^2\eta(4z)^2}, \qquad \theta_M(z) = \frac{\eta(z)^2}{\eta(2z)}, \qquad \theta_F(z) = 2\frac{\eta(4z)^2}{\eta(2z)}, \quad (33)$$

where $\eta(z)$ is the "Dedekind eta-function"

$$\eta(z) = \Delta(z)^{1/24} = q^{1/24}\prod_{n=1}^{\infty}\left(1 - q^n\right). \qquad (34)$$

The proof of (33) is immediate by the usual q-expansion principle: we multiply the identities out (writing, e.g., the first as $\theta(z)\eta(z)^2\eta(4z)^2 = \eta(2z)^5$) and then verify the equality of enough coefficients to account for all possible zeros of a modular form of the corresponding weight. More efficiently, we can use our knowledge of the transformation behavior of $\Delta(z)$ and hence of $\eta(z)$ under Γ_1 to see that the quotients on the right in (33) are finite at every cusp and hence, since they also have no poles in the upper half-plane, are holomorphic modular forms of weight $1/2$, after which the equality with the theta-functions on the left follows directly.

More generally, one can ask when a quotient of products of eta-functions is a holomorphic modular form. Since $\eta(z)$ is non-zero in $\mathfrak{H}$, such a quotient never has finite zeros, and the only issue is whether the numerator vanishes to at least the same order as the denominator at each cusp. Based on extensive numerical calculations, I formulated a general conjecture saying that there are essentially only finitely many such products of any given weight, and a second explicit conjecture giving the complete list for weight $1/2$ (i.e., when the number of η's in the numerator is one bigger than in the denominator). Both conjectures were proved by Gerd Mersmann in a brilliant Master's thesis.

For the weight $1/2$ result, the meaning of "essentially" is that the product should be primitive, i.e., it should have the form $\prod \eta(n_i z)^{a_i}$ where the n_i are positive integers with no common factor. (Otherwise one would obtain infinitely many examples by rescaling, e.g., one would have both $\theta_M(z) = \eta(z)^2/\eta(2z)$ and $\theta_M(2z) = \eta(2z)^2/\eta(4z)$ on the list.) The classification is then as follows:

Theorem (Mersmann). *There are precisely* 14 *primitive eta-products which are holomorphic modular forms of weight* $1/2$:

$$\eta(z), \quad \frac{\eta(z)^2}{\eta(2z)}, \quad \frac{\eta(2z)^2}{\eta(z)}, \quad \frac{\eta(z)\,\eta(4z)}{\eta(2z)}, \quad \frac{\eta(2z)^3}{\eta(z)\,\eta(4z)}, \quad \frac{\eta(2z)^5}{\eta(z)^2\eta(4z)^2},$$

$$\frac{\eta(z)^2\eta(6z)}{\eta(2z)\,\eta(3z)}, \quad \frac{\eta(2z)^2\eta(3z)}{\eta(z)\,\eta(6z)}, \quad \frac{\eta(2z)\,\eta(3z)^2}{\eta(z)\,\eta(6z)}, \quad \frac{\eta(z)\,\eta(6z)^2}{\eta(2z)\,\eta(3z)},$$

$$\frac{\eta(z)\,\eta(4z)\,\eta(6z)^2}{\eta(2z)\,\eta(3z)\,\eta(12z)}, \quad \frac{\eta(2z)^2\eta(3z)\,\eta(12z)}{\eta(z)\,\eta(4z)\,\eta(6z)}, \quad \frac{\eta(2z)^5\eta(3z)\,\eta(12z)}{\eta(z)^2\eta(4z)^2\eta(6z)^2},$$

$$\frac{\eta(z)\,\eta(4z)\,\eta(6z)^5}{\eta(2z)^2\eta(3z)^2\eta(12z)^2}.$$

Finally, we mention that $\eta(z)$ itself has the theta-series representation

$$\eta(z) = \sum_{n=1}^{\infty} \chi_{12}(n)\, q^{n^2/24} = q^{1/24} - q^{25/24} - q^{49/24} + q^{121/24} + \cdots$$

where $\chi_{12}(12m \pm 1) = 1$, $\chi_{12}(12m \pm 5) = -1$, and $\chi_{12}(n) = 0$ if n is divisible by 2 or 3. This identity was discovered numerically by Euler (in the simpler-looking but less enlightening version $\prod_{n=1}^{\infty}(1 - q^n) = \sum_{n=1}^{\infty}(-1)^n q^{(3n^2+n)/2}$) and proved by him only after several years of effort. From a modern point of view, his theorem is no longer surprising because one now knows the following beautiful general result, proved by J-P. Serre and H. Stark in 1976:

Theorem (Serre–Stark). *Every modular form of weight* $1/2$ *is a linear combination of unary theta series.*

Explicitly, this means that every modular form of weight $1/2$ with respect to any subgroup of finite index of $\mathrm{SL}(2, \mathbb{Z})$ is a linear combination of sums of the form $\sum_{n \in \mathbb{Z}} q^{a(n+c)^2}$ with $a \in \mathbb{Q}_{>0}$ and $c \in \mathbb{Q}$. Euler's formula for $\eta(z)$ is a typical case of this, and of course each of the other products given in Mersmann's theorem must also have a representation as a theta series. For instance, the last function on the list, $\eta(z)\eta(4z)\eta(6z)^5/\eta(2z)^2\eta(3z)^2\eta(12z)^2$, has the expansion $\sum_{n>0,\,(n,6)=1} \chi_8(n)q^{n^2/24}$, where $\chi_8(n)$ equals $+1$ for $n \equiv \pm 1 \pmod 8$ and -1 for $n \equiv \pm 3 \pmod 8$.

We end this subsection by mentioning one more application of the Jacobi theta series.

♠ The Kac–Wakimoto Conjecture

For any two natural numbers m and n, denote by $\Delta_m(n)$ the number of representations of n as a sum of m triangular numbers (numbers of the form $a(a-1)/2$ with a integral). Since $8a(a-1)/2+1 = (2a-1)^2$, this can also be written as the number $r_m^{\mathrm{odd}}(8n+m)$ of representations of $8n+m$ as a sum of m odd squares. As part of an investigation in the theory of affine superalgebras, Kac and Wakimoto were led to conjecture the formula

$$\Delta_{4s^2}(n) = \sum_{\substack{r_1, a_1, \ldots, r_s, a_s \in \mathbb{N}_{\mathrm{odd}} \\ r_1 a_1 + \cdots + r_s a_s = 2n + s^2}} P_s(a_1, \ldots, a_s) \tag{35}$$

for m of the form $4s^2$ (and a similar formula for m of the form $4s(s+1)$), where $\mathbb{N}_{\mathrm{odd}} = \{1, 3, 5, \ldots\}$ and P_s is the polynomial

$$P_s(a_1, \ldots, a_s) = \frac{\prod_i a_i \cdot \prod_{i<j}\left(a_i^2 - a_j^2\right)^2}{4^{s(s-1)}\, s!\, \prod_{j=1}^{2s-1} j!} .$$

Two proofs of this were subsequently given, one by S. Milne using elliptic functions and one by myself using modular forms. Milne's proof is very ingenious, with a number of other interesting identities appearing along the way, but is quite involved. The modular proof is much simpler. One first notes that, P_s being a homogeneous polynomial of degree $2s^2 - s$ and odd in each argument, the right-hand side of (35) is the coefficient of q^{2n+s^2} in a function $F(z)$ which is a linear combination of products $g_{h_1}(z) \cdots g_{h_s}(z)$ with $h_1 + \cdots + h_s = s^2$, where $g_h(z) = \sum_{r, a \in \mathbb{N}_{\mathrm{odd}}} a^{2h-1} q^{ra}$ ($h \geq 1$). Since g_h is a modular form (Eisenstein series) of weight $2h$ on $\Gamma_0(4)$, this function F is a modular form of weight $2s^2$ on the same group. Moreover, its Fourier expansion belongs to $q^{s^2} \mathbb{Q}[[q^2]]$ (because $P_s(a_1, \ldots, a_s)$ vanishes if any two a_i are equal, and the smallest value of $r_1 a_1 + \cdots + r_s a_s$ with all r_i and a_i in $\mathbb{N}_{\mathrm{odd}}$ and all a_i distinct is $1 + 3 + \cdots + 2s - 1 = s^2$), and from the formula given in §1 for the number of zeros of a modular form we find that this property characterizes $F(z)$ uniquely in $M_{2s^2}(\Gamma_0(4))$ up to a scalar factor. But $\theta_F(z)^{4s^2}$ has the same property, so the two functions must be proportional. This proves (35) up to a scalar factor, easily determined by setting $n = 0$. ♡

3.2 Theta Series in Many Variables

We now consider quadratic forms in an arbitrary number m of variables. Let $Q : \mathbb{Z}^m \to \mathbb{Z}$ be a positive definite quadratic form which takes integral values on $\mathbb{Z}^m$. We associate to Q the theta series

$$\Theta_Q(z) = \sum_{x_1, \ldots, x_m \in \mathbb{Z}} q^{Q(x_1, \ldots, x_m)} = \sum_{n=0}^{\infty} R_Q(n)\, q^n , \tag{36}$$

where of course $q = e^{2\pi i z}$ as usual and $R_Q(n) \in \mathbb{Z}_{\geq 0}$ denotes the number of representations of n by Q, i.e., the number of vectors $x \in \mathbb{Z}^m$ with $Q(x) = n$. The basic statement is that Θ_Q is always a modular form of weight $m/2$. In the case of even m we can be more precise about the modular transformation behavior, since then we are in the realm of modular forms of integral weight where we have given complete definitions of what modularity means. The quadratic form $Q(x)$ is a linear combination of products $x_i x_j$ with $1 \leq i, j \leq m$. Since $x_i x_j = x_j x_i$, we can write $Q(x)$ uniquely as

$$Q(x) = \frac{1}{2} x^t A x = \frac{1}{2} \sum_{i,j=1}^{m} a_{ij} x_i x_j, \tag{37}$$

where $A = (a_{ij})_{1 \leq i,j \leq m}$ is a symmetric $m \times m$ matrix and the factor $1/2$ has been inserted to avoid counting each term twice. The integrality of Q on $\mathbb{Z}^m$ is then equivalent to the statement that the symmetric matrix A has integral elements and that its diagonal elements a_{ii} are even. Such an A is called an *even integral matrix*. Since we want $Q(x) > 0$ for $x \neq 0$, the matrix A must be positive definite. This implies that $\det A > 0$. Hence A is non-singular and A^{-1} exists and belongs to $M_m(\mathbb{Q})$. The *level* of Q is then defined as the smallest positive integer $N = N_Q$ such that NA^{-1} is again an even integral matrix. We also have the *discriminant* $\Delta = \Delta_Q$ of A, defined as $(-1)^m \det A$. It is always congruent to 0 or 1 modulo 4, so there is an associated character (Kronecker symbol) χ_Δ, which is the unique Dirichlet character modulo N satisyfing $\chi_\Delta(p) = \left(\dfrac{\Delta}{p}\right)$ (Legendre symbol) for any odd prime $p \nmid N$. (The character χ_Δ in the special cases $\Delta = -4$, 12 and 8 already occurred in §2.2 (eq. (15)) and §3.1.) The precise description of the modular behavior of Θ_Q for $m \in 2\mathbb{Z}$ is then:

Theorem (Hecke, Schoenberg). *Let $Q : \mathbb{Z}^{2k} \to \mathbb{Z}$ be a positive definite integer-valued form in $2k$ variables of level N and discriminant Δ. Then Θ_Q is a modular form on $\Gamma_0(N)$ of weight k and character χ_Δ, i.e., we have $\Theta_Q\left(\frac{az+b}{cz+d}\right) = \chi_\Delta(a)(cz+d)^k \Theta_Q(z)$ for all $z \in \mathfrak{H}$ and $\left(\begin{smallmatrix} a & b \\ c & d \end{smallmatrix}\right) \in \Gamma_0(N)$.*

The proof, as in the unary case, relies essentially on the Poisson summation formula, which gives the identity $\Theta_Q(-1/Nz) = N^{k/2}(z/i)^k \Theta_{Q^*}(z)$, where $Q^*(x)$ is the quadratic form associated to NA^{-1}, but finding the precise modular behavior requires quite a lot of work. One can also in principle reduce the higher rank case to the one-variable case by using the fact that every quadratic form is diagonalizable over $\mathbb{Q}$, so that the sum in (36) can be broken up into finitely many sub-sums over sublattices or translated sublattices of $\mathbb{Z}^m$ on which $Q(x_1, \ldots, x_m)$ can be written as a linear combination of m squares.

There is another language for quadratic forms which is often more convenient, the language of lattices. From this point of view, a quadratic form is no longer a homogeneous quadratic polynomial in m variables, but a function Q

from a free $\mathbb{Z}$-module Λ of rank m to $\mathbb{Z}$ such that the associated scalar product $(x, y) = Q(x + y) - Q(x) - Q(y)$ $(x, y \in \Lambda)$ is bilinear. Of course we can always choose a $\mathbb{Z}$-basis of Λ, in which case Λ is identified with $\mathbb{Z}^m$ and Q is described in terms of a symmetric matrix A as in (37), the scalar product being given by $(x, y) = x^t A y$, but often the basis-free language is more convenient. In terms of the scalar product, we have a length function $\|x\|^2 = (x, x)$ (actually this is the square of the length, but one often says simply "length" for convenience) and $Q(x) = \frac{1}{2}\|x\|^2$, so that the integer-valued case we are considering corresponds to lattices in which all vectors have even length. One often chooses the lattice Λ inside the euclidean space $\mathbb{R}^m$ with its standard length function $(x, x) = \|x\|^2 = x_1^2 + \cdots + x_m^2$; in this case the square root of $\det A$ is equal to the volume of the quotient $\mathbb{R}^m/\Lambda$, i.e., to the volume of a fundamental domain for the action by translation of the lattice Λ on $\mathbb{R}^m$. In the case when this volume is 1, i.e., when $\Lambda \in \mathbb{R}^m$ has the same covolume as $\mathbb{Z}^m$, the lattice is called *unimodular*. Let us look at this case in more detail.

♠ Invariants of Even Unimodular Lattices

If the matrix A in (37) is even and unimodular, then the above theorem tells us that the theta series Θ_Q associated to Q is a modular form on the full modular group. This has many consequences.

Proposition 12. *Let $Q : \mathbb{Z}^m \to \mathbb{Z}$ be a positive definite even unimodular quadratic form in m variables. Then*

(i) the rank m is divisible by 8, and
(ii) the number of representations of $n \in \mathbb{N}$ by Q is given for large n by the formula

$$R_Q(n) = -\frac{2k}{B_k} \sigma_{k-1}(n) + O(n^{k/2}) \qquad (n \to \infty), \qquad (38)$$

where $m = 2k$ and B_k denotes the kth Bernoulli number.

Proof. For the first part it is enough to show that m cannot be an odd multiple of 4, since if m is either odd or twice an odd number then $4m$ or $2m$ is an odd multiple of 4 and we can apply this special case to the quadratic form $Q \oplus Q \oplus Q \oplus Q$ or $Q \oplus Q$, respectively. So we can assume that $m = 2k$ with k even and must show that k is divisible by 4 and that (38) holds. By the theorem above, the theta series Θ_Q is a modular form of weight k on the full modular group $\Gamma_1 = \mathrm{SL}(2, \mathbb{Z})$ (necessarily with trivial character, since there are no non-trivial Dirichlet characters modulo 1). By the results of Section 2, this modular form is a linear combination of $\mathbb{G}_k(z)$ and a cusp form of weight k, and from the Fourier expansion (13) we see that the coefficient of $\mathbb{G}_k$ in this decomposition equals $-2k/B_k$, since the constant term $R_Q(0)$ of Θ_Q equals 1. (The only vector of length 0 is the zero vector.) Now Proposition 8 implies the asymptotic formula (38), and the fact that k must be divisible by 4 also follows because if $k \equiv 2 \pmod 4$ then B_k is positive and therefore the right-hand side of (38) tends to $-\infty$ as $k \to \infty$, contradicting $R_Q(n) \geq 0$.

The first statement of Proposition 12 is purely algebraic, and purely algebraic proofs are known, but they are not as simple or as elegant as the modular proof just given. No non-modular proof of the asymptotic formula (38) is known.

Before continuing with the theory, we look at some examples, starting in rank 8. Define the lattice $\Lambda_8 \subset \mathbb{R}^8$ to be the set of vectors belonging to either $\mathbb{Z}^8$ or $(\mathbb{Z}+\frac{1}{2})^8$ for which the sum of the coordinates is even. This is unimodular because the lattice $\mathbb{Z}^8 \cup (\mathbb{Z} + \frac{1}{2})^8$ contains both it and $\mathbb{Z}^8$ with the same index 2, and is even because $x_i^2 \equiv x_i \pmod 2$ for $x_i \in \mathbb{Z}$ and $x_i^2 \equiv \frac{1}{4} \pmod 2$ for $x_i \in \mathbb{Z}+\frac{1}{2}$. The lattice Λ_8 is sometimes denoted E_8 because, if we choose the $\mathbb{Z}$-basis $u_i = e_i - e_{i+1}$ $(1 \leq i \leq 6)$, $u_7 = e_6 + e_7$, $u_8 = -\frac{1}{2}(e_1 + \cdots + e_8)$ of Λ_8, then every u_i has length 2 and (u_i, u_j) for $i \neq j$ equals -1 or 0 according whether the ith and jth vertices (in a standard numbering) of the "E_8" Dynkin diagram in the theory of Lie algebras are adjacent or not. The theta series of Λ_8 is a modular form of weight 4 on $\mathrm{SL}(2,\mathbb{Z})$ whose Fourier expansion begins with 1, so it is necessarily equal to $E_4(z)$, and we get "for free" the information that for every integer $n \geq 1$ there are exactly $240\,\sigma_3(n)$ vectors x in the E_8 lattice with $(x, x) = 2n$.

From the uniqueness of the modular form $E_4 \in M_4(\Gamma_1)$ we in fact get that $r_Q(n) = 240\sigma_3(n)$ for any even unimodular quadratic form or lattice of rank 8, but here this is not so interesting because the known classification in this rank says that Λ_8 is, in fact, the only such lattice up to isomorphism. However, in rank 16 one knows that there are two non-equivalent lattices: the direct sum $\Lambda_8 \oplus \Lambda_8$ and a second lattice Λ_{16} which is not decomposable. Since the theta series of both lattices are modular forms of weight 8 on the full modular group with Fourier expansions beginning with 1, they are both equal to the Eisenstein series $E_8(z)$, so we have $r_{\Lambda_8 \oplus \Lambda_8}(n) = r_{\Lambda_{16}}(n) = 480\,\sigma_7(n)$ for all $n \geq 1$, even though the two lattices in question are distinct. (Their distinctness, and a great deal of further information about the relative positions of vectors of various lengths in these or in any other lattices, can be obtained by using the theory of Jacobi forms which was mentioned briefly in §3.1 rather than just the theory of modular forms.)

In rank 24, things become more interesting, because now $\dim M_{12}(\Gamma_1) = 2$ and we no longer have uniqueness. The even unimodular lattices of this rank were classified completely by Niemeyer in 1973. There are exactly 24 of them up to isomorphism. Some of them have the same theta series and hence the same number of vectors of any given length (an obvious such pair of lattices being $\Lambda_8 \oplus \Lambda_8 \oplus \Lambda_8$ and $\Lambda_8 \oplus \Lambda_{16}$), but not all of them do. In particular, exactly one of the 24 lattices has the property that it has no vectors of length 2. This is the famous Leech lattice (famous among other reasons because it has a huge group of automorphisms, closely related to the monster group and other sporadic simple groups). Its theta series is the unique modular form of weight 12 on Γ_1 with Fourier expansion starting $1 + 0q + \cdots$, so it must equal $E_{12}(z) - \frac{21736}{691}\Delta(z)$, i.e., the number $r_{\text{Leech}}(n)$ of vectors of length $2n$ in the Leech lattice equals $\frac{21736}{691}\left(\sigma_{11}(n) - \tau(n)\right)$ for every positive integer n. This gives another proof and an interpretation of Ramanujan's congruence (28).

In rank 32, things become even more interesting: here the complete classification is not known, and we know that we cannot expect it very soon, because there are more than 80 million isomorphism classes! This, too, is a consequence of the theory of modular forms, but of a much more sophisticated part than we are presenting here. Specifically, there is a fundamental theorem of Siegel saying that the average value of the theta series associated to the quadratic forms in a single genus (we omit the definition) is always an Eisenstein series. Specialized to the particular case of even unimodular forms of rank $m = 2k \equiv 0 \pmod 8$, which form a single genus, this theorem says that there are only finitely many such forms up to equivalence for each k and that, if we number them $Q_1, \ldots, Q_I$, then we have the relation

$$\sum_{i=1}^{I} \frac{1}{w_i} \Theta_{Q_i}(z) = \mathsf{m}_k \, E_k(z), \tag{39}$$

where w_i is the number of automorphisms of the form Q_i (i.e., the number of matrices $\gamma \in \mathrm{SL}(m, \mathbb{Z})$ such that $Q_i(\gamma x) = Q_i(x)$ for all $x \in \mathbb{Z}^m$) and m_k is the positive rational number given by the formula

$$\mathsf{m}_k = \frac{B_k}{2k} \frac{B_2}{4} \frac{B_4}{8} \cdots \frac{B_{2k-2}}{4k-4},$$

where B_i denotes the ith Bernoulli number. In particular, by comparing the constant terms on the left- and right-hand sides of (39), we see that $\sum_{i=1}^{I} 1/w_i = \mathsf{m}_k$, the *Minkowski-Siegel mass formula*. The numbers $\mathsf{m}_4 \approx 1.44 \times 10^{-9}$, $\mathsf{m}_8 \approx 2.49 \times 10^{-18}$ and $\mathsf{m}_{12} \approx 7,94 \times 10^{-15}$ are small, but $\mathsf{m}_{16} \approx 4,03 \times 10^7$ (the next two values are $\mathsf{m}_{20} \approx 4.39 \times 10^{51}$ and $\mathsf{m}_{24} \approx 1.53 \times 10^{121}$), and since $w_i \geq 2$ for every i (one has at the very least the automorphisms $\pm \mathrm{Id}_m$), this shows that $I > 80000000$ for $m = 32$ as asserted.

A further consequence of the fact that $\Theta_Q \in M_k(\Gamma_1)$ for Q even and unimodular of rank $m = 2k$ is that the minimal value of $Q(x)$ for non-zero $x \in \Lambda$ is bounded by $r = \dim M_k(\Gamma_1) = [k/12] + 1$. The lattice L is called *extremal* if this bound is attained. The three lattices of rank 8 and 16 are extremal for trivial reasons. (Here $r = 1$.) For $m = 24$ we have $r = 2$ and the only extremal lattice is the Leech lattice. Extremal unimodular lattices are also known to exist for $m = 32, 40, 48, 56, 64$ and 80, while the case $m = 72$ is open. Surprisingly, however, there are no examples of large rank:

Theorem (Mallows–Odlyzko–Sloane). *There are only finitely many non-isomorphic extremal even unimodular lattices.*

We sketch the proof, which, not surprisingly, is completely modular. Since there are only finitely many non-isomorphic even unimodular lattices of any given rank, the theorem is equivalent to saying that there is an absolute bound on the value of the rank m for extremal lattices. For simplicity, let us suppose that $m = 24n$. (The cases $m = 24n + 8$ and $m = 24n + 16$ are similar.) The theta series of any extremal unimodular lattice of this rank must be equal to the unique modular form $f_n \in M_{12n}(\mathrm{SL}(2, \mathbb{Z}))$ whose q-development has the form $1 + O(q^{n+1})$. By an elementary argument which we omit but which the reader may want to look for, we find that this q-development has the form

$$f_n(z) = 1 + n \, a_n \, q^{n+1} + \left(\frac{nb_n}{2} - 24 \, n \, (n + 31) \, a_n \right) q^{n+2} + \cdots$$

where a_n and b_n are the coefficients of $\Delta(z)^n$ in the modular functions $j(z)$ and $j(z)^2$, respectively, when these are expressed (locally, for small q) as Laurent series in the modular form $\Delta(z) = q - 24q^2 + 252q^3 - \cdots$. It is not hard to show that a_n has the asymptotic behavior $a_n \sim An^{-3/2}C^n$ for some constants $A = 225153.793389\cdots$ and $C = 1/\Delta(z_0) = 69.1164201716\cdots$, where $z_0 = 0.52352170017992\cdots i$ is the unique zero on the imaginary axis of the function $E_2(z)$ defined in (17) (this is because $E_2(z)$ is the logarithmic derivative of $\Delta(z)$), while b_n has a similar expansion but with A replaced by $2\lambda A$ with $\lambda = j(z_0) - 720 = 163067.793145\cdots$. It follows that the coefficient $\frac{1}{2}nb_n - 24n(n+31)a_n$ of q^{n+2} in f_n is negative for n larger than roughly 6800, corresponding to $m \approx 163000$, and that therefore extremal lattices of rank larger than this cannot exist.   ♡

♠ Drums Whose Shape One Cannot Hear

Marc Kac asked a famous question, "Can one hear the shape of a drum?" Expressed more mathematically, this means: can there be two riemannian manifolds (in the case of real "drums" these would presumably be two-dimensional manifolds with boundary) which are not isometric but have the same spectra of eigenvalues of their Laplace operators? The first example of such a pair of manifolds to be found was given by Milnor, and involved 16-dimensional closed "drums." More drum-like examples consisting of domains in $\mathbb{R}^2$ with polygonal boundary are now also known, but they are difficult to construct, whereas Milnor's example is very easy. It goes as follows. As we already mentioned, there are two non-isomorphic even unimodular lattices $\Lambda_1 = \Gamma_8 \oplus \Gamma_8$ and $\Lambda_2 = \Gamma_{16}$ in dimension 16. The fact that they are non-isomorphic means that the two Riemannian manifolds $M_1 = \mathbb{R}^{16}/\Lambda_1$ and $M_2 = \mathbb{R}^{16}/\Lambda_2$, which are topologically both just tori $(S^1)^{16}$, are not isometric to each other. But the spectrum of the Laplace operator on any torus $\mathbb{R}^n/\Lambda$ is just the set of norms $\|\lambda\|^2$ ($\lambda \in \Lambda$), counted with multiplicities, and these spectra agree for M_1 and M_2 because the theta series $\sum_{\lambda \in \Lambda_1} q^{\|\lambda\|^2}$ and $\sum_{\lambda \in \Lambda_2} q^{\|\lambda\|^2}$ coincide.
♡

We should not leave this section without mentioning at least briefly that there is an important generalization of the theta series (36) in which each term $q^{Q(x_1,\ldots,x_m)}$ is weighted by a polynomial $P(x_1,\ldots,x_m)$. If this polynomial is homogeneous of degree d and is *spherical with respect to* Q (this means that $\Delta P = 0$, where Δ is the Laplace operator with respect to a system of coordinates in which $Q(x_1,\ldots,x_m)$ is simply $x_1^2 + \cdots + x_m^2$), then the theta series $\Theta_{Q,P}(z) = \sum_x P(x) q^{Q(x)}$ is a modular form of weight $m/2 + d$ (on the same group and with respect to the same character as in the case $P = 1$), and is a cusp form if d is strictly positive. The possibility of putting non-trivial weights into theta series in this way considerably enlarges their range of applications, both in coding theory and elsewhere.

4 Hecke Eigenforms and L-series

In this section we give a brief sketch of Hecke's fundamental discoveries that the space of modular forms is spanned by modular forms with multiplicative Fourier coefficients and that one can associate to these forms Dirichlet series which have Euler products and functional equations. These facts are at the basis of most of the higher developments of the theory: the relations of modular forms to arithmetic algebraic geometry and to the theory of motives, and the adelic theory of automorphic forms. The last two subsections describe some basic examples of these higher connections.

4.1 Hecke Theory

For each integer $m \geq 1$ there is a linear operator T_m, the mth *Hecke operator*, acting on modular forms of any given weight k. In terms of the description of modular forms as homogeneous functions on lattices which was given in §1.1, the definition of T_m is very simple: it sends a homogeneous function F of degree $-k$ on lattices $\Lambda \subset \mathbb{C}$ to the function $T_m F$ defined (up to a suitable normalizing constant) by $T_m F(\Lambda) = \sum F(\Lambda')$, where the sum runs over all sublattices $\Lambda' \subset \Lambda$ of index m. The sum is finite and obviously still homogeneous in Λ of the same degree $-k$. Translating from the language of lattices to that of functions in the upper half-plane by the usual formula $f(z) = F(\Lambda_z)$, we find that the action of T_m is given by

$$T_m f(z) = m^{k-1} \sum_{\left(\begin{smallmatrix} a & b \\ c & d \end{smallmatrix}\right) \in \Gamma_1 \backslash \mathcal{M}_m} (cz+d)^{-k} f\left(\frac{az+b}{cz+d}\right) \qquad (z \in \mathfrak{H}), \quad (40)$$

where $\mathcal{M}_m$ denotes the set of 2×2 integral matrices of determinant m and where the normalizing constant m^{k-1} has been introduced for later convenience (T_m normalized in this way will send forms with integral Fourier coefficients to forms with integral Fourier coefficients). The sum makes sense because the transformation law (2) of f implies that the summand associated to a matrix $M = \left(\begin{smallmatrix} a & b \\ c & d \end{smallmatrix}\right) \in \mathcal{M}_m$ is indeed unchanged if M is replaced by γM with $\gamma \in \Gamma_1$, and from (40) one also easily sees that $T_m f$ is holomorphic in $\mathfrak{H}$ and satisfies the same transformation law and growth properties as f, so T_m indeed maps $M_k(\Gamma_1)$ to $M_k(\Gamma_1)$. Finally, to calculate the effect of T_m on Fourier developments, we note that a set of representatives of $\Gamma_1 \backslash \mathcal{M}_m$ is given by the upper triangular matrices $\left(\begin{smallmatrix} a & b \\ 0 & d \end{smallmatrix}\right)$ with $ad = m$ and $0 \le b < d$ (this is an easy exercise), so

$$T_m f(z) = m^{k-1} \sum_{\substack{ad=m \\ a,d>0}} \frac{1}{d^k} \sum_{b \,(\mathrm{mod}\, d)} f\left(\frac{az+b}{d}\right). \tag{41}$$

If $f(z)$ has the Fourier development (3), then a further calculation with (41), again left to the reader, shows that the function $T_m f(z)$ has the Fourier expansion

$$T_m f(z) = \sum_{\substack{d|m \\ d>0}} (m/d)^{k-1} \sum_{\substack{n\ge 0 \\ d|n}} a_n \, q^{mn/d^2} = \sum_{n\ge 0} \left(\sum_{\substack{r|(m,n) \\ r>0}} r^{k-1} a_{mn/r^2} \right) q^n. \tag{42}$$

An easy but important consequence of this formula is that the operators T_m ($m \in \mathbb{N}$) all commute.

Let us consider some examples. The expansion (42) begins $\sigma_{k-1}(m)a_0 + a_m q + \cdots$, so if f is a cusp form (i.e., $a_0 = 0$), then so is $T_m f$. In particular, since the space $S_{12}(\Gamma_1)$ of cusp forms of weight 12 is 1-dimensional, spanned by $\Delta(z)$, it follows that $T_m \Delta$ is a multiple of Δ for every $m \ge 1$. Since the Fourier expansion of Δ begins $q + \cdots$ and that of $T_m \Delta$ begins $\tau(m)q + \cdots$, the eigenvalue is necessarily $\tau(m)$, so $T_m \Delta = \tau(m)\Delta$ and (42) gives

$$\tau(m)\tau(n) = \sum_{r|(m,n)} r^{11} \tau\left(\frac{mn}{r^2}\right) \qquad \text{for all } m, n \ge 1,$$

proving Ramanujan's multiplicativity observations mentioned in §2.4. By the same argument, if $f \in M_k(\Gamma_1)$ is any simultaneous eigenfunction of all of the T_m, with eigenvalues λ_m, then $a_m = \lambda_m a_1$ for all m. We therefore have $a_1 \ne 0$ if f is not identically 0, and if we normalize f by $a_1 = 1$ (such an f is called a *normalized Hecke eigenform*, or *Hecke form* for short) then we have

$$T_m f = a_m f, \qquad a_m a_n = \sum_{r|(m,n)} r^{k-1} a_{mn/r^2} \qquad (m, n \ge 1). \tag{43}$$

Examples of this besides $\Delta(z)$ are the unique normalized cusp forms $f(z) = \Delta(z)E_{k-12}(z)$ in the five further weights where $\dim S_k(\Gamma_1) = 1$ (viz. $k = 16$, 18, 20, 22 and 26) and the function $\mathbb{G}_k(z)$ for all $k \ge 4$, for which we have $T_m \mathbb{G}_k = \sigma_{k-1}(m)\mathbb{G}_k$, $\sigma_{k-1}(m)\sigma_{k-1}(n) = \sum_{r|(m,n)} r^{k-1}\sigma_{k-1}(mn/r^2)$. (This was the reason for the normalization of $\mathbb{G}_k$ chosen in §2.2.) In fact, a theorem of Hecke asserts that $M_k(\Gamma_1)$ has a basis of normalized simultaneous eigenforms for all k, and that this basis is unique. We omit the proof, though it is not difficult (one introduces a scalar product on the space of cusp forms of weight k, shows that the T_m are self-adjoint with respect to this scalar product, and appeals to a general result of linear algebra saying that commuting self-adjoint operators can always be simultaneously diagonalized), and content ourselves instead with one further example, also due to Hecke. Consider $k = 24$, the first weight where $\dim S_k(\Gamma_1)$ is greater than 1. Here S_k is 2-dimensional, spanned by $\Delta E_4^3 = q + 696q^2 + \cdots$ and $\Delta^2 = q^2 - 48q^3 + \ldots$. Computing the first two Fourier expansions of the images under T_2 of these two functions by (42), we find that $T_2(\Delta E_4^3) = 696\,\Delta E_4^3 + 20736000\,\Delta^2$ and $T_2(\Delta^2) = \Delta E_4^3 + 384\,\Delta^2$. The matrix $\left(\begin{smallmatrix} 696 & 20736000 \\ 1 & 384 \end{smallmatrix}\right)$ has distinct eigenvalues $\lambda_1 = 540 + 12\sqrt{144169}$ and $\lambda_2 = 540 - 12\sqrt{144169}$, so there are precisely two normalized eigenfunctions of T_2 in $S_{24}(\Gamma_1)$, namely the functions $f_1 = \Delta E_4^3 - (156 - 12\sqrt{144169})\Delta^2 = q + \lambda_1 q^2 + \cdots$ and $f_2 = \Delta E_4^3 - (156 + 12\sqrt{144169})\Delta^2 = q + \lambda_2 q^2 + \cdots$, with $T_2 f_i = \lambda_i f_i$ for $i = 1, 2$. The uniqueness of these eigenfunctions and the fact that T_m commutes with T_2 for all $m \geq 1$ then implies that $T_m f_i$ is a multiple of f_i for all $m \geq 1$, so $\mathbb{G}_{24}$, f_1 and f_2 give the desired unique basis of $M_{24}(\Gamma_1)$ consisting of normalized Hecke eigenforms.

Finally, we mention without giving any details that Hecke's theory generalizes to congruence groups of $\mathrm{SL}(2, \mathbb{Z})$ like the group $\Gamma_0(N)$ of matrices $\left(\begin{smallmatrix} a & b \\ c & d \end{smallmatrix}\right) \in \Gamma_1$ with $c \equiv 0 \pmod{N}$, the main differences being that the definition of T_m must be modified somewhat if m and N are not coprime and that the statement about the existence of a unique base of Hecke forms becomes more complicated: the space $M_k(\Gamma_0(N))$ is the direct sum of the space spanned by all functions $f(dz)$ where $f \in M_k(\Gamma_0(N'))$ for some proper divisor N' of N and d divides N/N' (the so-called "old forms") and a space of "new forms" which is again uniquely spanned by normalized eigenforms of all Hecke operators T_m with $(m, N) = 1$. The details can be found in any standard textbook.

4.2 L-series of Eigenforms

Let us return to the full modular group. We have seen that $M_k(\Gamma_1)$ contains, and is in fact spanned by, normalized Hecke eigenforms $f = \sum a_m q^m$ satisfying (43). Specializing this equation to the two cases when m and n are coprime and when $m = p^\nu$ and $n = p$ for some prime p gives the two equations (which together are equivalent to (43))

$$a_{mn} = a_m \, a_n \text{ if } (m, n) = 1, \quad a_{p^{\nu+1}} = a_p \, a_{p^\nu} - p^{k-1} a_{p^{\nu-1}} \quad (p \text{ prime}, \nu \geq 1).$$

The first says that the coefficients a_n are *multiplicative* and hence that the Dirichlet series $L(f, s) = \sum\limits_{n=1}^{\infty} \dfrac{a_n}{n^s}$, called the *Hecke L-series* of f, has an Euler product $L(f, s) = \prod_{p \text{ prime}} \left(1 + \dfrac{a_p}{p^s} + \dfrac{a_{p^2}}{p^{2s}} + \cdots\right)$, and the second tells us

that the power series $\sum_{\nu=0}^{\infty} a_{p^\nu} x^\nu$ for p prime equals $1/(1 - a_p x + p^{k-1} x^2)$. Combining these two statements gives Hecke's fundamental Euler product development

$$L(f, s) = \prod_{p \text{ prime}} \frac{1}{1 - a_p \, p^{-s} + p^{k-1-2s}} \tag{44}$$

for the L-series of a normalized Hecke eigenform $f \in M_k(\Gamma_1)$, a simple example being given by

$$L(\mathbb{G}_k, s) = \prod_p \frac{1}{1 - (p^{k-1} + 1)p^{-s} + p^{k-1-2s}} = \zeta(s)\,\zeta(s - k + 1).$$

For eigenforms on $\Gamma_0(N)$ there is a similar result except that the Euler factors for $p|N$ have to be modified suitably.

The L-series have another fundamental property, also discovered by Hecke, which is that they can be analytically continued in s and then satisfy functional equations. We again restrict to $\Gamma = \Gamma_1$ and also, for convenience, to cusp forms, though not any more just to eigenforms. (The method of proof extends to non-cusp forms but is messier there since $L(f, s)$ then has poles, and since M_k is spanned by cusp forms and by $\mathbb{G}_k$, whose L-series is completely known, there is no loss in making the latter restriction.) From the estimate $a_n = O(n^{k/2})$ proved in §2.4 we know that $L(f, s)$ converges absolutely in the half-plane $\Re(s) > 1 + k/2$. Take s in that half-plane and consider the Euler gamma function

$$\Gamma(s) = \int_0^\infty t^{s-1} e^{-t} \, dt.$$

Replacing t by λt in this integral gives $\Gamma(s) = \lambda^s \int_0^\infty t^{s-1} e^{-\lambda t} \, dt$ or $\lambda^{-s} = \Gamma(s)^{-1} \int_0^\infty t^{s-1} e^{-\lambda t} \, dt$ for any $\lambda > 0$. Applying this to $\lambda = 2\pi n$, multiplying by a_n, and summing over n, we obtain

$$(2\pi)^{-s}\,\Gamma(s)\,L(f, s) = \sum_{n=1}^{\infty} a_n \int_0^\infty t^{s-1} e^{-2\pi nt} \, dt = \int_0^\infty t^{s-1} f(it) \, dt$$

$$\left(\Re(s) > \frac{k}{2} + 1\right),$$

where the interchange of integration and summation is justified by the absolute convergence. Now the fact that $f(it)$ is exponentially small for $t \to \infty$ (because f is a cusp form) and for $t \to 0$ (because $f(-1/z) = z^k f(z)$) implies that the integral converges absolutely for all $s \in \mathbb{C}$ and hence that the function

$$L^*(f, s) := (2\pi)^{-s}\,\Gamma(s)\,L(f, s) = (2\pi)^{-s}\,\Gamma(s) \sum_{n=1}^{\infty} \frac{a_n}{n^s} \tag{45}$$

extends holomorphically from the half-plane $\Re(s) > 1 + k/2$ to the entire complex plane. The substitution $t \to 1/t$ together with the transformation equation $f(i/t) = (it)^k f(it)$ of f then gives the functional equation

$$L^*(f, k - s) = (-1)^{k/2} L^*(f, s) \tag{46}$$

of $L^*(f, s)$. We have proved:

Proposition 13. *Let* $f = \sum_{n=1}^{\infty} a_n q^n$ *be a cusp form of weight* k *on the full modular group. Then the L-series $L(f, s)$ extends to an entire function of s and satisfies the functional equation* (46), *where $L^*(f, s)$ is defined by equation* (45).

It is perhaps worth mentioning that, as Hecke also proved, the converse of Proposition 13 holds as well: if a_n $(n \geq 1)$ are complex numbers of polynomial growth and the function $L^*(f, s)$ defined by (45) continues analytically to the whole complex plane and satisfies the functional equation (46), then $f(z) = \sum_{n=1}^{\infty} a_n e^{2\pi i n z}$ is a cusp form of weight k on Γ_1.

4.3 Modular Forms and Algebraic Number Theory

In §3, we used the theta series $\theta(z)^2$ to determine the number of representations of any integer n as a sum of two squares. More generally, we can study the number $r(Q, n)$ of representations of n by a positive definite binary quadratic $Q(x, y) = ax^2 + bxy + cy^2$ with integer coefficients by considering the weight 1 theta series $\Theta_Q(z) = \sum_{x,y \in \mathbb{Z}} q^{Q(x,y)} = \sum_{n=0}^{\infty} r(Q, n) q^n$. This theta series depends only on the class $[Q]$ of Q up to equivalences $Q \sim Q \circ \gamma$ with $\gamma \in \Gamma_1$. We showed in §1.2 that for any $D < 0$ the number $h(D)$ of Γ_1-equivalence classes $[Q]$ of binary quadratic forms of discriminant $b^2 - 4ac = D$ is finite. If D is a fundamental discriminant (i.e., not representable as $D' r^2$ with D' congruent to 0 or 1 mod 4 and $r > 1$), then $h(D)$ equals the class number of the imaginary quadratic field $K = \mathbb{Q}(\sqrt{D})$ of discriminant D and there is a well-known bijection between the Γ_1-equivalence classes of binary quadratic forms of discriminant D and the ideal classes of K such that $r(Q, n)$ for any form Q equals w times the number $r(\mathcal{A}, n)$ of integral ideals $\mathfrak{a}$ of K of norm n belonging to the corresponding ideal class $\mathcal{A}$, where w is the number of roots of unity in K ($= 6$ or 4 if $D = -3$ or $D = -4$ and 2 otherwise). The L-series $L(\Theta_Q, s)$ of Θ_Q is therefore w times the "partial zeta-function" $\zeta_{K, \mathcal{A}}(s) = \sum_{\mathfrak{a} \in \mathcal{A}} N(\mathfrak{a})^{-s}$. The ideal classes of K form an abelian group. If χ is a homomorphism from this group to $\mathbb{C}^*$, then the L-series $L_K(s, \chi) = \sum_{\mathfrak{a}} \chi(\mathfrak{a}) / N(\mathfrak{a})^s$ (sum over all integral ideals of K) can be written as $\sum_{\mathcal{A}} \chi(\mathcal{A}) \zeta_{K, \mathcal{A}}(s)$ (sum over all ideal classes of K) and hence is the L-series of the weight 1 modular form $f_\chi(z) = w^{-1} \sum_{\mathcal{A}} \chi(\mathcal{A}) \Theta_{\mathcal{A}}(z)$. On the other hand, from the unique prime decomposition of ideals in K it follows that $L_K(s, \chi)$ has an Euler product. Hence f_χ is a Hecke eigenform. If $\chi = \chi_0$ is the trivial character, then $L_K(s, \chi) = \zeta_K(s)$, the Dedekind zeta function of K, which factors as $\zeta(s)L(s,\varepsilon_D)$, the product of the Riemann zeta-function and the Dirichlet L-series of the character $\varepsilon_D(n) = \left(\dfrac{D}{n}\right)$ (Kronecker symbol). Therefore in this case we get $\sum_{[Q]} r(Q,n) = w \sum_{d|n} \varepsilon_D(d)$ (an identity known to Gauss) and correspondingly

$$f_{\chi_0}(z) = \frac{1}{w} \sum_{[Q]} \Theta_Q(z) = \frac{h(D)}{2} + \sum_{n=1}^{\infty} \left(\sum_{d|n} \left(\frac{D}{d}\right)\right) q^n,$$

an Eisenstein series of weight 1. If the character χ has order 2, then it is a so-called genus character and one knows that $L_K(s,\chi)$ factors as $L(s,\varepsilon_{D_1})L(s,\varepsilon_{D_2})$ where D_1 and D_2 are two other discriminants with product D. In this case, too, $f_\chi(z)$ is an Eisenstein series. But in all other cases, f_χ is a cusp form and the theory of modular forms gives us non-trivial information about representations of numbers by quadratic forms.

♠ Binary Quadratic Forms of Discriminant −23

We discuss an explicit example, taken from a short and pretty article written by van der Blij in 1952. The class number of the discriminant $D = -23$ is 3, with the $SL(2,\mathbb{Z})$-equivalence classes of binary quadratic forms of this discriminant being represented by the three forms

$$\begin{aligned} Q_0(x,y) &= x^2 + xy + 6y^2, \\ Q_1(x,y) &= 2x^2 + xy + 3y^2, \\ Q_2(x,y) &= 2x^2 - xy + 3y^2. \end{aligned}$$

Since Q_1 and Q_2 represent the same integers, we get only two distinct theta series

$$\begin{aligned} \Theta_{Q_0}(z) &= 1 + 2q + 2q^4 + 4q^6 + 4q^8 + \cdots, \\ \Theta_{Q_1}(z) &= 1 + 2q^2 + 2q^3 + 2q^4 + 2q^6 + \cdots. \end{aligned}$$

The linear combination corresponding to the trivial character is the Eisenstein series

$$\begin{aligned} f_{\chi_0} &= \frac{1}{2}\left(\Theta_{Q_0} + 2\Theta_{Q_1}\right) = \frac{3}{2} + \sum_{n=1}^{\infty}\left(\sum_{d|n}\left(\frac{-23}{d}\right)\right)q^n \\ &= \frac{3}{2} + q + 2q^2 + 2q^3 + 3q^4 + \cdots, \end{aligned}$$

in accordance with the general identity $w^{-1}\sum_{[Q]} r(Q,n) = \sum_{d|n}\varepsilon_D(d)$ mentioned above. If χ is one of the two non-trivial characters, with values $e^{\pm 2\pi i/3} = \frac{1}{2}(-1 \pm i\sqrt{3})$ on Q_1 and Q_2, we have

$$f_\chi = \frac{1}{2}(\Theta_{Q_0} - \Theta_{Q_1}) = q - q^2 - q^3 + q^6 + \cdots.$$

This is a Hecke eigenform in the space $S_1(\Gamma_0(23), \varepsilon_{-23})$. Its L-series has the form

$$L(f_\chi, s) = \prod_p \frac{1}{1 - a_p\, p^{-s} + \varepsilon_{-23}(p)\, p^{-2s}}$$

where $\varepsilon_{-23}(p)$ equals the Legendre symbol $(p/23)$ by quadratic reciprocity and

$$a_p = \begin{cases} 1 & \text{if } p = 23, \\ 0 & \text{if } (p/23) = -1, \\ 2 & \text{if } (p/23) = 1 \text{ and } p \text{ is representable as } x^2 + xy + 6y^2, \\ -1 & \text{if } (p/23) = 1 \text{ and } p \text{ is representable as } 2x^2 + xy + 3y^2. \end{cases} \tag{47}$$

On the other hand, the space $S_1(\Gamma_0(23), \varepsilon_{-23})$ is one-dimensional, spanned by the function

$$\eta(z)\,\eta(23z) = q \prod_{n=1}^{\infty} (1 - q^n)(1 - q^{23n}). \tag{48}$$

We therefore obtain the explicit "reciprocity law"

Proposition 14 (van der Blij). *Let p be a prime. Then the number a_p defined in (47) is equal to the coefficient of q^p in the product (48).*

As an application of this, we observe that the q-expansion on the right-hand side of (48) is congruent modulo 23 to $\Delta(z) = q \prod (1 - q^n)^{24}$ and hence that $\tau(p)$ is congruent modulo 23 to the number a_p defined in (47) for every prime number p, a congruence for the Ramanujan function $\tau(n)$ of a somewhat different type than those already given in (27) and (28).   ♡

Proposition 14 gives a concrete example showing how the coefficients of a modular form – here $\eta(z)\eta(23z)$ – can answer a question of number theory – here, the question whether a given prime number which splits in $\mathbb{Q}(\sqrt{-23})$ splits into principal or non-principal ideals. But actually the connection goes much deeper. By elementary algebraic number theory we have that the L-series $L(s) = L_K(s, \chi)$ is the quotient $\zeta_F(s)/\zeta(s)$ of the Dedekind function of F by the Riemann zeta function, where $F = \mathbb{Q}(\alpha)$ ($\alpha^3 - \alpha - 1 = 0$) is the cubic field of discriminant -23. (The composite $K \cdot F$ is the Hilbert class field of K.) Hence the four cases in (47) also describe the splitting of p in F: 23 is ramified, quadratic non-residues of 23 split as $p = \mathfrak{p}_1\mathfrak{p}_2$ with $N(\mathfrak{p}_i) = p^i$, and quadratic residues of 23 are either split completely (as products of three prime ideals of norm p) or are inert (remain prime) in F, according whether they are represented by Q_0 or Q_1. Thus the modular form $\eta(z)\eta(23z)$ describes not only the algebraic number theory of the quadratic field K, but also the splitting of primes in the higher degree field F. This is the first non-trivial example of the connection found by Weil–Langlands and Deligne–Serre which relates modular forms of weight one to the arithmetic of number fields whose Galois groups admit non-trivial two-dimensional representations.

4.4 Modular Forms Associated to Elliptic Curves and Other Varieties

If X is a smooth projective algebraic variety defined over $\mathbb{Q}$, then for almost all primes p the equations defining X can be reduced modulo p to define a smooth variety X_p over the field $\mathbb{Z}/p\mathbb{Z} = \mathbb{F}_p$. We can then count the number of points in X_p over the finite field $\mathbb{F}_{p^r}$ for all $r \geq 1$ and, putting all this information together, define a "local zeta function" $Z(X_p, s) = \exp\left(\sum_{r=1}^{\infty} |X_p(\mathbb{F}_{p^r})| \, p^{-rs}/r\right)$ ($\Re(s) \gg 0$) and a "global zeta function" (Hasse–Weil zeta function) $Z(X/\mathbb{Q}, s) = \prod_p Z(X_p, s)$, where the product is over all primes. (The factors $Z_p(X, s)$ for the "bad" primes p, where the equations defining X yield a singular variety over $\overline{\mathbb{F}_p}$, are defined in a more complicated but completely explicit way and are again power series in p^{-s}.) Thanks to the work of Weil, Grothendieck, Dwork, Deligne and others, a great deal is known about the local zeta functions – in particular, that they are rational functions of p^{-s} and have all of their zeros and poles on the vertical lines $\Re(s) = 0, \frac{1}{2}, \ldots, n - \frac{1}{2}, n$ where n is the dimension of X – but the global zeta function remains mysterious. In particular, the general conjecture that $Z(X/\mathbb{Q}, s)$ can be meromorphically continued to all s is known only for very special classes of varieties.

In the case where $X = E$ is an elliptic curve, given, say, by a Weierstrass equation

$$y^2 = x^3 + Ax + B \qquad (A, B \in \mathbb{Z}, \quad \Delta := -4A^3 - 27B^2 \neq 0), \qquad (49)$$

the local factors can be made completely explicit and we find that $Z(E/\mathbb{Q}, s) = \dfrac{\zeta(s)\zeta(s-1)}{L(E/\mathbb{Q}, s)}$ where the L-series $L(E/\mathbb{Q}, s)$ is given for $\Re(s) \gg 0$ by an Euler product of the form

$$L(E/\mathbb{Q}, s) = \prod_{p \nmid \Delta} \frac{1}{1 - a_p(E)\, p^{-s} + p^{1-2s}} \cdot \prod_{p \mid \Delta} \frac{1}{(\text{polynomial of degree} \leq 2 \text{ in } p^{-s})} \qquad (50)$$

with $a_p(E)$ defined for $p \nmid \Delta$ as $p - \left|\left\{(x,y) \in (\mathbb{Z}/p\mathbb{Z})^2 \mid y^2 = x^3 + Ax + B\right\}\right|$. In the mid-1950's, Taniyama noticed the striking formal similarity between this Euler product expansion and the one in (44) when $k = 2$ and asked whether there might be cases of overlap between the two, i.e., cases where the L-function of an elliptic curve agrees with that of a Hecke eigenform of weight 2 having eigenvalues $a_p \in \mathbb{Z}$ (a necessary condition if they are to agree with the integers $a_p(E)$).

Numerical examples show that this at least sometimes happens. The simplest elliptic curve (if we order elliptic curves by their "conductor," an invariant in $\mathbb{N}$ which is divisible only by primes dividing the discriminant Δ in (49)) is the curve $Y^2 - Y = X^3 - X^2$ of conductor 11. (This can be put into the form (49) by setting $y = 216Y - 108$, $x = 36X - 12$, giving $A = -432$,

$B = 8208$, $\Delta = -2^8 \cdot 3^{12} \cdot 11$, but the equation in X and Y, the so-called "minimal model," has much smaller coefficients.) We can compute the numbers a_p by counting solutions of $Y^2 - Y = X^3 - X^2$ in $(\mathbb{Z}/p\mathbb{Z})^2$. (For $p > 3$ this is equivalent to the recipe given above because the equations relating (x, y) and (X, Y) are invertible in characteristic p, and for $p = 2$ or 3 the minimal model gives the correct answer.) For example, we have $a_5 = 5 - 4 = 1$ because the equation $Y^2 - Y = X^3 - X^2$ has the 4 solutions (0,0), (0,1), (1,0) and (1,1) in $(\mathbb{Z}/5\mathbb{Z})^2$. Then we have

$$
L(E/\mathbb{Q}, s) = \left(1 + \frac{2}{2^s} + \frac{2}{2^{2s}}\right)^{-1} \left(1 + \frac{1}{3^s} + \frac{3}{3^{2s}}\right)^{-1} \left(1 - \frac{1}{5^s} + \frac{5}{5^{2s}}\right)^{-1} \cdots
$$

$$
= \frac{1}{1^s} - \frac{2}{2^s} - \frac{1}{3^s} + \frac{2}{4^s} + \frac{1}{5^s} + \cdots = L(f, s),
$$

where $f \in S_2(\Gamma_0(11))$ is the modular form

$$
f(z) = \eta(z)^2 \eta(11z)^2 = q \prod_{n=1}^{\infty} \left(1 - q^n\right)^2 \left(1 - q^{11n}\right)^2 = x - 2q^2 - q^3 + 2q^4 + q^5 + \cdots .
$$

In the 1960's, one direction of the connection suggested by Taniyama was proved by M. Eichler and G. Shimura, whose work establishes the following theorem.

Theorem (Eichler–Shimura). *Let $f(z)$ be a Hecke eigenform in $S_2(\Gamma_0(N))$ for some $N \in \mathbb{N}$ with integral Fourier coefficients. Then there exists an elliptic curve $E/\mathbb{Q}$ such that $L(E/\mathbb{Q}, s) = L(f, s)$.*

Explicitly, this means that $a_p(E) = a_p(f)$ for all primes p, where $a_n(f)$ is the coefficient of q^n in the Fourier expansion of f. The proof of the theorem is in a sense quite explicit. The quotient of the upper half-plane by $\Gamma_0(N)$, compactified appropriately by adding a finite number of points (cusps), is a complex curve (Riemann surface), traditionally denoted by $X_0(N)$, such that the space of holomorphic 1-forms on $X_0(N)$ can be identified canonically (via $f(z) \mapsto f(z)dz$) with the space of cusp forms $S_2(\Gamma_0(N))$. One can also associate to $X_0(N)$ an abelian variety, called its *Jacobian*, whose tangent space at any point can be identified canonically with $S_2(\Gamma_0(N))$. The Hecke operators T_p introduced in 4.1 act not only on $S_2(\Gamma_0(N))$, but on the Jacobian variety itself, and if the Fourier coefficients $a_p = a_p(f)$ are in $\mathbb{Z}$, then so do the differences $T_p - a_p \cdot \mathrm{Id}$. The subvariety of the Jacobian annihilated by all of these differences (i.e., the set of points x in the Jacobian whose image under T_p equals a_p times x; this makes sense because an abelian variety has the structure of a group, so that we can multiply points by integers) is then precisely the sought-for elliptic curve E. Moreover, this construction shows that we have an even more intimate relationship between the curve E and the form f than the L-series equality $L(E/\mathbb{Q}, s) = L(f, s)$, namely, that there is an actual map from the modular curve $X_0(N)$ to the elliptic curve E which is induced by f. Specifically, if we define $\phi(z) = \sum_{n=1}^{\infty} \frac{a_n(f)}{n} e^{2\pi i n z}$, so that $\phi'(z) = 2\pi i f(z) dz$, then the fact that f is modular of weight 2 implies that the difference $\phi(\gamma(z)) - \phi(z)$ has zero derivative and hence is constant for all $\gamma \in \Gamma_0(N)$, say $\phi(\gamma(z)) - \phi(z) = C(\gamma)$. It is then easy to see that the map $C : \Gamma_0(N) \to \mathbb{C}$ is a homomorphism, and in our case (f an eigenform, eigenvalues in $\mathbb{Z}$), it turns out that its image is a lattice $\Lambda \subset \mathbb{C}$, and the quotient map $\mathbb{C}/\Lambda$ is isomorphic to the elliptic curve E. The fact that $\phi(\gamma(z)) - \phi(z) \in \Lambda$ then implies that the composite map $\mathfrak{H} \xrightarrow{\phi} \mathbb{C} \xrightarrow{\mathrm{pr}} \mathbb{C}/\Lambda$ factors through the projection $\mathfrak{H} \to \Gamma_0(N)\backslash\mathfrak{H}$, i.e., ϕ induces a map (over $\mathbb{C}$) from $X_0(N)$ to E. This map is in fact defined over $\mathbb{Q}$, i.e., there are modular functions $X(z)$ and $Y(z)$ with rational Fourier coefficients which are invariant under $\Gamma_0(N)$ and which identically satisfy the equation $Y(z)^2 = X(z)^3 + AX(z) + B$ (so that the map from $X_0(N)$ to E in its Weierstrass form is simply $z \mapsto (X(z), Y(z))$) as well as the equation $X'(z)/2Y(z) = 2\pi i f(z)$. (Here we are simplifying a little.)

Gradually the idea arose that perhaps the answer to Taniyama's original question might be yes in *all* cases, not just sometimes. The results of Eichler and Shimura showed this in one direction, and strong evidence in the other direction was provided by a theorem proved by A. Weil in 1967 which said that if the L-series of an elliptic curve $E/\mathbb{Q}$ and certain "twists" of it satisfied the conjectured analytic properties (holomorphic continuation and functional equation), then E really did correspond to a modular form in the above way. The conjecture that every E over $\mathbb{Q}$ is modular became famous (and was called according to taste by various subsets of the names Taniyama, Weil and Shimura, although none of these three people had ever stated the conjecture explicitly in print). It was finally proved at the end of the 1990's by Andrew Wiles and his collaborators and followers:

Theorem (Wiles–Taylor, Breuil–Conrad–Diamond–Taylor). *Every elliptic curve over $\mathbb{Q}$ can be parametrized by modular functions.*

The proof, which is extremely difficult and builds on almost the entire apparatus built up during the previous decades in algebraic geometry, representation theory and the theory of automorphic forms, is one of the pinnacles of mathematical achievement in the 20th century.

♠ Fermat's Last Theorem

In the 1970's, Y. Hellegouarch was led to consider the elliptic curve (49) in the special case when the roots of the cubic polynomial on the right were nth powers of rational integers for some prime number $n > 2$, i.e., if this cubic factors as $(x - a^n)(x - b^n)(x - c^n)$ where a, b, c satisfy the Fermat equation $a^n + b^n + c^n = 0$. A decade later, G. Frey studied the same elliptic curve and discovered that the associated Galois representation (we do not explain this here) had properties which contradicted the properties which Galois representations of elliptic curves were expected to satisfy. Precise conjectures about the modularity of certain Galois representations were then made by Serre which would fail for the representations attached to the Hellegouarch-Frey curve, so that the correctness of these conjectures would imply the insolubility of Fermat's equation. (Very roughly, the conjectures imply that, if the Galois representation associated to the above curve E is modular at all, then the corresponding cusp form would have to be congruent modulo n to a cusp form of weight 2 and level 1 or 2, and there aren't any.) In 1990, K. Ribet proved a special case of Serre's conjectures (the general case is now also known, thanks to recent work of Khare, Wintenberger, Dieulefait and Kisin) which was sufficient to yield the same implication. The proof by Wiles and Taylor of the Taniyama-Weil conjecture (still with some minor restrictions on E which were later lifted by the other authors cited above, but in sufficient generality to make Ribet's result applicable) thus sufficed to give the proof of the following theorem, first claimed by Fermat in 1637:

Theorem (Ribet, Wiles–Taylor). *If $n > 2$, there are no positive integers with $a^n + b^n = c^n$.*   ♡

Finally, we should mention that the connection between modularity and algebraic geometry does not apply only to elliptic curves. Without going into detail, we mention only that the Hasse–Weil zeta function $Z(X/\mathbb{Q}, s)$ of an arbitrary smooth projective variety X over $\mathbb{Q}$ splits into factors corresponding to the various cohomology groups of X, and that if any of these cohomology groups (or any piece of them under some canonical decomposition, say with respect to the action of a finite group of automorphisms of X) is two-dimensional, then the corresponding piece of the zeta function is conjectured to be the L-series of a Hecke eigenform of weight $i + 1$, where the cohomology group in question is in degree i. This of course includes the case when $X = E$ and $i = 1$, since the first cohomology group of a curve of genus 1 is 2-dimensional, but it also applies to many higher-dimensional varieties. Many examples are now known, an early one, due to R. Livné, being given by the cubic hypersurface $x_1^3 + \cdots + x_{10}^3 = 0$ in the projective space $\{x \in \mathbb{P}^9 \mid x_1 + \cdots + x_{10} = 0\}$, whose zeta-function equals $\prod_{j=0}^7 \zeta(s - i)^{m_i} \cdot L(s - 2, f)^{-1}$ where $(m_0, \ldots m_7) = (1, 1, 1, -83, 43, 1, 1, 1)$ and $f = q + 2q^2 - 8q^3 + 4q^4 + 5q^5 + \cdots$ is the unique new form of weight 4 on $\Gamma_0(10)$. Other examples arise from so-called "rigid Calabi-Yau 3-folds," which have been studied intensively in recent years in connection with the phenomenon, first discovered by mathematical physicists, called "mirror symmetry." We skip all further discussion, referring to the survey paper and book cited in the references at the end of these notes.

5 Modular Forms and Differential Operators

The starting point for this section is the observation that the derivative of
a modular form is not modular, but nearly is. Specifically, if f is a modular
form of weight k with the Fourier expansion (3), then by differentiating (2)
we see that the derivative

$$Df = f' := \frac{1}{2\pi i}\frac{df}{dz} = q\frac{df}{dq} = \sum_{n=1}^{\infty} n\, a_n\, q^n \qquad (51)$$

(where the factor $2\pi i$ has been included in order to preserve the rationality
properties of the Fourier coefficients) satisfies

$$f'\left(\frac{az+b}{cz+d}\right) = (cz+d)^{k+2} f'(z) + \frac{k}{2\pi i}\, c\,(cz+d)^{k+1} f(z). \qquad (52)$$

If we had only the first term, then f' would be a modular form of weight $k+2$.
The presence of the second term, far from being a problem, makes the theory
much richer. To deal with it, we will:

- modify the differentiation operator so that it preserves modularity;
- make combinations of derivatives of modular forms which are again modular;
- relax the notion of modularity to include functions satisfying equations
  like (52);
- differentiate with respect to $t(z)$ rather than z itself, where $t(z)$ is a modular function.

These four approaches will be discussed in the four subsections 5.1–5.4, respectively.

5.1 Derivatives of Modular Forms

As already stated, the first approach is to introduce modifications of the operator D which do preserve modularity. There are two ways to do this, one
holomorphic and one not. We begin with the holomorphic one. Comparing
the transformation equation (52) with equations (19) and (17), we find that
for any modular form $f \in M_k(\Gamma_1)$ the function

$$\vartheta_k f := f' - \frac{k}{12}\, E_2\, f\,, \qquad (53)$$

sometimes called the *Serre derivative*, belongs to $M_{k+2}(\Gamma_1)$. (We will often
drop the subscript k, since it must always be the weight of the form to which
the operator is applied.) A first consequence of this basic fact is the following.
We introduce the ring $\widetilde{M}_*(\Gamma_1) := M_*(\Gamma_1)[E_2] = \mathbb{C}[E_2, E_4, E_6]$, called the *ring
of quasimodular forms on* $SL(2, \mathbb{Z})$. (An intrinsic definition of the elements of
this ring, and a definition for other groups $\Gamma \subset G$, will be given in the next
subsection.) Then we have:

Proposition 15. *The ring* $\widetilde{M}_*(\Gamma_1)$ *is closed under differentiation. Specifically, we have*

$$E_2' = \frac{E_2^2 - E_4}{12}, \qquad E_4' = \frac{E_2 E_4 - E_6}{3}, \qquad E_6' = \frac{E_2 E_6 - E_4^2}{2}. \qquad (54)$$

Proof. Clearly ϑE_4 and ϑE_6, being holomorphic modular forms of weight 6 and 8 on Γ_1, respectively, must be proportional to E_6 and E_4^2, and by looking at the first terms in their Fourier expansion we find that the factors are $-1/3$ and $-1/2$. Similarly, by differentiating (19) we find the analogue of (53) for E_2, namely that the function $E_2' - \frac{1}{12} E_2^2$ belongs to $M_4(\Gamma)$. It must therefore be a multiple of E_4, and by looking at the first term in the Fourier expansion one sees that the factor is $-1/12$. $\qquad\square$

Proposition 15, first discovered by Ramanujan, has many applications. We describe two of them here. Another, in transcendence theory, will be mentioned in Section 6.

♠ Modular Forms Satisfy Non-Linear Differential Equations

An immediate consequence of Proposition 15 is the following:

Proposition 16. *Any modular form or quasi-modular form on* Γ_1 *satisfies a non-linear third order differential equation with constant coefficients.*

Proof. Since the ring $\widetilde{M}_*(\Gamma_1)$ has transcendence degree 3 and is closed under differentiation, the four functions f, f', f'' and f''' are algebraically dependent for any $f \in \widetilde{M}_*(\Gamma_1)$. $\qquad\square$

As an example, by applying (54) several times we find that the function E_2 satisfies the non-linear differential equation $f''' - ff'' + \frac{3}{2} f'^2 = 0$. This is called the *Chazy equation* and plays a role in the theory of Painlevé equations. We can now use modular/quasimodular ideas to describe a full set of solutions of this equation. First, define a "modified slash operator" $f \mapsto f\|_2 g$ by $(f\|_2 g)(z) = (cz+d)^{-2} f\left(\frac{az+b}{cz+d}\right) + \frac{\pi}{12} \frac{c}{cz+d}$ for $g = \left(\begin{smallmatrix} a & b \\ c & d \end{smallmatrix}\right)$. This is not linear in f (it is only affine), but it is nevertheless a group operation, as one checks easily, and, at least locally, it makes sense for any matrix $\left(\begin{smallmatrix} a & b \\ c & d \end{smallmatrix}\right) \in \mathrm{SL}(2, \mathbb{C})$. Now one checks by a direct, though tedious, computation that $\mathsf{Ch}[f\|_2 g] = \mathsf{Ch}[f]\|_8 g$ (where defined) for any $g \in \mathrm{SL}(2, \mathbb{C})$, where $\mathsf{Ch}[f] = f''' - ff'' + \frac{3}{2} f'^2$. (Again this is surprising, because the operator Ch is not linear.) Since E_2 is a solution of $\mathsf{Ch}[f] = 0$, it follows that $E_2\|_2 g$ is a solution of the same equation (in $g^{-1}\mathfrak{H} \subset \mathbb{P}^1(\mathbb{C})$) for every $g \in \mathrm{SL}(2, \mathbb{C})$, and since $E_2\|_2 \gamma = E_2$ for $\gamma \in \Gamma_1$ and $\|_2$ is a group operation, it follows that this function depends only on the class of g in $\Gamma_1\backslash\mathrm{SL}(2, \mathbb{C})$. But $\Gamma_1\backslash\mathrm{SL}(2, \mathbb{C})$ is 3-dimensional and a third-order differential equation generically has a 3-dimensional solution space (the values of f, f' and f'' at a point determine all higher derivatives recursively and hence fix the function uniquely in a neighborhood of any point where it is holomorphic), so that, at least generically, this describes all solutions of the non-linear differential equation $\mathsf{Ch}[f] = 0$ in modular terms.   ♡

Our second "application" of the ring $\widetilde{M}_*(\Gamma_1)$ describes an unexpected appearance of this ring in an elementary and apparently unrelated context.

♠ Moments of Periodic Functions

This very pretty application of the modular forms E_2, E_4, E_6 is due to P. Gallagher. Denote by $\mathfrak{P}$ the space of periodic real-valued functions on $\mathbb{R}$, i.e., functions $f : \mathbb{R} \to \mathbb{R}$ satisfying $f(x + 2\pi) = f(x)$. For each ∞-tuple $\mathbf{n} = (n_0, n_1, \dots)$ of non-negative integers (all but finitely many equal to 0) we define a "coordinate" $I_{\mathbf{n}}$ on the infinite-dimensional space $\mathfrak{P}$ by $I_{\mathbf{n}}[f] = \int_0^{2\pi} f(x)^{n_0} f'(x)^{n_1} \cdots dx$ (higher moments). Apart from the relations among these coming from integration by parts, like $\int f''(x)dx = 0$ or $\int f'(x)^2 dx = -\int f(x)f''(x)dx$, we also have various inequalities. The general problem, certainly too hard to be solved completely, would be to describe all equalities and inequalities among the $I_{\mathbf{n}}[f]$. As a special case we can ask for the complete list of inequalities satisfied by the four moments $(A[f], B[f], C[f], D[f]) :=$ $\left(\int_0^{2\pi} f, \int_0^{2\pi} f^2, \int_0^{2\pi} f^3, \int_0^{2\pi} f'^2 \right)$ as f ranges over $\mathfrak{P}$. Surprisingly enough, the answer involves quasimodular forms on $\mathrm{SL}(2, \mathbb{Z})$. First, by making a linear shift $f \mapsto \lambda f + \mu$ with $\lambda, \mu \in \mathbb{R}$ we suppose that $A[f] = 0$ and $D[f] = 1$. The problem is then to describe the subset $\mathfrak{X} \subset \mathbb{R}^2$ of pairs $(B[f], C[f]) = \left(\int_0^{2\pi} f(x)^2 dx, \int_0^{2\pi} f(x)^3 dx \right)$ where f ranges over functions in $\mathfrak{P}$ satisfying $\int_0^{2\pi} f(x)dx = 0$ and $\int_0^{2\pi} f'(x)^2 \, dx = 1$.

Theorem (Gallagher). *We have* $\mathfrak{X} = \left\{ (B, C) \in \mathbb{R}^2 \mid 0 < B \leq 1, \ C^2 \leq \Phi(B) \right\}$ *where the function* $\Phi : (0, 1] \to \mathbb{R}_{\geq 0}$ *is given parametrically by*

$$\Phi\left(\frac{\mathbb{G}_2'(it)}{\mathbb{G}_4'(it)} \right) = \frac{(\mathbb{G}_4'(it) - \mathbb{G}_2''(it))^2}{2 \, \mathbb{G}_4'(it)^3} \qquad (0 < t \leq \infty).$$

The idea of the proof is as follows. First, from $A = 0$ and $D = 1$ we deduce $0 < B[f] \leq 1$ by an inequality of Wirtinger (just look at the Fourier expansion of f). Now let $f \in \mathfrak{P}$ be a function – but one must prove that it exists! – which maximizes $C = C[f]$ for given values of A, B and D. By a standard calculus-of-variations-type argument (replace f by $f + \varepsilon g$ where $g \in \mathfrak{P}$ is orthogonal to 1, f and f'', so that A, B and D do not change to first order, and then use that C also cannot change to first order since otherwise its value could not be extremal, so that g must also be orthogonal to f^2), we show that the four functions 1, f, f^2 and f'' are linearly dependent. From this it follows by integrating once that the five functions 1, f, f^2, f^3 and f'^2 are also linearly dependent. After a rescaling $f \mapsto \lambda f + \mu$, we can write this dependency as $f'(x)^2 = 4f(x)^3 - g_2 f(x) - g_3$ for some constants g_2 and g_3. But this is the famous differential equation of the Weierstrass $\wp$-function, so $f(2\pi x)$ is the restriction to $\mathbb{R}$ of the function $\wp(x, \mathbb{Z}\tau + \mathbb{Z})$ for some $\tau \in \mathfrak{H}$, necessarily of the form $\tau = it$ with $t > 0$ because everything is real. Now the coefficients g_2 and g_3, by the classical Weierstrass theory, are simple multiples of $\mathbb{G}_4(it)$ and $\mathbb{G}_6(it)$, and for this function f the value of $A(f) = \int_0^{2\pi} f(x)dx$ is known to be a multiple of $\mathbb{G}_2(it)$. Working out all the details, and then rescaling f again to get $A = 0$ and $D = 1$, one finds the result stated in the theorem. $\heartsuit$

We now turn to the second modification of the differentiation operator which preserves modularity, this time, however, at the expense of sacrificing holomorphy. For $f \in M_k(\Gamma)$ (we now no longer require that Γ be the full modular group Γ_1) we define

$$\partial_k f(z) = f'(z) - \frac{k}{4\pi y} f(z), \tag{55}$$

where y denotes the imaginary part of z. Clearly this is no longer holomorphic, but from the calculation

$$\frac{1}{\Im(\gamma z)} = \frac{|cz + d|^2}{y} = \frac{(cz + d)^2}{y} - 2ic(cz + d) \qquad \left(\gamma = \begin{pmatrix} a & b \\ c & d \end{pmatrix} \in \mathrm{SL}(2, \mathbb{R}) \right)$$

and (52) one easily sees that it transforms like a modular form of weight $k + 2$, i.e., that $(\partial_k f)|_{k+2}\gamma = \partial_k f$ for all $\gamma \in \Gamma$. Moreover, this remains true even if f is modular but not holomorphic, if we interpret f' as $\frac{1}{2\pi i} \frac{\partial f}{\partial z}$. This means that we can apply $\partial = \partial_k$ repeatedly to get non-holomorphic modular forms $\partial^n f$ of weight $k + 2n$ for all $n \geq 0$. (Here, as with ϑ_k, we can drop the subscript k because ∂_k will only be applied to forms of weight k; this is convenient because we can then write $\partial^n f$ instead of the more correct $\partial_{k+2n-2} \cdots \partial_{k+2}\partial_k f$.) For example, for $f \in M_k(\Gamma)$ we find

$$\begin{aligned} \partial^2 f &= \left(\frac{1}{2\pi i} \frac{\partial}{\partial z} - \frac{k+2}{4\pi y} \right) \left(f' - \frac{k}{4\pi y} f \right) \\ &= f'' - \frac{k}{4\pi y} f' - \frac{k}{16\pi^2 y^2} f - \frac{k+2}{4\pi y} f' + \frac{k(k+2)}{16\pi^2 y^2} f \\ &= f'' - \frac{k+1}{2\pi y} f' + \frac{k(k+1)}{16\pi^2 y^2} f \end{aligned}$$

and more generally, as one sees by an easy induction,

$$\partial^n f = \sum_{r=0}^{n} (-1)^{n-r} \binom{n}{r} \frac{(k+r)_{n-r}}{(4\pi y)^{n-r}} D^r f, \tag{56}$$

where $(a)_m = a(a+1)\cdots(a+m-1)$ is the Pochhammer symbol. The inversion of (56) is

$$D^n f = \sum_{r=0}^{n} \binom{n}{r} \frac{(k+r)_{n-r}}{(4\pi y)^{n-r}} \partial^r f, \tag{57}$$

and describes the decomposition of the holomorphic but non-modular form $f^{(n)} = D^n f$ into non-holomorphic but modular pieces: the function $y^{r-n}\partial^r f$ is multiplied by $(cz + d)^{k+n+r}(c\bar{z} + d)^{n-r}$ when z is replaced by $\frac{az+b}{cz+d}$ with $\left(\begin{smallmatrix} a & b \\ c & d \end{smallmatrix}\right) \in \Gamma$.

Formula (56) has a consequence which will be important in §6. The usual way to write down modular forms is via their Fourier expansions, i.e., as power series in the quantity $q = e^{2\pi i z}$ which is a local coordinate at infinity for the modular curve $\Gamma \backslash \mathfrak{H}$. But since modular forms are holomorphic functions in the upper half-plane, they also have Taylor series expansions in the neighborhood of any point $z = x + iy \in \mathfrak{H}$. The "straight" Taylor series expansion, giving $f(z + w)$ as a power series in w, converges only in the disk $|w| < y$ centered at z and tangent to the real line, which is unnatural since the domain of holomorphy of f is the whole upper half-plane, not just this disk. Instead, we should remember that we can map $\mathfrak{H}$ isomorphically to the unit disk, with z mapping to 0, by sending $z' \in \mathfrak{H}$ to $w = \frac{z'-z}{z'-\bar{z}}$. The inverse of this map is given by $z' = \frac{z-\bar{z}w}{1-w}$, and then if f is a modular form of weight k we should also include the automorphy factor $(1 - w)^{-k}$ corresponding to this fractional linear transformation (even though it belongs to $\mathrm{PSL}(2, \mathbb{C})$ and not Γ). The most natural way to study f near z is therefore to expand $(1 - w)^{-k} f\left(\frac{z-\bar{z}w}{1-w}\right)$ in powers of w. The following proposition describes the coefficients of this expansion in terms of the operator (55).

Proposition 17. *Let f be a modular form of weight k and $z = x + iy$ a point of $\mathfrak{H}$. Then*

$$\left(1 - w\right)^{-k} f\left(\frac{z - \bar{z}w}{1 - w}\right) = \sum_{n=0}^{\infty} \partial^n f(z) \frac{(4\pi y w)^n}{n!} \qquad (|w| < 1). \tag{58}$$

Proof. From the usual Taylor expansion, we find

$$\left(1 - w\right)^{-k} f\left(\frac{z - \bar{z}w}{1 - w}\right) = \left(1 - w\right)^{-k} f\left(z + \frac{2iyw}{1 - w}\right)$$

$$= \left(1 - w\right)^{-k} \sum_{r=0}^{\infty} \frac{D^r f(z)}{r!} \left(\frac{-4\pi y w}{1 - w}\right)^r,$$

and now expanding $(1 - w)^{-k-r}$ by the binomial theorem and using (56) we obtain (58).

Proposition 17 is useful because, as we will see in §6, the expansion (58), after some renormalizing, often has algebraic coefficients that contain interesting arithmetic information.

5.2 Rankin–Cohen Brackets and Cohen–Kuznetsov Series

Let us return to equation (52) describing the near-modularity of the derivative of a modular form $f \in M_k(\Gamma)$. If $g \in M_\ell(\Gamma)$ is a second modular form on the same group, of weight ℓ, then this formula shows that the non-modularity of $f'(z)g(z)$ is given by an additive correction term $(2\pi i)^{-1}kc\,(cz+d)^{k+\ell+1}f(z)\,g(z)$. This correction term, multiplied by ℓ, is symmetric in f and g, so the difference $[f,g] = kfg' - \ell f'g$ is a modular form of weight $k + \ell + 2$ on Γ. One checks easily that the bracket $[\,\cdot\,,\,\cdot\,]$ defined in this way is anti-symmetric and satisfies the Jacobi identity, making $M_*(\Gamma)$ into a graded Lie algebra (with grading given by the weight $+\,2$). Furthermore, the bracket $g \mapsto [f,g]$ with a fixed modular form f is a derivation with respect to the usual multiplication, so that $M_*(\Gamma)$ even acquires the structure of a so-called Poisson algebra.

We can continue this construction to find combinations of higher derivatives of f and g which are modular, setting $[f,g]_0 = fg$, $[f,g]_1 = [f,g] = kfg' - \ell f'g$,

$$[f,g]_2 = \frac{k(k+1)}{2}fg'' - (k+1)(\ell+1)f'g' + \frac{\ell(\ell+1)}{2}f''g\,,$$

and in general

$$[f,g]_n = \sum_{\substack{r,\,s \geq 0 \\ r+s=n}} (-1)^r \binom{k+n-1}{s}\binom{\ell+n-1}{r} D^r f\, D^s g \qquad (n \geq 0), \quad (59)$$

the nth *Rankin–Cohen bracket* of f and g.

Proposition 18. *For $f \in M_k(\Gamma)$ and $g \in M_\ell(\Gamma)$ and for every $n \geq 0$, the function $[f,g]_n$ defined by (59) belongs to $M_{k+\ell+2n}(\Gamma)$.*

There are several ways to prove this. We will do it using *Cohen–Kuznetsov series*. If $f \in M_k(\Gamma)$, then the Cohen–Kuznetsov series of f is defined by

$$\widetilde{f}_D(z,X) = \sum_{n=0}^\infty \frac{D^n f(z)}{n!\,(k)_n}\, X^n \ \in \ \mathrm{Hol}_0(\mathfrak{H})[[X]]\,, \qquad (60)$$

where $(k)_n = (k+n-1)!/(k-1)! = k(k+1)\cdots(k+n-1)$ is the Pochhammer symbol already used above and $\mathrm{Hol}_0(\mathfrak{H})$ denotes the space of holomorphic functions in the upper half-plane of subexponential growth (see §1.1). This series converges for all $X \in \mathbb{C}$ (although for our purposes only its properties as a formal power series in X will be needed). Its key property is given by:

Proposition 19. *If $f \in M_k(\Gamma)$, then the Cohen–Kuznetsov series defined by (60) satisfies the modular transformation equation*

$$\widetilde{f}_D\left(\frac{az+b}{cz+d},\, \frac{X}{(cz+d)^2}\right) = (cz+d)^k \exp\left(\frac{c}{cz+d}\frac{X}{2\pi i}\right)\widetilde{f}_D(z,X)\,. \quad (61)$$

for all $z \in \mathfrak{H}$, $X \in \mathbb{C}$, and $\gamma = \left(\begin{smallmatrix} a & b \\ c & d \end{smallmatrix}\right) \in \Gamma$.

Proof. This can be proved in several different ways. One way is direct: one shows by induction on n that the derivative $D^n f(z)$ transforms under Γ by

$$D^n f\left(\frac{az+b}{cz+d}\right) = \sum_{r=0}^{n} \binom{n}{r} \frac{(k+r)_{n-r}}{(2\pi i)^{n-r}} c^{n-r}(cz+d)^{k+n+r} D^r f(z)$$

for all $n \geq 0$ (equation (52) is the case $n = 1$ of this), from which the claim follows easily. Another, more elegant, method is to use formula (56) or (57) to establish the relationship

$$\tilde{f}_D(z, X) = e^{X/4\pi y}\, \tilde{f}_\partial(z, X) \qquad (z = x + iy \in \mathfrak{H},\ X \in \mathbb{C}) \qquad (62)$$

between $\tilde{f}_D(z, X)$ and the modified Cohen–Kuznetsov series

$$\tilde{f}_\partial(z, X) = \sum_{n=0}^{\infty} \frac{\partial^n f(z)}{n!\, (k)_n} X^n \ \in \operatorname{Hol}_0(\mathfrak{H})[[X]]. \qquad (63)$$

The fact that each function $\partial^n f(z)$ transforms like a modular form of weight $k + 2n$ on Γ implies that $\tilde{f}_\partial(z, X)$ is multiplied by $(cz + d)^k$ when z and X are replaced by $\frac{az+b}{cz+d}$ and $\frac{X}{(cz+d)^2}$, and using (62) one easily deduces from this the transformation formula (61). Yet a third way is to observe that $\tilde{f}_D(z, X)$ is the unique solution of the differential equation $\left(X\frac{\partial^2}{\partial X^2} + k\frac{\partial}{\partial X} - D\right)\tilde{f}_D = 0$ with the initial condition $\tilde{f}_D(z, 0) = f(z)$ and that $(cz + d)^{-k} e^{-cX/2\pi i(cz+d)} \tilde{f}_D\left(\frac{az+b}{cz+d}, \frac{X}{(cz+d)^2}\right)$ satisfies the same differential equation with the same initial condition.

Now to deduce Proposition 18 we simply look at the product of $\tilde{f}_D(z, -X)$ with $\tilde{g}_D(z, X)$. Proposition 19 implies that this product is multiplied by $(cz + d)^{k+\ell}$ when z and X are replaced by $\frac{az+b}{cz+d}$ and $\frac{X}{(cz+d)^2}$ (the factors involving an exponential in X cancel), and this means that the coefficient of X^n in the product, which is equal to $\frac{[f,g]_n}{(k)_n (\ell)_n}$, is modular of weight $k + \ell + 2n$ for every $n \geq 0$.

Rankin–Cohen brackets have many applications in the theory of modular forms. We will describe two – one very straightforward and one more subtle – at the end of this subsection, and another one in §5.4. First, however, we make a further comment about the Cohen–Kuznetsov series attached to a modular form. We have already introduced two such series: the series $\tilde{f}_D(z, X)$ defined by (60), with coefficients proportional to $D^n f(z)$, and the series $\tilde{f}_\partial(z, X)$ defined by (63), with coefficients proportional to $\partial^n f(z)$. But, at least when Γ is the full modular group Γ_1, we had defined a third differentiation operator besides D and ∂, namely the operator ϑ defined in (53), and it is natural to ask whether there is a corresponding Cohen–Kuznetsov series $\tilde{f}_\vartheta$ here also. The answer is yes, but this series is not simply given by $\sum_{n\geq0} \vartheta^n f(z) X^n/n! (k)_n$. Instead, for $f \in M_k(\Gamma_1)$, we define a sequence of modified derivatives $\vartheta^{[n]}f \in M_{k+2n}(\Gamma_1)$ for $n \geq 0$ by

$$\vartheta^{[0]}f = f, \quad \vartheta^{[1]}f = \vartheta f, \quad \vartheta^{[r+1]}f = \vartheta(\vartheta^{[r]}f) - r(k+r-1)\frac{E_4}{144}\vartheta^{[r-1]}f \text{ for } r \geq 1$$
$$(64)$$

(the last formula also holds for $r = 0$ with $f^{[-1]}$ defined as 0 or in any other way), and set

$$\tilde{f}_\vartheta(z, X) = \sum_{n=0}^{\infty} \frac{\vartheta^{[n]}f(z)}{n! (k)_n} X^n.$$

Using the first equation in (54) we find by induction on n that

$$D^n f = \sum_{r=0}^{n} \binom{n}{r} (k + r)_{n-r} \left(\frac{E_2}{12}\right)^{n-r} \vartheta^{[r]}f \quad (n = 0, 1, \ldots), \quad (65)$$

(together with similar formulas for $\partial^n f$ in terms of $\vartheta^{[n]}f$ and for $\vartheta^{[n]}f$ in terms of $D^n f$ or $\partial^n f$), the explicit version of the expansion of $D^n f$ as a polynomial in E_2 with modular coefficients whose existence is guaranteed by Proposition 15. This formula together with (62) gives us the relations

$$\tilde{f}_\vartheta(z, X) = e^{-XE_2(z)/12} \tilde{f}_D(z, X) = e^{-XE_2^*(z)/12} \tilde{f}_\partial(z, X) \quad (66)$$

between the new series and the old ones, where $E_2^*(z)$ is the non-holomorphic Eisenstein series defined in (21), which transforms like a modular form of weight 2 on Γ_1. More generally, if we are on any discrete subgroup Γ of $SL(2, \mathbb{R})$, we choose a holomorphic or meromorphic function ϕ in $\mathfrak{H}$ such that the function $\phi^*(z) = \phi(z) - \frac{1}{4\pi y}$ transforms like a modular form of weight 2 on Γ, or equivalently such that $\phi(\frac{az+b}{cz+d}) = (cz + d)^2\phi(z) + \frac{1}{2\pi i}c(cz + d)$ for all $\left(\begin{smallmatrix} a & b \\ c & d \end{smallmatrix}\right) \in \Gamma$. (Such a ϕ always exists, and if Γ is commensurable with Γ_1 is simply the sum of $\frac{1}{12}E_2(z)$ and a holomorphic or meromorphic modular form of weight 2 on $\Gamma \cap \Gamma_1$.) Then, just as in the special case $\phi = E_2/12$, the operator ϑ_ϕ defined by $\vartheta_\phi f := Df - k\phi f$ for $f \in M_k(\Gamma)$ sends $M_k(\Gamma)$ to $M_{k+2}(\Gamma)$, the function $\omega := \phi' - \phi^2$ belongs to $M_4(\Gamma)$ (generalizing the first equation in (54)), and if we generalize the above definition by introducing operators $\vartheta_\phi^{[n]} : M_k(\Gamma) \to M_{k+2n}(\Gamma)$ $(n = 0, 1, \ldots)$ by

$$\vartheta_\phi^{[0]}f = f, \quad \vartheta_\phi^{[r+1]}f = \vartheta_\phi^{[r]}f + r(k+r-1)\omega\,\vartheta_\phi^{[r-1]}f \quad \text{for } r \geq 0, \quad (67)$$

then (65) holds with $\frac{1}{12}E_2$ replaced by ϕ, and (66) is replaced by

$$\tilde{f}_{\vartheta_\phi}(z, X) := \sum_{n=0}^{\infty} \frac{\vartheta_\phi^{[n]}f(z)\,X^n}{n! (k)_n} = e^{-\phi(z)X} \tilde{f}_D(z, X) = e^{-\phi^*(z)X} \tilde{f}_\partial(z, X).$$
$$(68)$$

These formulas will be used again in §§6.3–6.4.

We now give the two promised applications of Rankin–Cohen brackets.

♦ Further Identities for Sums of Powers of Divisors

In §2.2 we gave identities among the divisor power sums $\sigma_\nu(n)$ as one of our first applications of modular forms and of identities like $E_4^2 = E_8$. By including the quasimodular form E_2 we get many more identities of the same type, e.g., the relationship $E_2^2 = E_4 + 12E_2'$ gives the identity $\sum_{m=1}^{n-1} \sigma_1(m)\sigma_1(n-m) = \frac{1}{12}\big(5\sigma_3(n) - (6n-1)\sigma_1(n)\big)$ and similarly for the other two formulas in (54). Using the Rankin–Cohen brackets we get yet more. For instance, the first Rankin–Cohen bracket of $E_4(z)$ and $E_6(z)$ is a cusp form of weight 12 on Γ_1, so it must be a multiple of $\Delta(z)$, and this leads to the formula $\tau(n) = n\big(\frac{7}{12}\sigma_5(n) + \frac{5}{12}\sigma_3(n)\big) - 70\sum_{m=1}^{n-1}(5m - 2n)\sigma_3(m)\sigma_5(n-m)$ for the coefficient $\tau(n)$ of q^n in Δ. (This can also be expressed completely in terms of elementary functions, without mentioning $\Delta(z)$ or $\tau(n)$, by writing Δ as a linear combination of any two of the three functions E_{12}, E_4E_8 and E_6^2.) As in the case of the identities mentioned in §2.2, all of these identities also have combinatorial proofs, but these involve much more work and more thought than the (quasi)modular ones.   ♡

♠ Exotic Multiplications of Modular Forms

A construction which is familiar both in symplectic geometrys and in quantum theory (Moyal brackets) is that of the deformation of the multiplication in an algebra. If A is an algebra over some field k, say with a commutative and associative multiplication $\mu : A \otimes A \to A$, then we can look at deformations μ_ε of μ given by formal power series $\mu_\varepsilon(x, y) = \mu_0(x, y) + \mu_1(x, y)\varepsilon + \mu_2(x, y)\varepsilon^2 + \cdots$ with $\mu_0 = \mu$ which are still associative but are no longer necessarily commutative. (To be more precise, μ_ε is a multiplication on A if there is a topology and the series above is convergent and otherwise, after being extended $k[[\varepsilon]]$-linearly, a multiplication on $A[[\varepsilon]]$.) The linear term $\mu_1(x, y)$ in the expansion of μ_ε is then anti-symmetric and satisfies the Jacobi identity, making A (or $A[[\varepsilon]]$) into a Lie algebra, and the μ_1-product with a fixed element of A is a derivation with respect to the original multiplication μ, giving the structure of a Poisson algebra. All of this is very reminiscent of the zeroth and first Rankin–Cohen brackets, so one can ask whether these two brackets arise as the beginning of the expansion of some deformation of the ordinary multiplication of modular forms. Surprisingly, this is not only true, but there is in fact a *two-parameter* family of such deformed multiplications:

Theorem. *Let u and v be two formal variables and Γ any subgroup of $SL(2, \mathbb{R})$. Then the multiplication $\mu_{u,v}$ on $\prod_{k=0}^{\infty} M_k(\Gamma)$ defined by*

$$\mu_{u,v}(f, g) = \sum_{n=0}^{\infty} t_n(k, \ell; u, v)\,[f, g]_n \qquad (f \in M_k(\Gamma),\ g \in M_\ell(\Gamma)), \qquad (69)$$

where the coefficients $t_n(k, \ell; u, v) \in \mathbb{Q}(k, l)[u, v]$ are given by

$$t_n(k, \ell; u, v) = v^n \sum_{0 \le j \le n/2} \binom{n}{2j} \frac{\binom{-\frac{1}{2}}{j}\binom{\frac{u}{v} - 1}{j}\binom{-\frac{u}{v}}{j}}{\binom{-k - \frac{1}{2}}{j}\binom{-\ell - \frac{1}{2}}{j}\binom{n + k + l - \frac{3}{2}}{j}},$$

$$(70)$$

is associative.

Of course the multiplication given by $(u, v) = (0, 0)$ is just the usual multiplication of modular forms, and the multiplications associated to (u, v) and $(\lambda u, \lambda v)$ are isomorphic by rescaling $f \mapsto \lambda^k f$ for $f \in M_k$, so the set of new multiplications obtained this way is parametrized by a projective line. Two of the multiplications obtained are noteworthy: the one corresponding to $(u : v) = (1 : 0)$ because it is the only commutative one, and the one corresponding to $(u, v) = (0, 1)$ because it is so simple, being given just by $f * g = \sum_n [f, g]_n$.

These deformed multiplications were found, at approximately the same time, in two independent investigations and as consequences of two quite different theories. On the one hand, Y. Manin, P. Cohen and I studied Γ-invariant and twisted Γ-invariant pseudodifferential operators in the upper half-plane. These are formal power series $\Psi(z) = \sum_{n \ge h} f_n(z) D^{-n}$, with D as in (52), transforming under Γ by $\Psi\left(\frac{az+b}{cz+d}\right) = \Psi(z)$ or by $\Psi\left(\frac{az+b}{cz+d}\right) = (cz + d)^\kappa \Psi(z)(cz + d)^{-\kappa}$, respectively, where κ is some complex parameter. (Notice that the second formula is different from the first because multiplication by a non-constant function of z does not commute with the differentiation operator D, and is well-defined even for non-integral κ because the ambiguity of argument involved in choosing a branch of $(cz + d)^\kappa$ cancels out when one divides by $(cz + d)^\kappa$ on the other side.) Using that D transforms under the action of $\left(\begin{smallmatrix} a & b \\ c & d \end{smallmatrix}\right)$ to $(cz + d)^2 D$, one finds that the leading coefficient f_h of F is then a modular form on Γ of weight $2h$ and that the higher coefficients f_{h+j} $(j \ge 0)$ are specific linear combinations (depending on h, j and κ) of $D^j g_0$, $D^{j-1} g_1, \ldots, g_j$ for some modular forms $g_i \in M_{2h+2i}(G)$, so that (assuming that Γ contains -1 and hence has only modular forms of even weight) we can canonically identify the space of invariant or twisted invariant pseudodifferential operators with $\prod_{k=0}^{\infty} M_k(\Gamma)$. On the other hand, pseudo-differential operators can be multiplied just by multiplying out the formal series defining them using Leibnitz's rule, this multiplication clearly being associative, and this then leads to the family of non-trivial multiplications of modular forms given in the theorem, with $u/v = \kappa - 1/2$. The other paper in which the same multiplications arise is one by A. and J. Untenberger which is based on a certain realization of modular forms as Hilbert-Schmidt operators on $L^2(\mathbb{R})$. In both constructions, the coefficients $t_n(k, \ell; u, v)$ come out in a different and less symmetric form than (70); the equivalence with the form given above is a very complicated combinatorial identity. $\heartsuit$

5.3 Quasimodular Forms

We now turn to the definition of the ring $\widetilde{M}_*(\Gamma)$ of quasimodular forms on Γ. In §5.1 we defined this ring for the case $\Gamma = \Gamma_1$ as $\mathbb{C}[E_2, E_4, E_6]$. We now give an intrinsic definition which applies also to other discrete groups Γ.

We start by defining the ring of *almost holomorphic modular forms* on Γ. By definition, such a form is a function in $\mathfrak{H}$ which transforms like a modular form but, instead of being holomorphic, is a polynomial in $1/y$ (with $y = \Im(z)$ as usual) with holomorphic coefficients, the motivating examples being the non-holomorphic Eisenstein series $E_2^*(z)$ and the non-holomorphic derivative $\partial f(z)$ of a holomorphic modular form as defined in equations (21) and (55), respectively. More precisely, an almost holomorphic modular form of *weight* k and *depth* $\leq p$ on Γ is a function of the form $F(z) = \sum_{r=0}^{p} f_r(z)(-4\pi y)^{-r}$ with each $f_r \in \mathrm{Hol}_0(\mathfrak{H})$ (holomorphic functions of moderate growth) satisfying $F|_k \gamma = F$ for all $\gamma \in \Gamma$. We denote by $\widehat{M}_k^{(\leq p)} = \widehat{M}_k^{(\leq p)}(\Gamma)$ the space of such forms and by $\widehat{M}_* = \bigoplus_k \widehat{M}_k$, $\widehat{M}_k = \cup_p \widehat{M}_k^{(\leq p)}$ the graded and filtered ring of all almost holomorphic modular forms, usually omitting Γ from the notations. For the two basic examples $E_2^* \in \widehat{M}_2^{(\leq 1)}(\Gamma_1)$ and $\partial_k f \in \widehat{M}_{k+2}^{(\leq 1)}(\Gamma)$ (where $f \in M_k(\Gamma)$) we have $f_0 = E_2$, $f_1 = 12$ and $f_0 = Df$, $f_1 = kf$, respectively.

We now define the space $\widetilde{M}_k^{(\leq p)} = \widetilde{M}_k^{(\leq p)}(\Gamma)$ of quasimodular forms of weight k and depth $\leq p$ on Γ as the space of "constant terms" $f_0(z)$ of $F(z)$ as F runs over $\widehat{M}_k^{(\leq p)}$. It is not hard to see that the almost holomorphic modular form F is uniquely determined by its constant term f_0, so the ring $\widetilde{M}_* = \bigoplus_k \widetilde{M}_k$ $(\widetilde{M}_k = \cup_p \widetilde{M}_k^{(\leq p)})$ of quasimodular forms on Γ is canonically isomorphic to the ring $\widehat{M}_*$ of almost holomorphic modular forms on Γ. One can also define quasimodular forms directly, as was pointed out to me by W. Nahm: a quasimodular form of weight k and depth $\leq p$ on Γ is a function $f \in \mathrm{Hol}_0(\mathfrak{H})$ such that, for fixed $z \in \mathfrak{H}$ and variable $\gamma = \left(\begin{smallmatrix} a & b \\ c & d \end{smallmatrix}\right) \in \Gamma$, the function $(f|_k\gamma)(z)$ is a polynomial of degree $\leq p$ in $\frac{c}{cz+d}$. Indeed, if $f(z) = f_0(z) \in \widetilde{M}_k$ corresponds to $F(z) = \sum_r f_r(z)(-4\pi y)^{-r} \in \widehat{M}_k$, then the modularity of F implies the identity $(f|_k\gamma)(z) = \sum_r f_r(z)\left(\frac{c}{cz+d}\right)^r$ with the same coefficients $f_r(z)$, and conversely.

The basic facts about quasimodular forms are summarized in the following proposition, in which Γ is a non-cocompact discrete subgroup of $\mathrm{SL}(2, \mathbb{R})$ and $\phi \in \widetilde{M}_2(\Gamma)$ is a quasimodular form of weight 2 on Γ which is not modular, e.g., $\phi = E_2$ if Γ is a subgroup of Γ_1. For $\Gamma = \Gamma_1$ part (i) of the proposition reduces to Proposition 15 above, while part (ii) shows that the general definition of quasimodular forms given here agrees with the ad hoc one given in §5.1 in this case.

Proposition 20. (i) *The space of quasimodular forms on Γ is closed under differentiation. More precisely, we have $D\big(\widetilde{M}_k^{(\leq p)}\big) \subseteq \widetilde{M}_{k+2}^{(\leq p+1)}$ for all $k, p \geq 0$.*
(ii) *Every quasimodular form on Γ is a polynomial in ϕ with modular coefficients. More precisely, we have $\widetilde{M}_k^{(\leq p)}(\Gamma) = \bigoplus_{r=0}^{p} M_{k-2r}(\Gamma) \cdot \phi^r$ for all $k, p \geq 0$.*
(iii) *Every quasimodular form on Γ can be written uniquely as a linear combination of derivatives of modular forms and of ϕ. More precisely, for all $k, p \geq 0$ we have*

$$
\widetilde{M}_k^{(\leq p)}(\Gamma) = \begin{cases} \bigoplus_{r=0}^{p} D^r\big(M_{k-2r}(\Gamma)\big) & \text{if } p < k/2, \\ \bigoplus_{r=0}^{k/2-1} D^r\big(M_{k-2r}(\Gamma)\big) \oplus \mathbb{C} \cdot D^{k/2-1}\phi & \text{if } p \geq k/2. \end{cases}
$$

Proof. Let $F = \sum f_r(-4\pi y)^{-r} \in \widehat{M}_k$ correspond to $f = f_0 \in \widetilde{M}_k$. The almost holomorphic form $\partial_k F \in \widehat{M}_{k+2}$ then has the expansion $\partial_k F = \sum [D(f_r) + (k - r + 1)f_{r-1}](-4\pi y)^{-r}$, with constant term Df. This proves the first statement. (One can also prove it in terms of the direct definition of quasimodular forms by differentiating the formula expressing the transformation behavior of f under Γ.) Next, one checks easily that if F belongs to $\widehat{M}_k^{(\leq p)}$, then the last coefficient $f_p(z)$ in its expansion is a modular form of weight $k - 2p$. It follows that $p \leq k/2$ (if $f_p \neq 0$, i.e., if F has depth exactly p) and also, since the almost holomorphic modular form ϕ^* corresponding to ϕ is the sum of ϕ and a non-zero multiple of $1/y$, that F is a linear combination of $f_p \phi^{*p}$ and an almost holomorphic modular form of depth strictly smaller than p, from which statement (ii) for almost holomorphic modular forms (and therefore also for quasimodular forms) follows by induction on p. Statement (iii) is proved exactly the same way, by subtracting from F a multiple of $\partial^p f_p$ if $p < k/2$ and a multiple of $\partial^{k/2-1}\phi$ if $p = k/2$ to prove by induction on p the corresponding statement with "quasimodular" and D replaced by "almost holomorphic modular" and ∂, and then again using the isomorphism between $\widehat{M}_*$ and $\widetilde{M}_*$.

There is one more important element of the structure of the ring of quasimodular (or almost holomorphic modular) forms. Let $F = \sum f_r(-4\pi y)^{-r}$ and $f = f_0$ be an almost holomorphic modular form of weight k and depth $\leq p$ and the quasimodular form which corresponds to it. One sees easily using the properties above that each coefficient f_r is quasimodular (of weight $k - 2r$ and depth $\leq p - r$) and that, if $\delta : \widetilde{M}_* \to \widetilde{M}_*$ is the map which sends f to f_1, then $f_r = \delta^r f / r!$ for all $r \geq 0$ (and $\delta^r f = 0$ for $r > p$), so that the expansion of $F(z)$ in powers of $-1/4\pi y$ is a kind of Taylor expansion formula. This gives us three operators from $\widetilde{M}_*$ to itself: the differentiation operator D, the operator E

which multiplies a quasimodular form of weight k by k, and the operator δ. Each of these operators is a derivation on the ring of quasimodular forms, and they satisfy the three commutation relations

$$[E, D] \;=\; 2\,D\,, \qquad [D, \delta] \;=\; -2\,\delta\,, \qquad [D, \delta] \;=\; E$$

(of which the first two just say that D and δ raise and lower the weight of a quasimodular form by 2, respectively), giving to $\widetilde{M}_*$ the structure of an $\mathfrak{sl}_2(\mathbb{C})$-module. Of course the ring $\widehat{M}_*$, being isomorphic to $\widetilde{M}_*$, also becomes an $\mathfrak{sl}_2(\mathbb{C})$-module, the corresponding operators being ϑ ($= \vartheta_k$ on $\widehat{M}_k$), E ($=$ multiplication by k on $\widehat{M}_k$) and δ^* ($=$ "derivation with respect to $-1/4\pi y$"). From this point of view, the subspace M_* of $\widetilde{M}_*$ or $\widehat{M}_*$ appears simply as the kernel of the lowering operator δ (or δ^*). Using this $\mathfrak{sl}_2(\mathbb{C})$-module structure makes many calculations with quasimodular or almost holomorphic modular forms simpler and more transparent.

♠ Counting Ramified Coverings of the Torns

We end this subsection by describing very briefly a beautiful and unexpected context in which quasimodular forms occur. Define

$$\Theta(X, z, \zeta) \;=\; \prod_{n>0}(1 - q^n) \prod_{\substack{n>0 \\ n\text{ odd}}} \left(1 - e^{n^2 X/8}\, q^{n/2}\, \zeta\right)\left(1 - e^{-n^2 X/8}\, q^{n/2}\, \zeta^{-1}\right),$$

expand $\Theta(X, z, \zeta)$ as a Laurent series $\sum_{n\in\mathbb{Z}} \Theta_n(X, z)\, \zeta^n$, and expand $\Theta_0(X, z)$ as a Taylor series $\sum_{r=0}^{\infty} A_r(z)\, X^{2r}$. Then a somewhat intricate calculation involving Eisenstein series, theta series and quasimodular forms on $\Gamma(2)$ shows that each A_r is a quasimodular form of weight $6r$ on $\mathrm{SL}(2, \mathbb{Z})$, i.e., a weighted homogeneous polynomial in E_2, E_4, and E_6. In particular, $A_0(z) = 1$ (this is a consequence of the Jacobi triple product identity, which says that $\Theta_n(0, z) = (-1)^n q^{n^2/2}$), so we can also expand $\log \Theta_0(X, z)$ as $\sum_{r=1}^{\infty} F_r(z)\, X^{2r}$ and $F_r(z)$ is again quasimodular of weight $6r$. These functions arose in the study of the "one-dimensional analogue of mirror symmetry": the coefficient of q^m in F_r counts the generically ramified coverings of degree m of a curve of genus 1 by a curve of genus $r + 1$. ("Generic" means that each point has $\geq m - 1$ preimages.) We thus obtain:

Theorem. *The generating function of generically ramified coverings of a torus by a surface of genus $g > 1$ is a quasimodular form of weight $6g-6$ on $\mathrm{SL}(2, \mathbb{Z})$.*

As an example, we have $F_1 = A_1 = \frac{1}{103680}(10E_2^3 - 6E_2 E_4 - 4E_6) = q^2 + 8q^3 + 30q^4 + 80q^5 + 180q^6 + \cdots$. In this case (but for no higher genus), the function $F_1 = \frac{1}{1440}(E_4' + 10E_2'')$ is also a linear combination of derivatives of Eisenstein series and we get a simple explicit formula $n(\sigma_3(n) - n\sigma_1(n))/6$ for the (correctly counted) number of degree n coverings of a torus by a surface of genus 2. ♡

5.4 Linear Differential Equations and Modular Forms

The statement of Proposition 16 is very simple, but not terribly useful, because it is difficult to derive properties of a function from a non-linear differential equation. We now prove a much more useful fact: if we express a modular form as a function, not of z, but of a modular function of z (i.e., a meromorphic modular form of weight zero), then it always satisfies a *linear* differentiable equation of finite order with algebraic coefficients. Of course, a modular form cannot be written as a single-valued function of a modular function, since the latter is invariant under modular transformations and the former transforms with a non-trivial automorphy factor, but in can be expressed locally as such a function, and the global non-uniqueness then simply corresponds to the monodromy of the differential equation.

The fact that we have mentioned is by no means new – it is at the heart of the original discovery of modular forms by Gauss and of the later work of Fricke and Klein and others, and appears in modern literature as the theory of Picard–Fuchs differential equations or of the Gauss–Manin connection – but it is not nearly as well known as it ought to be. Here is a precise statement.

Proposition 21. *Let $f(z)$ be a (holomorphic or meromorphic) modular form of positive weight k on some group Γ and $t(z)$ a modular function with respect to Γ. Express $f(z)$ (locally) as $\Phi(t(z))$. Then the function $\Phi(t)$ satisfies a linear differential equation of order $k + 1$ with algebraic coefficients, or with polynomial coefficients if $\Gamma \backslash \mathfrak{H}$ has genus 0 and $t(z)$ generates the field of modular functions on Γ.*

This proposition is perhaps the single most important source of applications of modular forms in other branches of mathematics, so with no apology we sketch three different proofs, each one giving us different information about the differential equation in question.

Proof 1: We want to find a linear relation among the derivatives of f with respect to t. Since $f(z)$ is not defined in $\Gamma \backslash \mathfrak{H}$, we must replace d/dt by the operator $D_t = t'(z)^{-1} d/dz$, which makes sense in $\mathfrak{H}$. We wish to show that the functions $D_t^n f$ $(n = 0, 1, \ldots, k + 1)$ are linearly dependent over the field of modular functions on Γ, since such functions are algebraic functions of $t(z)$ in general and rational functions in the special cases when $\Gamma \backslash \mathfrak{H}$ has genus 0 and $t(z)$ is a "Hauptmodul" (i.e., $t : \Gamma \backslash \mathfrak{H} \to \mathbb{P}_1(\mathbb{C})$ is an isomorphism). The difficulty is that, as seen in §5.1, differentiating the transformation equation (2) produces undesired extra terms as in (52). This is because the automorphy factor $(cz+d)^k$ in (2) is non-constant and hence contributes non-trivially when we differentiate. To get around this, we replace $f(z)$ by the *vector-valued* function $F : \mathfrak{H} \to \mathbb{C}^{k+1}$ whose mth component $(0 \leq m \leq k)$ is $F_m(z) = z^{k-m} f(z)$. Then

$$F_m\left(\frac{az+b}{cz+d}\right) = (az+b)^{k-m}(cz+d)^m f(z) = \sum_{n=0}^{m} M_{mn} F_n(z), \quad (71)$$

for all $\gamma = \left(\begin{smallmatrix} a & b \\ c & d \end{smallmatrix}\right) \in \Gamma$, where $M = S^k(\gamma) \in \mathrm{SL}(k+1, \mathbb{Z})$ denotes the kth symmetric power of γ. In other words, we have $F(\gamma z) = M F(z)$ where the new "automorphy factor" M, although it is more complicated than the automorphy factor in (2) because it is a matrix rather than a scalar, is now independent of z, so that when we differentiate we get simply $(cz+d)^{-2} F'(\gamma z) = M F'(z)$. Of course, this equation contains $(cz+d)^{-2}$, so differentiating again with respect to z would again produce unwanted terms. But the Γ-invariance of t implies that $t'\left(\frac{az+b}{cz+d}\right) = (cz+d)^2 t'(z)$, so the factor $(cz+d)^{-2}$ is cancelled if we replace d/dz by $D_t = d/dt$. Thus (71) implies $D_t F(\gamma z) = M D_t F(z)$, and by induction $D_t^r F(\gamma z) = M D_t^r F(z)$ for all $r \geq 0$. Now consider the $(k+2) \times (k+2)$ matrix

$$\begin{pmatrix} f & D_t f & \cdots & D_t^{k+1} f \\ F & D_t F & \cdots & D_t^{k+1} F \end{pmatrix}.$$

The top and bottom rows of this matrix are identical, so its determinant is 0. Expanding by the top row, we find $0 = \sum_{n=0}^{k+1} (-1)^n \det(A_n(z)) D_t^n f(z)$, where $A_n(z)$ is the $(k+1) \times (k+1)$ matrix $\left(F \; D_t F \; \cdots \; \widehat{D_t^n F} \; \cdots \; D_t^{k+1} F\right)$. From $D_t^r F(\gamma z) = M D_t^r F(z)$ $(\forall r)$ we get $A_n(\gamma z) = M A_n(z)$ and hence, since $\det(M) = 1$, that $A_n(z)$ is a Γ-invariant function. Since $A_n(z)$ is also meromorphic (including at the cusps), it is a modular function on Γ and hence an algebraic or rational function of $t(z)$, as desired. The advantage of this proof is that it gives us all $k+1$ linearly independent solutions of the differential equation satisfied by $f(z)$: they are simply the functions $f(z)$, $zf(z), \ldots, z^k f(z)$.

Proof 2 (following a suggestion of Ouled Azaiez): We again use the differentiation operator D_t, but this time work with quasimodular rather than vector-valued modular forms. As in §5.2, choose a quasimodular form $\phi(z)$ of weight 2 on Γ with $\delta(\phi) = 1$ (e.g., $\frac{1}{12} E_2$ if $\Gamma = \Gamma_1$), and write each $D_t^n f(z)$ $(n = 0, 1, 2, \ldots)$ as a polynomial in $\phi(z)$ with (meromorphic) modular coefficients. For instance, for $n = 1$ we find $D_t f = k \dfrac{f}{t'} \phi + \dfrac{\vartheta_\phi f}{t'}$ where ϑ_ϕ denotes the Serre derivative with respect to ϕ. Using that $\phi' - \phi^2 \in M_4$, one finds by induction that each $D_t^n f$ is the sum of $k(k-1) \cdots (k-n+1) (\phi/t')^n f$ and a polynomial of degree $< n$ in ϕ. It follows that the $k+2$ functions $\{D_t^n(f)/f\}_{0 \leq n \leq k+1}$ are linear combinations of the $k+1$ functions $\{(\phi/t')^n\}_{0 \leq n \leq k}$ with modular functions as coefficients, and hence that they are linearly dependent over the field of modular functions on Γ.

Proof 3: The third proof will give an explicit differential equation satisfied by f. Consider first the case when f has weight 1. The funcion $t' = D(t)$ is a (meromorphic) modular form of weight 2, so we can form the Rankin–Cohen brackets $[f, t']_1$ and $[f, f]_2$ of weights 5 and 6, respectively, and the quotients $A = \frac{[f, t']_1}{ft'^2}$ and $B = -\frac{[f, f]_2}{2f^2 t'^2}$ of weight 0. Then

$$D_t^2 f + A \, D_t f + B f = \frac{1}{t'} \left(\frac{f'}{t'}\right)' + \frac{ft'' - 2f't'}{ft'^2} \frac{f'}{t'} - \frac{ff'' - 2f'^2}{t'^2 f^2} f = 0.$$

Since A and B are modular functions, they are rational (if t is a Hauptmodul) or algebraic (in any case) functions of t, say $A(z) = a(t(z))$ and $B(z) = b(t(z))$, and then the function $\Phi(t)$ defined locally by $f(z) = \Phi(t(z))$ satisfies $\Phi''(t) + a(t)\Phi'(t) + b(t)\Phi(t) = 0$.

Now let f have arbitrary (integral) weight $k > 0$ and apply the above construction to the function $h = f^{1/k}$, which is formally a modular form of weight 1. Of course h is not really a modular form, since in general it is not even a well-defined function in the upper half-plane (it changes by a kth root of unity when we go around a zero of f). But it is defined locally, and the functions $A = \frac{[h,t']_1}{ht'^2}$ and $B = -\frac{[h,h]_2}{2h^2t'^2}$ are well-defined, because they are both homogeneous of degree 0 in h, so that the kth roots of unity occurring when we go around a zero of f cancel out. (In fact, a short calculation shows that they can be written directly in terms of Rankin–Cohen brackets of f and t', namely $A = \frac{[f,t']_1}{kft'^2}$ and $B = -\frac{[f,f]_2}{k^2(k+1)f^2t'^2}$.) Now just as before $A(z) = a(t(z))$, $B(z) = b(t(z))$ for some rational or algebraic functions $a(t)$ and $b(t)$. Then $\Phi(t)^{1/k}$ is annihilated by the second order differential operator $L = d^2/dt^2 + a(t)d/dt + b(t)$ and $\Phi(t)$ itself is annihilated by the $(k+1)$st order differential operator $\mathrm{Sym}^k(L)$ whose solutions are the kth powers or k-fold products of solutions of the differential equation $L\Psi = 0$. The coefficients of this operator can be given by explicit expressions in terms of a and b and their derivatives (for instance, for $k = 2$ we find $\mathrm{Sym}^2(L) = d^3/dt^3 + 3a\,d^2/dt^2 + (a' + 2a^2 + 4b)\,d/dt + 2(b' + 2ab))$, and these in turn can be written as weight 0 quotients of appropriate Rankin–Cohen brackets.

Here are two classical examples of Proposition 21. For the first, we take $\Gamma = \Gamma(2)$, $f(z) = \theta_3(z)^2$ (with $\theta_3(z) = \sum q^{n^2/2}$ as in (32)) and $t(z) = \lambda(z)$, where $\lambda(z)$ is the Legendre modular function

$$\lambda(z) = 16\frac{\eta(z/2)^8\,\eta(2z)^{16}}{\eta(z)^{24}} = 1 - \frac{\eta(z/2)^{16}\,\eta(2z)^8}{\eta(z)^{24}} = \left(\frac{\theta_2(z)}{\theta_3(z)}\right)^4, \quad (72)$$

which is known to be a Hauptmodul for the group $\Gamma(2)$. Then

$$\theta_3(z)^2 = \sum_{n=0}^{\infty}\binom{2n}{n}\left(\frac{\lambda(z)}{16}\right)^n = F\left(\frac{1}{2},\frac{1}{2};1;\lambda(z)\right), \quad (73)$$

where $F(a,b;c;x) = \sum_{n=0}^{\infty}\frac{(a)_n(b)_n}{n!\,(c)_n}x^n$ with $(a)_n$ as in eq. (56) denotes the Gauss hypergeometric function, which satisfies the second order differential equation $x(x-1)\,y'' + ((a+b+1)x-c)\,y' + aby = 0$. For the second example we take $\Gamma = \Gamma_1$, $t(z) = 1728/j(z)$, and $f(z) = E_4(z)$. Since f is a modular form of weight 4, it should satisfy a fifth order linear differential equation with respect to $t(z)$, but by the third proof above one should even have that the fourth root of f satisfies a second order differential equation, and indeed one finds

$$\sqrt[4]{E_4(z)} = F\left(\tfrac{1}{12}, \tfrac{5}{12}; 1; t(z)\right) = 1 + \frac{1 \cdot 5}{1 \cdot 1} \frac{12}{j(z)} + \frac{1 \cdot 5 \cdot 13 \cdot 17}{1 \cdot 1 \cdot 2 \cdot 2} \frac{12^2}{j(z)^2} + \cdots,$$
(74)

a classical identity which can be found in the works of Fricke and Klein.

♠ The Irrationality of $\zeta(3)$

In 1978, Roger Apéry created a sensation by proving:

Theorem. *The number $\zeta(3) = \sum_{n \geq 1} n^{-3}$ is irrational.*

What he actually showed was that, if we define two sequences $\{a_n\} = \{1, 5, 73, 1445, \dots\}$ and $\{b_n\} = \{0, 6, 351/4, 62531/36, \dots\}$ as the solutions of the recursion

$$(n+1)^3 u_{n+1} = (34n^3 + 51n^2 + 27n + 5) u_n - n^3 u_{n-1} \qquad (n \geq 1) \quad (75)$$

with initial conditions $a_0 = 1$, $a_1 = 5$, $b_0 = 0$, $b_1 = 6$, then we have

$$a_n \in \mathbb{Z} \ (\forall n \geq 0), \qquad D_n^3 b_n \in \mathbb{Z} \ (\forall n \geq 0), \qquad \lim_{n \to \infty} \frac{b_n}{a_n} = \zeta(3), \quad (76)$$

where D_n denotes the least common multiple of $1, 2, \dots, n$. These three assertions together imply the theorem. Indeed, both a_n and b_n grow like C^n by the recursion, where $C = 33.97\dots$ is the larger root of $C^2 - 34C + 1 = 0$. On the other hand, $a_{n-1}b_n - a_n b_{n-1} = 6/n^3$ by the recursion and induction, so the difference between b_n/a_n and its limiting value $\zeta(3)$ decreases like C^{-2n}. Hence the quantity $x_n = D_n^3(b_n - a_n\zeta(3))$ grows like by $D_n^3/(C+o(1))^n$, which tends to 0 as $n \to \infty$ since $C > e^3$ and $D_n^3 = (e^3 + o(1))^n$ by the prime number theorem. (Chebyshev's weaker elementary estimate of D_n would suffice here.) But if $\zeta(3)$ were rational then the first two statements in (76) would imply that the x_n are rational numbers with bounded denominators, and this is a contradiction.

Apéry's own proof of the three properties (76), which involved complicated explicit formulas for the numbers a_n and b_n as sums of binomial coefficients, was very ingenious but did not give any feeling for why any of these three properties hold. Subsequently, two more enlightening proofs were found by Frits Beukers, one using representations of a_n and b_n as multiple integrals involving Legendre polynomials and the other based on modularity. We give a brief sketch of the latter one. Let $\Gamma = \Gamma_0^+(6)$ as in §3.1 be the group $\Gamma_0(6) \cup \Gamma_0(6)W$, where $W = W_6 = \frac{1}{\sqrt{6}}\begin{pmatrix} 0 & -1 \\ 6 & 0 \end{pmatrix}$. This group has genus 0 and the Hauptmodul

$$t(z) = \left(\frac{\eta(z)\,\eta(6z)}{\eta(2z)\,\eta(3z)}\right)^{12} = q - 12\,q^2 + 66\,q^3 - 220\,q^4 + \cdots,$$

where $\eta(z)$ is the Dedekind eta-function defined in (34). For $f(z)$ we take the function

$$f(z) \;=\; \frac{\bigl(\eta(2z)\,\eta(3z)\bigr)^7}{\bigl(\eta(z)\,\eta(6z)\bigr)^5} \;=\; 1 + 5\,q + 13\,q^2 + 23\,q^3 + 29\,q^4 + \cdots,$$

which is a modular form of weight 2 on Γ (and in fact an Eisenstein series, namely $5\mathbb{G}_2(z) - 2\mathbb{G}_2(2z) + 3\mathbb{G}_2(3z) - 30\mathbb{G}_2(6z)$). Proposition 21 then implies that if we expand $f(z)$ as a power series $1 + 5\,t(z) + 73\,t(z)^2 + 1445\,t(z)^3 + \cdots$ in $t(z)$, then the coefficients of the expansion satisfy a linear recursion with polynomial coefficients (this is equivalent to the statement that the power series itself satisfies a differential equation with polynomial coefficients), and if we go through the proof of the proposition to calculate the differential equation explicitly, we find that the recursion is exactly the one defining a_n. Hence $f(z) = \sum_{n=0}^{\infty} a_n\, t(z)^n$, and the integrality of the coefficients a_n follows immediately since $f(z)$ and $t(z)$ have integral coefficients and t has leading coefficient 1.

To get the properties of the second sequence $\{b_n\}$ is a little harder. Define

$$\begin{aligned}
g(z) \;&=\; \mathbb{G}_4(z) - 28\,\mathbb{G}_4(2z) + 63\,\mathbb{G}_4(3z) - 36\,\mathbb{G}_4(6z) \\
&=\; q - 14\,q^2 + 91\,q^3 - 179\,q^4 + \cdots,
\end{aligned}$$

an Eisenstein series of weight 4 on Γ. Write the Fourier expansion as $g(z) = \sum_{n=1}^{\infty} c_n\, q^n$ and define $\tilde{g}(z) = \sum_{n=1}^{\infty} n^{-3} c_n\, q^n$, so that $\tilde{g}''' = g$. This is the so-called Eichler integral associated to g and inherits certain modular properties from the modularity of g. (Specifically, the difference $\tilde{g}|_{-2}\,\gamma - \tilde{g}$ is a polynomial of degree ≤ 2 for every $\gamma \in \Gamma$, where $|_{-2}$ is the slash operator defined in (8).) Using this (we skip the details), one finds by an argument analogous to the proof of Proposition 21 that if we expand the product $f(z)\tilde{g}(z)$ as a power series $t(z) + \frac{117}{8}\,t(z)^2 + \frac{62531}{216}\,t(z)^3 + \cdots$ in $t(z)$, then this power series again satisfies a differential equation. This equation turns out to be the same one as the one satisfied by f (but with right-hand side 1 instead of 0), so the coefficients of the new expansion satisfy the same recursion and hence (since they begin 0, 1) must be one-sixth of Apéry's coefficients b_n, i.e., we have $6f(z)\tilde{g}(z) = \sum_{n=0}^{\infty} b_n\, t(z)^n$. The integrality of $D_n^3 b_n$ (indeed, even of $D_n^3 b_n/6$) follows immediately: the coefficients c_n are integral, so D_n^3 is a common denominator for the first n terms of $f\tilde{g}$ as a power series in q and hence also as a power series in $t(z)$. Finally, from the definition (13) of $\mathbb{G}_4(z)$ we find that the Fourier coefficients of g are given by the Dirichlet series identity

$$\sum_{n=1}^{\infty} \frac{c_n}{n^s} \;=\; \left(1 - \frac{28}{2^s} + \frac{63}{3^s} - \frac{36}{6^s}\right) \zeta(s)\,\zeta(s-3),$$

so the limiting value of b_n/a_n (which must exist because $\{a_n\}$ and $\{b_n\}$ satisfy the same recursion) is given by

$$\frac{1}{6} \lim_{n \to \infty} \frac{b_n}{a_n} = \tilde{g}(z)\Big|_{t(z)=1/C} = \sum_{n=1}^{\infty} \frac{c_n}{n^3} q^n\Big|_{q=1} = \sum_{n=1}^{\infty} \frac{c_n}{n^s}\Big|_{s=3} = \frac{1}{6}\zeta(3),$$

proving also the last assertion in (76).   ♡

♠ An Example Coming from Percolation Theory

Imagine a rectangle R of width $r > 0$ and height 1 on which has been drawn a fine square grid (of size roughly $rN \times N$ for some integer N going to infinity). Each edge of the grid is colored black or white with probability $1/2$ (the critical probability for this problem). The (horizontal) *crossing probability* $\Pi(r)$ is then defined as the limiting value, as $N \to \infty$, of the probability that there exists a path of black edges connecting the left and right sides of the rectangle. A simple combinatorial argument shows that there is always either a vertex-to-vertex path from left to right passing only through black edges or else a square-to-square path from top to bottom passing only through white edges, but never both. This implies that $\Pi(r) + \Pi(1/r) = 1$. The hypothesis of conformality which, though not proved, is universally believed and is at the basis of the modern theory of percolation, says that the corresponding problem, with R replaced by any (nice) open domain in the plane, is unchanged under conformal (biholomorphic) mappings, and this can be used to compute the crossing probability as the solution of a differential equation. The result, due to J. Cardy, is the formula $\Pi(r) = 2\pi\sqrt{3}\,\Gamma(\frac{1}{3})^{-3}\,t^{1/3}\,F(\frac{1}{3}, \frac{2}{3}; \frac{4}{3}; t)$, where t is the cross-ratio of the images of the four vertices of the rectangle R when it is mapped biholomorphically onto the unit disk by the Riemann uniformization theorem. This cross-ratio is known to be given by $t = \lambda(ir)$ with $\lambda(z)$ as in (72). In modular terms, using Proposition 21, we find that this translates to the formula $\Pi(r) = -2^{7/3}\,3^{-1/2}\,\pi^2\,\Gamma(\frac{1}{3})^{-3}\int_r^{\infty} \eta(iy)^4 dy$; i.e., the derivative of $\Pi(r)$ is essentially the restriction to the imaginary axis of the modular form $\eta(z)^4$ of weight 2. Conversely, an easy argument using modular forms on $SL(2, \mathbb{Z})$ shows that Cardy's function is the *unique* function satisfying the functional equation $\Pi(r) + \Pi(1/r) = 1$ and having an expansion of the form $e^{-2\pi\alpha r}$ times a power series in $e^{-2\pi r}$ for some $\alpha \in \mathbb{R}$ (which is then given by $\alpha = 1/6$). Unfortunately, there seems to be no physical argument implying *a priori* that the crossing probability has the latter property, so one cannot (yet?) use this very simple modular characterization to obtain a new proof of Cardy's famous formula.   ♡

6 Singular Moduli and Complex Multiplication

The theory of complex multiplication, the last topic which we will treat in detail in these notes, is by any standards one of the most beautiful chapters in all of number theory. To describe it fully one needs to combine themes relating to elliptic curves, modular forms, and algebraic number theory. Given the emphasis of these notes, we will discuss mostly the modular forms side, but in this introduction we briefly explain the notion of complex multiplication in the language of elliptic curves.

An elliptic curve over $\mathbb{C}$, as discussed in §1, can be represented by a quotient $E = \mathbb{C}/\Lambda$, where Λ is a lattice in $\mathbb{C}$. If $E' = \mathbb{C}/\Lambda'$ is another curve and λ a complex number with $\lambda\Lambda \subseteq \Lambda'$, then multiplication by λ induces an algebraic map from E to E'. In particular, if $\lambda\Lambda \subseteq \Lambda$, then we get a map from E to itself. Of course, we can always achieve this with $\lambda \in \mathbb{Z}$, since Λ is a $\mathbb{Z}$-module. These are the only possible real values of λ, and for generic lattices also the only possible complex values. Elliptic curves $E = \mathbb{C}/\Lambda$ where $\lambda\Lambda \subseteq \Lambda$ for some non-real value of λ are said to *admit complex multiplication*.

As we have seen in these notes, there are two completely different ways in which elliptic curves are related to modular forms. On the one hand, the moduli space of elliptic curves is precisely the domain of definition $\Gamma_1 \backslash \mathfrak{H}$ of modular functions on the full modular group, via the map $\Gamma_1 z \leftrightarrow [\mathbb{C}/\Lambda_z]$, where $\Lambda_z = \mathbb{Z}z + \mathbb{Z} \subset \mathbb{C}$. On the other hand, elliptic curves over $\mathbb{Q}$ are supposed (and now finally known) to have parametrizations by modular functions and to have Hasse-Weil L-functions that coincide with the Hecke L-series of certain cusp forms of weight 2. The elliptic curves with complex multiplication are of special interest from both points of view. If we think of $\mathfrak{H}$ as parametrizing elliptic curves, then the points in $\mathfrak{H}$ corresponding to elliptic curves with complex multiplication (usually called *CM points* for short) are simply the numbers $\mathfrak{z} \in \mathfrak{H}$ which satisfy a quadratic equation over $\mathbb{Z}$. The basic fact is that the value of $j(\mathfrak{z})$ (or of any other modular function with algebraic coefficients evaluated at $\mathfrak{z}$) is then an algebraic number; this says that an elliptic curve with complex multiplication is always defined over $\overline{\mathbb{Q}}$ (i.e., has a Weierstrass equation with algebraic coefficients). Moreover, these special algebraic numbers $j(\mathfrak{z})$, classically called *singular moduli*, have remarkable properties: they give explicit generators of the class fields of imaginary quadratic fields, the differences between them factor into small prime factors, their traces are themselves the coefficients of modular forms, etc. This will be the theme of the first two subsections. If on the other hand we consider the L-function of a CM elliptic curve and the associated cusp form, then again both have very special properties: the former belongs to two important classes of number-theoretical Dirichlet series (Epstein zeta functions and L-series of grossencharacters) and the latter is a theta series with spherical coefficients associated to a binary quadratic form. This will lead to several applications which are treated in the final two subsections.

6.1 Algebraicity of Singular Moduli

In this subsection we will discuss the proof, refinements, and applications of the following basic statement:

Proposition 22. *Let $\mathfrak{z} \in \mathfrak{H}$ be a CM point. Then $j(\mathfrak{z})$ is an algebraic number.*

Proof. By definition, $\mathfrak{z}$ satisfies a quadratic equation over $\mathbb{Z}$, say $A\mathfrak{z}^2 + B\mathfrak{z} + C = 0$. There is then always a matrix $M \in M(2, \mathbb{Z})$, not proportional to the identity, which fixes $\mathfrak{z}$. (For instance, we can take $M = \left(\begin{smallmatrix} B & C \\ -A & 0 \end{smallmatrix}\right)$.) This matrix has a positive determinant, so it acts on the upper half-plane in the usual way. The two functions $j(z)$ and $j(Mz)$ are both modular functions on the subgroup $\Gamma_1 \cap M^{-1}\Gamma_1 M$ of finite index in Γ_1, so they are algebraically dependent, i.e., there is a non-zero polynomial $P(X, Y)$ in two variables such that $P(j(Mz), j(z))$ vanishes identically. By looking at the Fourier expansion at ∞ we can see that the polynomial P can be chosen to have coefficients in $\mathbb{Q}$. (We omit the details, since we will prove a more precise statement below.) We can also assume that the polynomial $P(X, X)$ is not identically zero. (If some power of $X - Y$ divides P then we can remove it without affecting the validity of the relation $P(j(Mz), j(z)) \equiv 0$, since $j(Mz)$ is not identically equal to $j(z)$.) The number $j(\mathfrak{z})$ is then a root of the non-trivial polynomial $P(X, X) \in \mathbb{Q}[X]$, so it is an algebraic number.

More generally, if $f(z)$ is any modular function (say, with respect to a subgroup of finite index of $\mathrm{SL}(2, \mathbb{Z})$) with algebraic Fourier coefficients in its q-expansion at infinity, then $f(\mathfrak{z}) \in \overline{\mathbb{Q}}$, as one can see either by showing that $f(Mz)$ and $f(z)$ are algebraically dependent over $\overline{\mathbb{Q}}$ for any $M \in M(2, \mathbb{Z})$ with $\det M > 0$ or by showing that $f(z)$ and $j(z)$ are algebraically dependent over $\overline{\mathbb{Q}}$ and using Proposition 22. The full theory of complex multiplication describes precisely the number field in which these numbers $f(\mathfrak{z})$ lie and the way that $\mathrm{Gal}(\overline{\mathbb{Q}}/\mathbb{Q})$ acts on them. (Roughly speaking, any Galois conjugate of any $f(\mathfrak{z})$ has the form $f^*(\mathfrak{z}^*)$ for some other modular form with algebraic coefficients and CM point $\mathfrak{z}^*$, and there is a recipe to compute both.) We will not explain anything about this in these notes, except for a few words in the third application below, but refer the reader to the texts listed in the references.

We now return to the j-function and the proof above. The key point was the algebraic relation between $j(z)$ and $j(Mz)$, where M was a matrix in $M(2, \mathbb{Z})$ of positive determinant m fixing the point $\mathfrak{z}$. We claim that the polynomial P relating $j(Mz)$ and $j(z)$ can be chosen to depend only on m. More precisely, we have:

Proposition 23. *For each $m \in \mathbb{N}$ there is a polynomial $\Psi_m(X, Y) \in \mathbb{Z}[X, Y]$, symmetric up to sign in its two arguments and of degree $\sigma_1(m)$ with respect to either one, such that $\Psi_m(j(Mz), j(z)) \equiv 0$ for every matrix $M \in M(2, \mathbb{Z})$ of determinant m.*

Proof. Denote by $\mathcal{M}_m$ the set of matrices in $M(2, \mathbb{Z})$ of determinant m. The group Γ_1 acts on $\mathcal{M}_m$ by right and left multiplication, with finitely many orbits. More precisely, an easy and standard argument shows that the finite set

$$\mathcal{M}_m^* = \left\{ \begin{pmatrix} a & b \\ 0 & d \end{pmatrix} \mid a, b, d \in \mathbb{Z}, \ ad = m, \ 0 \le b < d \right\} \subset \mathcal{M}_m \qquad (77)$$

is a full set of representatives for $\Gamma_1 \backslash \mathcal{M}_m$; in particular, we have

$$\left| \Gamma_1 \backslash \mathcal{M}_m \right| = \left| \mathcal{M}_m^* \right| = \sum_{ad=m} d = \sigma_1(m). \tag{78}$$

We claim that we have an identity

$$\prod_{M \in \Gamma_1 \backslash \mathcal{M}_m} (X - j(Mz)) = \Psi_m(X, j(z)) \qquad (z \in \mathfrak{H},\ X \in \mathbb{C}) \tag{79}$$

for some polynomial $\Psi_m(X, Y)$. Indeed, the left-hand side of (79) is well-defined because $j(Mz)$ depends only on the class of M in $\Gamma_1 \backslash \mathcal{M}_m$, and it is Γ_1-invariant because $\mathcal{M}_m$ is invariant under right multiplication by elements of Γ_1. Furthermore, it is a polynomial in X (of degree $\sigma_1(m)$, by (78)) each of whose coefficients is a holomorphic function of z and of at most exponential growth at infinity, since each $j(Mz)$ has these properties and each coefficient of the product is a polynomial in the $j(Mz)$. But a Γ_1-invariant holomorphic function in the upper half-plane with at most exponential growth at infinity is a polynomial in $j(z)$, so that we indeed have (79) for some polynomial $\Psi_m(X, Y) \in \mathbb{C}[X, Y]$. To see that the coefficients of Ψ_m are in $\mathbb{Z}$, we use the set of representatives (77) and the Fourier expansion of j, which has the form $j(z) = \sum_{n=-1}^{\infty} c_n q^n$ with $c_n \in \mathbb{Z}$ ($c_{-1} = 1$, $c_0 = 744$, $c_1 = 196884$ etc.). Thus

$$\Psi_m(X, j(z)) = \prod_{\substack{ad=m \\ d>0}} \prod_{b=0}^{d-1} \left(X - j \left(\frac{az+b}{d} \right) \right)$$

$$= \prod_{\substack{ad=m \\ d>0}} \prod_{b \ (\mathrm{mod}\ d)} \left(X - \sum_{n=-1}^{\infty} c_n \zeta_d^{bn} q^{an/d} \right),$$

where q^α for $\alpha \in \mathbb{Q}$ denotes $e^{2\pi i \alpha z}$ and $\zeta_d = e^{2\pi i/d}$. The expression in parentheses belongs to the ring $\mathbb{Z}[\zeta_d][X][q^{-1/d}, q^{1/d}]]$ of Laurent series in $q^{1/d}$ with coefficients in $\mathbb{Z}[\zeta_d]$, but applying a Galois conjugation $\zeta_d \mapsto \zeta_d^r$ with $r \in (\mathbb{Z}/d\mathbb{Z})^*$ just replaces b in the inner product by br, which runs over the same set $\mathbb{Z}/d\mathbb{Z}$, so the inner product has coefficients in $\mathbb{Z}$. The fractional powers of q go away at the same time (because the product over b is invariant under $z \mapsto z+1$), so each inner product, and hence $\Psi_m(X, j(z))$ itself, belongs to $\mathbb{Z}[X][q^{-1}, q]]$. Now the fact that it is a polynomial in $j(z)$ and that $j(z)$ has a Fourier expansion with integral coefficients and leading coefficient q^{-1} imlies that $\Psi_m(X, j(z)) \in \mathbb{Z}[X, j(z)]$. Finally, the symmetry of $\Psi_m(X, Y)$ up to sign follows because $z' = Mz$ with $M = \left(\begin{smallmatrix} a & b \\ c & d \end{smallmatrix} \right) \in \mathcal{M}_m$ is equivalent to $z = M'z'$ with $M' = \left(\begin{smallmatrix} d & -b \\ -c & a \end{smallmatrix} \right) \in \mathcal{M}_m$.

An example will make all of this clearer. For $m = 2$ we have

$$\prod_{M \in \Gamma_1 \backslash \mathcal{M}_m} (X - j(z)) = \left(X - j\left(\frac{z}{2}\right)\right)\left(X - j\left(\frac{z+1}{2}\right)\right)(X - j(2z))$$

by (77). Write this as $X^3 - A(z)X^2 + B(z)X - C(z)$. Then

$$A(z) = j\left(\frac{z}{2}\right) + j\left(\frac{z+1}{2}\right) + j(2z)$$

$$= \left(q^{-1/2} + 744 + o(q)\right) + \left(-q^{-1/2} + 744 + o(q)\right) + \left(q^{-2} + 744 + o(q)\right)$$

$$= q^{-2} + 0\,q^{-1} + 2232 + o(1)$$

$$= j(z)^2 - 1488\,j(z) + 16200 + o(1)$$

as $z \to i\infty$, and since $A(z)$ is holomorphic and Γ_1-invariant this implies that $A = j^2 - 1488j + 16200$. A similar calculation gives $B = 1488j^2 + 40773375j + 8748000000$ and $C = -j^3 + 162000j^2 - 8748000000j + 157464000000000$, so

$$\Psi_2(X, Y) = -X^2Y^2 + X^3 + 1488X^2Y + 1488XY^2 + Y^3 - 162000X^2$$

$$+ 40773375XY - 162000Y^2 + 8748000000X + 8748000000Y$$

$$- 157464000000000. \tag{80}$$

Remark. The polynomial $\Psi_m(X, Y)$ is not in general irreducible: if m is not square-free, then it factors into the product of all $\Phi_{m/r^2}(X, Y)$ with $r \in \mathbb{N}$ such that $r^2 | m$, where $\Phi_m(X, Y)$ is defined exactly like $\Psi_m(X, Y)$ but with $\mathcal{M}_m$ replaced by the set $\mathcal{M}_m^0$ of primitive matrices of determinant m. The polynomial $\Phi_m(X, Y)$ is always irreducible.

To obtain the algebraicity of $j(\mathfrak{z})$, we used that this value was a root of $P(X, X)$. We therefore should look at the restriction of the polynomial $\Psi_m(X, Y)$ to the diagonal $X = Y$. In the example (80) just considered, two properties of this restriction are noteworthy. First, it is (up to sign) monic, of degree 4. Second, it has a striking factorization:

$$\Psi_2(X, X) = -(X - 8000) \cdot (X + 3375)^2 \cdot (X - 1728). \tag{81}$$

We consider each of these properties in turn. We assume that m is not a square, since otherwise $\Psi_m(X, X)$ is identically zero because $\Psi_m(X, Y)$ contains the factor $\Psi_1(X, Y) = X - Y$.

Proposition 24. *For m not a perfect square, the polynomial $\Psi_m(X, X)$ is, up to sign, monic of degree* $\sigma_1^+(m) := \sum_{d|m} \max(d, m/d)$.

Proof. Using the identity $\prod_{b \pmod d}(x - \zeta_d^b y) = x^d - y^d$ we find

$$
\begin{aligned}
\Psi_m(j(z), j(z)) &= \prod_{ad=m} \prod_{b \pmod d} \left(j(z) - j\left(\frac{az+b}{d}\right) \right) \\
&= \prod_{ad=m} \prod_{b \pmod d} \left(q^{-1} - \zeta_d^{-b} q^{-a/d} + o(1) \right) \\
&= \prod_{ad=m} \left(q^{-d} - q^{-a} + \text{(lower order terms)} \right) \sim \pm q^{-\sigma_1^+(m)}
\end{aligned}
$$

as $\Im(z) \to \infty$, and this proves the proposition since $j(z) \sim q^{-1}$.

Corollary. *Singular moduli are algebraic integers.* □

Now we consider the factors of $\Psi_m(X, X)$. First let us identify the three roots in the factorization for $m = 2$ just given. The three CM points i, $(1 + i\sqrt{7})/2$ and $i\sqrt{2}$ are fixed, respectively, by the three matrices $\left(\begin{smallmatrix} 0 & -1 \\ 1 & 0 \end{smallmatrix}\right)$, $\left(\begin{smallmatrix} 1 & 1 \\ -1 & 1 \end{smallmatrix}\right)$ and $\left(\begin{smallmatrix} 0 & -1 \\ 2 & 0 \end{smallmatrix}\right)$ of determinant 2, so each of the corresponding j-values must be a root of the polynomial (81). Computing these values numerically to low accuracy to see which is which, we find

$$
j(i) = 1728, \qquad j\left(\frac{1+i\sqrt{7}}{2}\right) = -3375, \qquad j(i\sqrt{2}) = 8000.
$$

(Another way to distinguish the roots, without doing any transcendental calculations, would be to observe, for instance, that i and $i\sqrt{2}$ are fixed by matrices of determinant 3 but $(1+i\sqrt{7})/2$ is not, and that $X + 3375$ does not occur as a factor of $\Psi_3(X, X)$ but both $X - 1728$ and $X - 8000$ do.)

The same method can be used for any other CM point. Here is a table of the first few values of $j(\mathfrak{z}_D)$, where $\mathfrak{z}_D$ equals $\frac{1}{2}\sqrt{D}$ for D even and $\frac{1}{2}(1+\sqrt{D})$ for D odd:

D	-3	-4	-7	-8	-11	-12	-15	-16	-19
$j(\mathfrak{z}_D)$	0	1728	-3375	8000	-32768	54000	$-\frac{191025+85995\sqrt{5}}{2}$	287496	-884736

We can make this more precise. For each discriminant (integer congruent to 0 or 1 mod 4) $D < 0$ we consider, as at the end of §1.2, the set $\mathfrak{Q}_D$ of primitive positive definite binary quadratic forms of discriminant D, i.e., functions $Q(x, y) = Ax^2 + Bxy + Cy^2$ with $A, B, C \in \mathbb{Z}$, $A > 0$, $\gcd(A, B, C) = 1$ and $B^2 - 4AC = D$. To each $Q \in \mathfrak{Q}_D$ we associate the root $\mathfrak{z}_Q$ of $Q(\mathfrak{z}, 1) = 0$ in $\mathfrak{H}$. This gives a Γ_1-equivariant bijection between $\mathfrak{Q}_D$ and the set $\mathfrak{z}_D \subset \mathfrak{H}$ of CM points of discriminant D. (The discriminant of a CM point is the smallest discriminant of a quadratic polynomial over $\mathbb{Z}$ of which it is a root.) In particular, the cardinality of $\Gamma_1 \backslash \mathfrak{z}_D$ is $h(D)$, the class number of D. We choose a set of representatives $\{\mathfrak{z}_{D,i}\}_{1 \leq i \leq h(D)}$ for $\Gamma_1 \backslash \mathfrak{z}_D$ (e.g., the points of

$\mathfrak{z}_D \cap \widetilde{\mathcal{F}}_1$, where $\widetilde{\mathcal{F}}_1$ is the fundamental domain constructed in §1.2, corresponding to the set $\mathfrak{Q}_D^{\mathrm{red}}$ in (5)), with $\mathfrak{z}_{D,1} = \mathfrak{z}_D$. We now form the *class polynomial*

$$H_D(X) = \prod_{\mathfrak{z} \in \Gamma_1 \backslash \mathfrak{z}_D} (X - j(\mathfrak{z})) = \prod_{1 \le i \le h(D)} (X - j(\mathfrak{z}_{D,i})). \qquad (82)$$

Proposition 25. *The polynomial $H_D(X)$ belongs to $\mathbb{Z}[X]$ and is irreducible. In particular, the number $j(\mathfrak{z}_D)$ is algebraic of degree exactly $h(D)$ over $\mathbb{Q}$, with conjugates $j(\mathfrak{z}_{D,i})$ $(1 \le i \le h(D))$.*

Proof. We indicate only the main ideas of the proof. We have already proved that $j(\mathfrak{z})$ for any CM point $\mathfrak{z}$ is a root of the equation $\Psi_m(X, X) = 0$ whenever m is the determinant of a matrix $M \in M(2, \mathbb{Z})$ fixing $\mathfrak{z}$. The main point is that the set of these m's depends only on the discriminant D of $\mathfrak{z}$, not on $\mathfrak{z}$ itself, so that the different numbers $j(\mathfrak{z}_{D,i})$ are roots of the same equations and hence are conjugates. Let $A\mathfrak{z}^2 + B\mathfrak{z} + C = 0$ $(A > 0)$ be the minimal equation of $\mathfrak{z}$ and suppose that $M = \left(\begin{smallmatrix} a & b \\ c & d \end{smallmatrix}\right) \in \mathcal{M}_m$ fixes $\mathfrak{z}$. Then since $cz^2 + (d - a)z - b = 0$ we must have $(c, d - a, -b) = u(A, B, C)$ for some $u \in \mathbb{Z}$. This gives

$$M = \begin{pmatrix} \frac{1}{2}(t - Bu) & -Cu \\ Au & \frac{1}{2}(t + Bu) \end{pmatrix}, \qquad \det M = \frac{t^2 - Du^2}{4}, \qquad (83)$$

where $t = \mathrm{tr}\, M$. Convesely, if t and u are any integers with $t^2 - Du^2 = 4m$, then (83) gives a matrix $M \in \mathcal{M}_m$ fixing $\mathfrak{z}$. Thus the set of integers $m = \det M$ with $M\mathfrak{z} = \mathfrak{z}$ is the set of numbers $\frac{1}{4}(t^2 - Du^2)$ with $t \equiv Du$ (mod 2) or, more invariantly, the set of norms of elements of the quadratic order

$$\mathcal{O}_D = \mathbb{Z}[\mathfrak{z}_D] = \left\{ \frac{t + u\sqrt{D}}{2} \,\middle|\, t, u \in \mathbb{Z}, \quad t \equiv Du \pmod 2 \right\} \qquad (84)$$

of discriminant D, and this indeed depends only on D, not on $\mathfrak{z}$. We can then obtain $H_D(X)$, or at least its square, as the g.c.d. of finitely many polymials $\Psi_m(X, X)$, just as we obtained $H_{-7}(X)^2 = (X + 3375)^2$ as the g.c.d. of $\Psi_2(X, X)$ and $\Psi_3(X, X)$ in the example above. (Start, for example, with a prime m_1 which is the norm of an element of $\mathcal{O}_D$ – there are known to be infinitely many – and then choose finitely many further m's which are also norms of elements in $\mathcal{O}_D$ but not any of the finitely many other quadratic orders in which m_1 is a norm.)

This argument only shows that $H_D(X)$ has rational coefficients, not that it is irreducible. The latter fact is proved most naturally by studying the arithmetic of the corresponding elliptic curves with complex multiplication (roughly speaking, the condition of having complex multiplication by a given order $\mathcal{O}_D$ is purely algebraic and hence is preserved by Galois conjugation), but since the emphasis in these notes is on modular methods and their applications, we omit the details.

The proof just given actually yields the formula

$$\Psi_m(X, X) = \pm \prod_{D<0} H_D(X)^{r_D(m)/w(D)} \qquad (m \in \mathbb{N},\ m \neq \text{square}), \qquad (85)$$

due to Kronecker. Here $r_D(m) = \left| \{ (t, u) \in \mathbb{Z}^2 \mid t^2 - Du^2 = 4m \} \right| = \left| \{ \lambda \in \mathcal{O}_D \mid N(\lambda) = m \} \right|$ and $w(D)$ is the number of units in $\mathcal{O}_D$, which is equal to 6 or 4 for $D = -3$ or -4, respectively, and to 2 otherwise. The product in (85) is finite since $r_D(m) \neq 0 \Rightarrow 4m = t^2 - u^2 D \geq |D|$ because u can't be equal to 0 if m is not a square.

There is another form of formula (85) which will be used in the second application below. As well as the usual class number $h(D)$, one has the *Hurwitz class number* $h^*(D)$ (the traditional notation is $H(|D|)$), defined as the number of *all* Γ_1-equivalence classes of positive definite binary quadratic forms of discriminant D, not just the primitive ones, counted with multiplicity equal to one over the order of their stabilizer in Γ_1 (which is 2 or 3 if the corresponding point in the fundamental domain for $\Gamma_1 \backslash \mathfrak{H}$ is at i or ρ and is 1 otherwise). In formulas, $h^*(D) = \sum_{r^2 \mid D} h'(D/r^2)$, where the sum is over all $r \in \mathbb{N}$ for which $r^2 \mid D$ (and for which D/r^2 is still congruent to 0 or 1 mod 4, since otherwise $h'(D/r^2)$ will be 0) and $h'(D) = h(D)/\frac{1}{2}w(D)$ with $w(D)$ as above. Similarly, we can define a modified class "polynomial" $H_D^*(X)$, of "degree" $h^*(D)$, by

$$H_D^*(X) = \prod_{r^2 \mid D} H_{D/r^2}(X)^{2/w(D)},$$

e.g., $H_{-12}^*(X) = X^{1/3}(X - 54000)$. (These are actual polynomials unless $|D|$ or $3|D|$ is a square.) Then (85) can be written in the following considerably simpler form:

$$\Psi_m(X, X) = \pm \prod_{t^2<4m} H_{t^2-4m}^*(X) \qquad (m \in \mathbb{N},\ m \neq \text{square}). \qquad (86)$$

This completes our long discussion of the algebraicity of singular moduli. We now describe some of the many applications.

♠ Strange Approximations to π

We start with an application that is more fun than serious. The discriminant $D = -163$ has class number one (and is in fact known to be the smallest such discriminant), so Proposition 25 implies that $j(\mathfrak{z}_D)$ is a rational integer. Moreover, it is large (in absolute value) because $j(z) \approx q^{-1}$ and the $q = e^{2\pi i z}$ corresponding to $z = \mathfrak{z}_{163}$ is roughly -4×10^{-18}. But then from $j(z) = q^{-1} + 744 + O(q)$ we find that q^{-1} is extremely close to an integer, giving the formula

$$e^{\pi\sqrt{163}} = 262537412640768743.99999999999925007259\cdots$$

which would be very startling if one did not know about complex multiplication. By taking logarithms, one gets an extremely good approximation to π:

$$\pi = \frac{1}{\sqrt{163}} \log(262537412640768744) - (2.237 \cdots \times 10^{-31}).$$

(A poorer but simpler approximation, based on the fact that $j(\mathfrak{z}_{163})$ is a perfect cube, is $\pi \approx \frac{3}{\sqrt{163}} \log(640320)$, with an error of about 10^{-16}.) Many further identities of this type were found, still using the theory of complex multiplication, by Ramanujan and later by Dan Shanks, the most spectactular one being

$$\pi - \frac{6}{\sqrt{3502}} \log\left[2\,\varepsilon(x_1)\,\varepsilon(x_2)\,\varepsilon(x_3)\,\varepsilon(x_4)\right] \approx 7.4 \times 10^{-82}$$

where $x_1 = 429 + 304\sqrt{2}$, $x_2 = \frac{627}{2} + 221\sqrt{2}$, $x_3 = \frac{1071}{2} + 92\sqrt{34}$, $x_4 = \frac{1553}{2} + 133\sqrt{34}$, and $\varepsilon(x) = x + \sqrt{x^2 - 1}$. Of course these formulas are more curiosities than useful ways to compute π, since the logarithms are no easier to compute numerically than π is and in any case, if we allow complex numbers, then Euler's formula $\pi = \frac{1}{\sqrt{-1}} \log(-1)$ is an exact formula of the same kind!
♡

♠ Computing Class Numbers

By comparing the degrees on both sides of (85) or (86) and using Proposition 24, we obtain the famous *Hurwitz-Kronecker class number relations*

$$\sigma_1^+(m) = \sum_{D<0} \frac{h(D)}{w(D)} r_D(m) = \sum_{t^2 < 4m} h^*(t^2 - 4m) \quad (m \in \mathbb{N},\ m \neq \text{square}).$$

(87)

(In fact the equality of the first and last terms is true also for m square if we replace the summation condition by $t^2 \leq 4m$ and define $h^*(0) = -\frac{1}{12}$, as one shows by a small modification of the proof given here.) We mention that this formula has a geometric interpretation in terms of intersection numbers. Let X denote the modular curve $\Gamma_1 \backslash \mathfrak{H}$. For each $m \geq 1$ there is a curve $T_m \subset X \times X$, the *Hecke correspondence*, corresponding to the mth Hecke operator T_m introduced in §4.1. (The preimage of this curve in $\mathfrak{H} \times \mathfrak{H}$ consists of all pairs (z, Mz) with $z \in \mathfrak{H}$ and $M \in \mathcal{M}_m$.) For $m = 1$, this curve is just the diagonal. Now the middle or right-hand term of (87) counts the "physical" intersection points of T_m and T_1 in $X \times X$ (with appropriate multiplicities if the intersections are not transversal), while the left-hand term computes the same number homologically, by first compactifying X to $\bar{X} = X \cup \{\infty\}$ (which is isomorphic to $\mathbb{P}^1(\mathbb{C})$ via $z \mapsto j(z)$) and T_m and T_1 to their closures $\bar{T}_m$ and $\bar{T}_1$ in $\bar{X} \times \bar{X}$ and then computing the intersection number of the homology classes of $\bar{T}_m$ and $\bar{T}_1$ in $H_2(\bar{X} \times \bar{X}) \cong \mathbb{Z}^2$ and correcting this by the contribution coming from intersections at infinity of the compactified curves.

Equation (87) gives a formula for $h^*(-4m)$ in terms of $h^*(D)$ with $|D| < 4m$. This does not quite suffice to calculate class numbers recursively since only half of all discriminants are multiples of 4. But by a quite similar type of argument (cf. the discussion in §6.2 below) one can prove a second class number relation, namely

$$\sum_{t^2 \leq 4m} (m - t^2) \, h^*(t^2 - 4m) \; = \; \sum_{d|m} \min(d, m/d)^3 \qquad (m \in \mathbb{N}) \qquad (88)$$

(again with the convention that $h^*(0) = -\frac{1}{12}$), and this together with (87) *does* suffice to determine $h^*(D)$ for all D recursively, since together they express $h^*(-4m) + 2h^*(1 - 4m)$ and $mh^*(1 - 4m) + 2(m - 1)h^*(1 - 4m)$ as linear combinations of $h^*(D)$ with $|D| < 4m - 1$, and every negative discriminant has the form $-4m$ or $1 - 4m$ for some $m \in \mathbb{N}$. This method is quite reasonable computationally if one wants to compute a table of class numbers $h^*(D)$ for $-X < D < 0$, with about the same running time (viz., $O(X^{3/2})$ operations) as the more direct method of counting all reduced quadratic forms with discriminant in this range.   ♡

♠ Explicit Class Field Theory for Imaginary Quadratic Fields

Class field theory, which is the pinnacle of classical algebraic number theory, gives a complete classification of all abelian extensions of a given number field K. In particular, it says that the unramified abelian extensions (we omit all definitions) are the subfields of a certain finite extension H/K, the *Hilbert class field*, whose degree over K is equal to the class number of K and whose Galois group over K is canonically isomorphic to the class group of K, while the ramified abelian extensions have a similar description in terms of more complicated partititons of the ideals of $\mathcal{O}_K$ into finitely many classes. However, this theory, beautiful though it is, gives no method to actually *construct* the abelian extensions, and in fact it is only known how to do this explicitly in two cases: $\mathbb{Q}$ and imaginary quadratic fields. If $K = \mathbb{Q}$ then the Hilbert class field is trivial, since the class number is 1, and the ramified abelian extensions are just the subfields of $\mathbb{Q}(e^{2\pi i/N})$ ($N \in \mathbb{N}$) by the Kronecker-Weber theorem. For imaginary quadratic fields, the result (in the unramified case) is as follows.

Theorem. *Let K be an imaginary quadratic field, with discriminant D and Hilbert class field H. Then the $h(D)$ singular moduli $j(\mathfrak{z}_{D,i})$ are conjugate to one another over K (not just over $\mathbb{Q}$), any one of them generates H over K, and the Galois group of H over K permutes them transitively. More precisely, if we label the CM points of discriminant D by the ideal classes of K and identify $\mathrm{Gal}(H/K)$ with the class group of K by the fundamental isomorphism of class field theory, then for any two ideal classes $\mathcal{A}, \mathcal{B}$ of K the element $\sigma_{\mathcal{A}}$ of $\mathrm{Gal}(H/K)$ sends $j(\mathfrak{z}_{\mathcal{B}})$ to $j(\mathfrak{z}_{\mathcal{AB}})$.*

The ramified abelian extensions can also be described completely by complex multiplication theory, but then one has to use the values of other modular functions (not just of $j(z)$) evaluated at all points of $K \cap \mathfrak{H}$ (not just those of discriminant D). Actually, even for the Hilbert class fields it can be advantageous to use modular functions other than $j(z)$. For instance, for $D = -23$, the first discriminant with class number not a power of 2 (and therefore the first non-trivial example, since Gauss's theory of genera describes the Hilbert class field of D when the class group has exponent 2 as a composite quadratic extension, e.g., H for $K = \mathbb{Q}(\sqrt{-15})$ is $\mathbb{Q}(\sqrt{-3}, \sqrt{5})$), the Hilbert class field is generated over K by the real root α of the polynomial $X^3 - X - 1$. The singular modulus $j(3_D)$, which also generates this field, is equal to $-5^3\alpha^{12}(2\alpha - 1)^3(3\alpha + 2)^3$ and is a root of the much more complicated irreducible polynomial $H_{-23}(X) = X^3 + 3491750X^2 - 5151296875X + 12771880859375$, but using more detailed results from the theory of complex multiplication one can show that the number $2^{-1/2}e^{\pi i/24}\eta(3_D)/\eta(2 \cdot 3_D)$ also generates H, and this number turns out to be α itself! The improvement is even more dramatic for larger values of $|D|$ and is important in situations where one actually wants to compute a class field explicitly, as in the applications to factorization and primality testing mentioned in §6.4 below. ♡

♠ Solutions of Diophantine Equations

In §4.4 we discussed that one can parametrize an elliptic curve E over $\mathbb{Q}$ by modular functions $X(z)$, $Y(z)$, i.e., functions which are invariant under $\Gamma_0(N)$ for some N (the conductor of E) and identically satisfy the Weierstrass equation $Y^2 = X^3 + AX + B$ defining E. If K is an imaginary quadratic field with discriminant D prime to N and congruent to a square modulo $4N$ (this is equivalent to requiring that $\mathcal{O}_K = \mathcal{O}_D$ contains an ideal $\mathfrak{n}$ with $\mathcal{O}_D/\mathfrak{n} \cong \mathbb{Z}/N\mathbb{Z}$), then there is a canonical way to lift the $h(D)$ points of $\Gamma_1\backslash\mathfrak{H}$ of discriminant D to $h(D)$ points in the covering $\Gamma_0(N)\backslash\mathfrak{H}$, the so-called *Heegner points*. (The way is quite simple: if $D \equiv r^2 \pmod{4N}$, then one looks at points 3_Q with $Q(x,y) = Ax^2 + Bxy + Cy^2 \in \mathfrak{Q}_D$ satisfying $A \equiv 0 \pmod{N}$ and $B \equiv r \pmod{2N}$; there is exactly one $\Gamma_0(N)$-equivalence class of such points in each Γ_1-equivalence class of points of 3_D.) One can then show that the values of $X(z)$ and $Y(z)$ at the Heegner points behave exactly like the values of $j(z)$ at the points of 3_D, viz., these values all lie in the Hilbert class field H of K and are permuted simply transitively by the Galois group of H over K. It follows that we get $h(D)$ points $P_i = (X(3_i), Y(3_i))$ in E with coordinates in H which are permuted by $\mathrm{Gal}(H/K)$, and therefore that the sum $P_K = P_1 + \cdots + P_{h(D)}$ has coordinates in K (and in many cases even in $\mathbb{Q}$). This method of constructing potentially non-trivial rational solutions of Diophantine equations of genus 1 (the question of their actual non-triviality will be discussed briefly in §6.2) was invented by Heegner as part of his proof of the fact, mentioned above, that -163 is the smallest quadratic discriminant with class number 1 (his proof, which was rather sketchy at many points, was not accepted at the time by the mathematical community, but was later shown by Stark to

be correct in all essentials) and has been used to construct non-trivial solutions of several classical Diophantine equations, e.g., to show that every
prime number congruent to 5 or 7 (mod 8) is a "congruent number" (= the
area of a right triangle with rational sides) and that every prime number
congruent to 4 or 7 (mod 9) is a sum of two rational cubes (Sylvester's
problem). ♡

6.2 Norms and Traces of Singular Moduli

A striking property of singular moduli is that they always are highly factoriz-
able numbers. For instance, the value of $j(\mathfrak{z}_{-163})$ used in the first application
in §6.1 is $-640320^3 = -2^{18}3^35^323^329^3$, and the value of $j(\mathfrak{z}_{-11}) = -32768$
is the 15th power of -2. As part of our investigation about the heights of
Heegner points (see below), B. Gross and I were led to a formula which explains and generalizes this phenomenon. It turns out, in fact, that the right
numbers to look at are not the values of the singular moduli themselves, but
their *differences*. This is actually quite natural, since the definition of the
modular function $j(z)$ involves an arbitrary choice of additive constant: any
function $j(z) + C$ with $C \in \mathbb{Z}$ would have the same analytic and arithmetic
properties as $j(z)$. The factorization of $j(\mathfrak{z})$, however, would obviously change
completely if we replaced $j(z)$ by, say, $j(z) + 1$ or $j(z) - 744 = q^{-1} + O(q)$
(the so-called "normalized Hauptmodul" of Γ_1), but this replacement would
have no effect on the difference $j(\mathfrak{z}_1) - j(\mathfrak{z}_2)$ of two singular moduli. That the
original singular moduli $j(\mathfrak{z})$ do nevertheless have nice factorizations is then
due to the accidental fact that 0 itself is a singular modulus, namely $j(\mathfrak{z}_{-3})$,
so that we can write them as differences $j(\mathfrak{z}) - j(\mathfrak{z}_{-3})$. (And the fact that the
values of $j(\mathfrak{z})$ tend to be perfect cubes is then related to the fact that $\mathfrak{z}_{-3}$ is
a fixed-point of order 3 of the action of Γ_1 on $\mathfrak{H}$.) Secondly, since the singular moduli are in general algebraic rather than rational integers, we should
not speak only of their differences, but of the norms of their differences, and
these norms will then also be highly factored. (For instance, the norms of
the singular moduli $j(\mathfrak{z}_{-15})$ and $j(\mathfrak{z}_{-23})$, which are algebraic integers of degree 2 and 3 whose values were given in §6.1, are $-3^65^311^3$ and $-5^911^317^3$,
respectively.)

   If we restrict ourselves for simplicity to the case that the discriminants D_1
and D_2 of $\mathfrak{z}_1$ and $\mathfrak{z}_2$ are coprime, then the norm of $j(\mathfrak{z}_1) - j(\mathfrak{z}_2)$ depends only
on D_1 and D_2 and is given by

$$J(D_1, D_2) = \prod_{\mathfrak{z}_1 \in \Gamma_1 \backslash \mathfrak{z}_{D_1}} \prod_{\mathfrak{z}_2 \in \Gamma_1 \backslash \mathfrak{z}_{D_2}} \left(j(\mathfrak{z}_1) - j(\mathfrak{z}_2) \right). \qquad (89)$$

(If $h(D_1) = 1$ then this formula reduces simply to $H_{D_2}(\mathfrak{z}_{D_1})$, while in general
$J(D_1, D_2)$ is equal, up to sign, to the resultant of the two irreducible polynomials $H_{D_1}(X)$ and $H_{D_2}(X)$.) These are therefore the numbers which we want
to study. We then have:

Theorem. *Let D_1 and D_2 be coprime negative discriminants. Then all prime factors of $J(D_1, D_2)$ are $\leq \frac{1}{4} D_1 D_2$. More precisely, any prime factors of $J(D_1, D_2)$ must divide $\frac{1}{4}(D_1 D_2 - x^2)$ for some $x \in \mathbb{Z}$ with $|x| < \sqrt{D_1 D_2}$ and $x^2 \equiv D_1 D_2 \pmod 4$.*

There are in fact two proofs of this theorem, one analytic and one arithmetic. We give some brief indications of what their ingredients are, without defining all the terms occurring. In the analytic proof, one looks at the Hilbert modular group $\mathrm{SL}(2, \mathcal{O}_F)$ (see the notes by J. Bruinier in this volume) associated to the real quadratic field $F = \mathbb{Q}(\sqrt{D_1 D_2})$ and constructs a certain Eisenstein series for this group, of weight 1 and with respect to the "genus character" associated to the decomposition of the discriminant of F as $D_1 \cdot D_2$. Then one restricts this form to the diagonal $z_1 = z_2$ (the story here is actually more complicated: the Eisenstein series in question is non-holomorphic and one has to take the holomorphic projection of its restriction) and makes use of the fact that there are no holomorphic modular forms of weight 2 on Γ_1. In the arithmetic proof, one uses that $p \mid J(D_1, D_2)$ if and only the CM elliptic curves with j-invariants $j(\mathfrak{z}_{D_1})$ and $j(\mathfrak{z}_{D_2})$ become isomorphic over $\overline{\mathbb{F}}_p$. Now it is known that the ring of endomorphisms of any elliptic curve over $\overline{\mathbb{F}}_p$ is isomorphic either to an order in a quadratic field or to an order in the (unique) quaternion algebra $B_{p,\infty}$ over $\mathbb{Q}$ ramified at p and at infinity. For the elliptic curve $\overline{E}$ which is the common reduction of the curves with complex multiplication by $\mathcal{O}_{D_1}$ and $\mathcal{O}_{D_2}$ the first alternative cannot occur, since a quadratic order cannot contain two quadratic orders coming from different quadratic fields, so there must be an order in $B_{p,\infty}$ which contains two elements α_1 and α_2 with square D_1 and D_2, respectively. Then the element $\alpha = \alpha_1 \alpha_2$ also belongs to this order, and if x is its trace then $x \in \mathbb{Z}$ (because α is in an order and hence integral), $x^2 < N(\alpha) = D_1 D_2$ (because $B_{p,\infty}$ is ramified at infinity), and $x^2 \equiv D_1 D_2 \pmod p$ (because $B_{p,\infty}$ is ramified at p). This proves the theorem, except that we have lost a factor "4" because the elements α_i actually belong to the smaller orders $\mathcal{O}_{4D_i}$ and we should have worked with the elements $\alpha_i/2$ or $(1 + \alpha_i)/2$ (depending on the parity of D_i) in $\mathcal{O}_{D_i}$ instead.

The theorem stated above is actually only the qualitative version of the full result, which gives a complete formula for the prime factorization of $J(D_1, D_2)$. Assume for simplicity that D_1 and D_2 are fundamental. For each positive integer n of the form $\frac{1}{4}(D_1 D_2 - x^2)$, we define a function ε from the set of divisors of n to $\{\pm 1\}$ by the requirement that ε is completely multiplicative (i.e., $\varepsilon(p_1^{r_1} \cdots p_s^{r_s}) = \varepsilon(p_1)^{r_1} \cdots \varepsilon(p_s)^{r_s}$ for any divisor $p_1^{r_1} \cdots p_s^{r_s}$ of n) and is given on primes $p \mid n$ by $\varepsilon(p) = \chi_{D_1}(p)$ if $p \nmid D_1$ and by $\varepsilon(p) = \chi_{D_2}(p)$ if $p \nmid D_2$, where χ_D is the Dirichlet character modulo D introduced at the beginning of §3.2. Notice that this makes sense: at least one of the two alternatives must hold, since $(D_1, D_2) = 1$ and p is prime, and if they both hold then the two definitions agree because $D_1 D_2$ is then congruent to a non-zero square modulo p if p is odd and to an odd square modulo 8 if $p = 2$, so $\chi_{D_1}(p)\chi_{D_2}(p) = 1$. We then define $F(n)$ (still for n of the form

$\frac{1}{4}(D_1 D_2 - x^2))$ by

$$F(n) = \prod_{d|n} d^{\varepsilon(n/d)}.$$

This number, which is a priori only rational since $\varepsilon(n/d)$ can be positive or negative, is actually integral and in fact *is always a power of a single prime number*: one can show easily that $\varepsilon(n) = -1$, so n contains an odd number of primes p with $\varepsilon(p) = -1$ and $2 \nmid \mathrm{ord}_p(n)$, and if we write

$$n = p_1^{2\alpha_1+1} \cdots p_r^{2\alpha_r+1} p_{r+1}^{2\beta_1} \cdots p_{r+s}^{2\beta_s} q_1^{\gamma_1} \cdots q_t^{\gamma_t}$$

with r odd and $\varepsilon(p_i) = -1$, $\varepsilon(q_j) = +1$, $\alpha_i, \beta_i, \gamma_j \geq 0$, then

$$F(n) = \begin{cases} p^{(\alpha_1+1)(\gamma_1+1)\cdots(\gamma_t+1)} & \text{if } r = 1,\ p_1 = p, \\ 1 & \text{if } r \geq 3. \end{cases} \tag{90}$$

The complete formula for $J(D_1, D_2)$ is then

$$J(D_1, D_2)^{8/w(D_1)w(D_2)} = \prod_{\substack{x^2 < D_1 D_2 \\ x^2 \equiv D_1 D_2 \,(\mathrm{mod}\ 4)}} F\left(\frac{D_1 D_2 - x^2}{4}\right). \tag{91}$$

As an example, for $D_1 = -7$, $D_2 = -43$ (both with class number one) this formula gives

$$J(-7, -43) = \prod_{\substack{1 \leq x \leq 17 \\ x \text{ odd}}} F\left(\frac{301 - x^2}{4}\right)$$

$$= F(75)F(73)F(69)F(63)F(55)F(45)F(33)F(19)F(3)$$

$$= 3 \cdot 73 \cdot 3^2 \cdot 7 \cdot 5^2 \cdot 5 \cdot 3^2 \cdot 19 \cdot 3,$$

and indeed $j(\jmath_{-7}) - j(\jmath_{-43}) = -3375 + 884736000 = 3^6 \cdot 5^3 \cdot 7 \cdot 19 \cdot 73$. This is the only instance I know of in mathematics where the prime factorization of a number (other than numbers like $n!$ which are defined as products) can be described in closed form.

♠ Heights of Heegner Points

This "application" is actually not an application of the result just described, but of the methods used to prove it. As already mentioned, the above theorem about differences of singular moduli was found in connection with the study of the height of Heegner points on elliptic curves. In the last "application" in §6.1 we explained what Heegner points are, first as points on the modular curve $X_0(N) = \Gamma_0(N)\backslash\mathfrak{H} \cup \{\text{cusps}\}$ and then via the modular parametrization as points on an elliptic curve E of conductor N. In fact we do not have to pass to an elliptic curve; the $h(D)$ Heegner points on $X_0(N)$ corresponding to complex multiplication by the order $\mathcal{O}_D$ are defined over the Hilbert class field H of $K = \mathbb{Q}(\sqrt{D})$ and we can add these points on the Jacobian $J_0(N)$ of $X_0(N)$ (rather than adding their images in E as before). The fact that they are permuted by $\mathrm{Gal}(H/K)$ means that this sum is a point $P_K \in J_0(N)$ defined over K (and sometimes even over $\mathbb{Q}$). The main question is whether this is a torsion point or not; if it is not, then we have an interesting solution of a Diophantine equation (e.g., a non-trivial point on an elliptic curve over $\mathbb{Q}$). In general, whether a point P on an elliptic curve or on an abelian variety (like $J_0(N)$) is torsion or not is measured by an invariant called the (global) *height* of P, which is always ≥ 0 and which vanishes if and only if P has finite order. This height is defined as the sum of *local heights*, some of which are "archimedean" (i.e., associated to the complex geometry of the variety and the point) and can be calculated as transcendental expressions involving Green's functions, and some of which are "non-archimedean" (i.e., associated to the geometry of the variety and the point over the p-adic numbers for some prime number p) and can be calculated arithmetically as the product of $\log p$ with an integer which measures certain geometric intersection numbers in characteristic p. Actually, the height of a point is the value of a certain positive definite quadratic form on the Mordell-Weil group of the variety, and we can also consider the associated bilinear form (the "height pairing"), which must then be evaluated for a pair of Heegner points. If $N = 1$, then the Jacobian $J_0(N)$ is trivial and therefore all global heights are automatically zero. It turns out that the archimedean heights are essentially the logarithms of the absolute values of the individual terms in the right-hand side of (89), while the p-adic heights are the logarithms of the various factors $F(n)$ on the right-hand side of (91) which are given by formula (90) as powers of the given prime number p. The fact that the global height vanishes is therefore equivalent to formula (91) in this case. If $N > 1$, then in general (namely, whenever $X_0(N)$ has positive genus) the Jacobian is non-trivial and the heights do not have to vanish identically. The famous conjecture of Birch and Swinnerton-Dyer, one of the seven million-dollar Clay Millennium Problems, says that the heights of points on an abelian variety are related to the values or derivatives of a certain L-function; more concretely, in the case of an elliptic curve $E/\mathbb{Q}$, the conjecture predicts that the order to which the L-series $L(E, s)$ vanishes at $s = 1$ is equal to the rank of the Mordell-Weil group $E(\mathbb{Q})$ and that the value of the first non-zero derivative of $L(E, s)$ at $s = 1$ is equal to a certain explicit expression involving the height pairings of a system of generators of $E(\mathbb{Q})$ with one another. The same kind of calculations as in the case $N = 1$ permitted a verification of this prediction in the case of Heegner points, the relevant derivative of the L-series being the first one:

Theorem. *Let E be an elliptic curve defined over $\mathbb{Q}$ whose L-series vanishes at $s = 1$. Then the height of any Heegner point is an explicit (and in general non-zero) multiple of the derivative $L'(E/\mathbb{Q}, 1)$.*

The phrase "in general non-zero" in this theorem means that for any given elliptic curve E with $L(E,1) = 0$ but $L'(E,1) \neq 0$ there are Heegner points whose height is non-zero and which therefore have infinite order in the Mordell-Weil group $E(\mathbb{Q})$. We thus get the following (very) partial statement in the direction of the full BSD conjecture:

Corollary. *If $E/\mathbb{Q}$ is an elliptic curve over $\mathbb{Q}$ whose L-series has a simple zero at $s = 1$, then the rank of $E(\mathbb{Q})$ is at least one.*

(Thanks to subsequent work of Kolyvagin using his method of "Euler systems" we in fact know that the rank is exactly equal to one in this case.) In the opposite direction, there are elliptic curves $E/\mathbb{Q}$ and Heegner points P in $E(\mathbb{Q})$ for which we know that the multiple occurring in the theorem is non-zero but where P can be checked directly to be a torsion point. In that case the theorem says that $L'(E/\mathbb{Q},1)$ must vanish and hence, if the L-series is known to have a functional equation with a minus sign (the sign of the functional equation can be checked algorithmically), that $L(E/\mathbb{Q},s)$ has a zero of order at least 3 at $s = 1$. This is important because it is exactly the hypothesis needed to apply an earlier theorem of Goldfeld which, assuming that such a curve is known, proves that the class numbers of $h(D)$ go to infinity in an effective way as $D \to -\infty$. Thus modular methods and the theory of Heegner points suffice to solve the nearly 200-year problem, due to Gauss, of showing that the set of discriminants $D < 0$ with a given class number is finite and can be determined explicitly, just as in Heegner's hands they had already sufficed to solve the special case when the given value of the class number was one.   ♡

So far in this subsection we have discussed the *norms* of singular moduli (or more generally, the norms of their differences), but algebraic numbers also have *traces*, and we can consider these too. Now the normalization of j does matter; it turns out to be best to choose the normalized Hauptmodul $j_0(z) = j(z) - 744$. For every discriminant $D < 0$ we therefore define $T(D) \in \mathbb{Z}$ to be the trace of $j_0(3_D)$. This is the sum of the $h(D)$ singular moduli $j_0(3_{D,i})$, but just as in the case of the Hurwitz class numbers it is better to use the modified trace $T^*(d)$ which is defined as the sum of the values of $j_0(z)$ at all points $z \in \Gamma_1 \backslash \mathfrak{H}$ satisfying a quadratic equation, primitive or not, of discriminant D (and with the points i and ρ, if they occur at all, being counted with multiplicity $1/2$ or $1/3$ as usual), i.e., $T^*(D) = \sum_{r^2|D} T(D/r^2)/(\frac{1}{2}w(D/r^2))$ with the same conventions as in the definition of $h^*(D)$. The result, quite different from (and much easier to prove than) the formula for the norms, is that these numbers $T^*(D)$ are the coefficients of a modular form. Specifically, if we denote by $g(z)$ the meromorphic modular form $\theta_M(z)E_4(4z)/\eta(4z)^6$ of weight $3/2$, where $\theta_M(z)$ is the Jacobi theta function $\sum_{n \in \mathbb{Z}}(-1)^n q^{n^2}$ as in §3.1, then we have:

Theorem. *The Fourier expansion of $g(z)$ is given by*

$$g(z) = q^{-1} - 2 - \sum_{d>0} T^*(-d)\,q^d. \tag{92}$$

We can check the first few coefficients of this by hand: the Fourier expansion of g begins $q^{-1} - 2 + 248\,q^3 - 492\,q^4 + 4119\,q^7 - \cdots$ and indeed we have $T^*(-3) = \frac{1}{3}(0 - 744) = -248$, $T^*(-4) = \frac{1}{2}(1728 - 744) = 492$, and $T^*(-7) = -3375 - 744 = -4119$.

The proof of this theorem, though somewhat too long to be given here, is fairly elementary and is essentially a refinement of the method used to prove the Hurwitz-Kronecker class number relations (87) and (88). More precisely, (87) was proved by comparing the degrees on both sides of (86), and these in turn were computed by looking at the most negative exponent occurring in the q-expansion of the two sides when X is replaced by $j(z)$; in particular, the number $\sigma_1^+(m)$ came from the calculation in Proposition 24 of the most negative power of q in $\Psi_m(j(z), j(z))$. If we look instead at the *next* coefficient (i.e., that of $q^{-\sigma_1^+(m)+1}$), then we find $-\sum_{t^2<4m} T^*(t^2 - 4m)$ for the right-hand side of (86), because the modular functions $H_D(j(z))$ and $H_D^*(j(z))$ have Fourier expansions beginning $q^{-h(D)}(1 - T(D)q + O(q^2))$ and $q^{-h^*(D)}(1 - T^*(D)q + O(q^2))$, respectively, while by looking carefully at the "lower order terms" in the proof of Proposition 24 we find that the corresponding coefficient for the left-hand side of (86) vanishes for all m unless m or $4m+1$ is a square, when it equals $+4$ or -2 instead. The resulting identity can then be stated uniformly for all m as

$$\sum_{t\in\mathbb{Z}} T^*(t^2 - 4m) = 0 \qquad (m \geq 0), \tag{93}$$

where we have artificially defined $T^*(0) = 2$, $T^*(1) = -1$, and $T^*(D) = 0$ for $D > 1$ (a definition made plausible by the result in the theorem we want to prove). By a somewhat more complicated argument involving modular forms of higher weight, we prove a second relation, analogous to (88):

$$\sum_{t\in\mathbb{Z}} (m - t^2) T^*(t^2 - 4m) = \begin{cases} 1 & \text{if } m = 0, \\ 240\sigma_3(n) & \text{if } m \geq 1. \end{cases} \tag{94}$$

(Of course, the factor $m - t^2$ here could be replaced simply by $-t^2$, in view of (93), but (94) is more natural, both from the proof and by analogy with (88).) Now, just as in the discussion of the Hurwitz-Kronecker class number relations, the two formulas (93) and (94) suffice to determine all $T^*(D)$ by recursion. But in fact we can solve these equations directly, rather than recursively (though this was not done in the original paper). Write the right-hand side of (92) as $t_0(z) + t_1(z)$ where $t_0(z) = \sum_{m=0}^{\infty} T^*(-4m)\,q^{4m}$ and $t_1(z) = \sum_{m=0}^{\infty} T^*(1 - 4m)\,q^{4m-1}$, and define two unary theta series $\theta_0(z)$ and $\theta_1(z)$ ($= \theta(4z)$ and $\theta_F(4z)$ in the notation of §3.1) as the sums $\sum q^{t^2}$ with t ranging over the even or odd integers, respectively. Then (93) says that $t_0\theta_0 + t_1\theta_1 = 0$ and (94) says that $[t_0, \theta_0] + [t_1, \theta_1] = 4E_4(4z)$, where $[t_i, \theta_i] = \frac{3}{2}t_i\theta_i' - \frac{1}{2}t_i'\theta_i$ is the first Rankin–Cohen bracket of t_i (in weight 3/2) and θ_i. By taking a linear combination of the second equation with the derivative of the first, we deduce that $\begin{pmatrix} \theta_0(z) & \theta_1(z) \\ \theta_0'(z) & \theta_1'(z) \end{pmatrix} \begin{pmatrix} t_0(z) \\ t_1(z) \end{pmatrix} = 2 \begin{pmatrix} 0 \\ E_4(4z) \end{pmatrix}$, and from this we can immediately solve for $t_0(z)$ and $t_1(z)$ and hence for their sum, proving the theorem.

♠ The Borcherds Product Formula

This is certainly not an "application" of the above theorem in any reasonable sense, since Borcherds's product formula is much deeper and more general and was proved earlier, but it turns out that there is a very close link and that one can even use this to give an elementary proof of Borcherds's formula in a special case. This special case is the beautiful product expansion

$$H_D^*(j(z)) = q^{-h^*(D)} \prod_{n=1}^{\infty} (1 - q^n)^{A_D(n^2)}, \tag{95}$$

where $A_D(m)$ denotes the mth Fourier coefficient of a certain meromorphic modular form $f_D(z)$ (with Fourier expansion beginning $q^D + O(q)$) of weight $1/2$. The link is that one can prove in an elementary way a "duality formula" $A_D(m) = -B_m(|D|)$, where $B_m(n)$ is the coefficient of q^n in the Fourier expansion of a certain other meromorphic modular form $g_m(z)$ (with Fourier expansion beginning $q^{-m} + O(1)$) of weight $3/2$. For $m = 1$ the function g_m coincides with the g of the theorem above, and since (95) immediately implies that $T^*(D) = A_D(1)$ this and the duality imply (92). Conversely, by applying Hecke operators (in half-integral weight) in a suitable way, one can give a generalization of (92) to all functions g_{n^2}, and this together with the duality formula gives the complete formula (95), not just its subleading coefficient.   ♡

6.3 Periods and Taylor Expansions of Modular Forms

In §6.1 we showed that the value of any modular function (with rational or algebraic Fourier coefficients; we will not always repeat this) at a CM point $\mathfrak{z}$ is algebraic. This is equivalent to saying that for any modular form $f(z)$, of weight k, the value of $f(\mathfrak{z})$ is an algebraic multiple of $\Omega_{\mathfrak{z}}^k$, where $\Omega_{\mathfrak{z}}$ depends on $\mathfrak{z}$ only, not on f or on k. Indeed, the second statement implies the first by specializing to $k = 0$, and the first implies the second by observing that if $f \in M_k$ and $g \in M_\ell$ then f^ℓ / g^k has weight 0 and is therefore algebraic at $\mathfrak{z}$, so that $f(\mathfrak{z})^{1/k}$ and $g(\mathfrak{z})^{1/\ell}$ are algebraically proportional. Furthermore, the number $\Omega_{\mathfrak{z}}$ is unchanged (at least up to an algebraic number, but it is only defined up to an algebraic number) if we replace $\mathfrak{z}$ by $M\mathfrak{z}$ for any $M \in M(2, \mathbb{Z})$ with positive determinant, because $f(Mz)/f(z)$ is a modular function, and since any two CM points which generate the same imaginary quadratic field are related in this way, this proves:

Proposition 26. *For each imaginary quadratic field K there is a number $\Omega_K \in \mathbb{C}^*$ such that $f(\mathfrak{z}) \in \overline{\mathbb{Q}} \cdot \Omega_K^k$ for all $\mathfrak{z} \in K \cap \mathfrak{H}$, all $k \in \mathbb{Z}$, and all modular forms f of weight k with algebraic Fourier coefficients.* □

To find Ω_K, we should compute $f(\mathfrak{z})$ for some special modular form f (of non-zero weight!) and some point or points $\mathfrak{z} \in K \cap \mathfrak{H}$. A natural choice for the modular form is $\Delta(z)$, since it never vanishes. Even better, to achieve weight 1, is its 12th root $\eta(z)^2$, and better yet is the function $\Phi(z) = \Im(z)|\eta(z)|^4$ (which at $\mathfrak{z} \in K \cap \mathfrak{H}$ is an algebraic multiple of Ω_K^2), since it is Γ_1-invariant. As for the choice of $\mathfrak{z}$, we can look at the CM points of discriminant D (= discriminant of K), but since there are $h(D)$ of them and none should be preferred over the others (their j-invariants are conjugate algebraic numbers), the only reasonable choice is to multiply them all together and take the $h(D)$-th root – or rather the $h'(D)$-th root (where $h'(D)$ as previously denotes $h(D)/\frac{1}{2}w(D)$, i.e., $h'(D) = \frac{1}{3}, \frac{1}{2}$ or $h(K)$ for $D = -3$, $D = -4$, or $D < -4$), because the elliptic fixed points ρ and i of $\Gamma_1\backslash\mathfrak{H}$ are always to be counted with multiplicity $\frac{1}{3}$ and $\frac{1}{2}$, respectively. Surprisingly enough, the product of the invariants $\Phi(\mathfrak{z}_{D,i})$ $(i = 1, \ldots, h(D))$ can be evaluated in closed form:

Theorem. *Let K be an imaginary quadratic field of discriminant D. Then*

$$\prod_{\mathfrak{z} \in \Gamma_1 \backslash \mathfrak{z}_D} \left(4\pi\sqrt{|D|}\,\Phi(\mathfrak{z})\right)^{2/w(D)} = \prod_{j=1}^{|D|-1} \Gamma(j/|D|)^{\chi_D(j)}, \tag{96}$$

where χ_D is the quadratic character associated to K and $\Gamma(x)$ is the Euler gamma function.

Corollary. *The number Ω_K in Proposition 26 can be chosen to be*

$$\Omega_K = \frac{1}{\sqrt{2\pi|D|}} \left(\prod_{j=1}^{|D|-1} \Gamma\left(\frac{j}{|D|}\right)^{\chi_D(j)} \right)^{1/2h'(D)}. \tag{97}$$

Formula (96), usually called the Chowla–Selberg formula, is contained in a paper published by S. Chowla and A. Selberg in 1949, but it was later noticed that it already appears in a paper of Lerch from 1897. We cannot give the complete proof here, but we describe the main idea, which is quite simple. The Dedekind zeta function $\zeta_K(s) = \sum N(\mathfrak{a})^{-s}$ (sum over all non-zero integral ideals of K) has two decompositions: an additive one as $\sum_{\mathcal{A}} \zeta_{K,\mathcal{A}}(s)$, where $\mathcal{A}$ runs over the ideal classes of K and $\zeta_{K,\mathcal{A}}(s)$ is the associated "partial zeta function" $(= \sum N(\mathfrak{a})^{-s}$ with $\mathfrak{a}$ running over the ideals in $\mathcal{A})$, and a multiplicative one as $\zeta(s)L(s, \chi_D)$, where $\zeta(s)$ denotes the Riemann zeta function and $L(s, \chi_D) = \sum_{n=1}^{\infty} \chi_D(n)n^{-s}$. Using these two decompositions, one can compute in two different ways the two leading terms of the Laurent expansion $\zeta_K(s) = \frac{A}{s-1} + B + O(s-1)$ as $s \to 1$. The residue at $s = 1$ of $\zeta_{K,\mathcal{A}}(s)$ is independent of $\mathcal{A}$ and equals $\pi/\frac{1}{2}w(D)\sqrt{|D|}$, leading to Dirichlet's class number formula $L(1, \chi_D) = \pi \, h'(D)/\sqrt{|D|}$. (This is, of course, precisely the method Dirichlet used.) The constant term in the Laurent expansion of $\zeta_{K,\mathcal{A}}(s)$ at $s = 1$ is given by the famous *Kronecker limit formula* and is (up to some normalizing constants) simply the value of $\log(\Phi(\mathfrak{z}))$, where $\mathfrak{z} \in K \cap \mathfrak{H}$ is the CM point corresponding to the ideal class $\mathcal{A}$. The Riemann zeta function has the expansion $(s - 1)^{-1} - \gamma + O(s - 1)$ near $s = 1$ ($\gamma = $ Euler's constant), and $L'(1, \chi_D)$ can be computed by a relatively elementary analytic argument and turns out to be a simple multiple of $\sum_j \chi_D(j) \log \Gamma(j/|D|)$. Combining everything, one obtains (96).

As our first "application," we mention two famous problems of transcendence theory which were solved by modular methods, one using the Chowla–Selberg formula and one using quasimodular forms.

♠ Two Transcendence Results

In 1976, G.V. Chudnovsky proved that for any $z \in \mathfrak{H}$, at least two of the three numbers $E_2(z)$, $E_4(z)$ and $E_6(z)$ are algebraically independent. (Equivalently, the field generated by all $f(z)$ with $f \in \widetilde{M}_*(\Gamma_1)^{\mathbb{Q}} = \mathbb{Q}[E_2, E_4, E_6]$ has transcendence degree at least 2.) Applying this to $z = i$, for which $E_2(z) = 3/\pi$, $E_4(z) = 3\,\Gamma(\frac{1}{4})^8/(2\pi)^6$ and $E_6(z) = 0$, one deduces immediately that $\Gamma(\frac{1}{4})$ is transcendental (and in fact algebraically independent of π). Twenty years later, Nesterenko, building on earlier work of Barré-Sirieix, Diaz, Gramain and Philibert, improved this result dramatically by showing that for any $z \in \mathfrak{H}$ at least three of the four numbers $e^{2\pi i z}$, $E_2(z)$, $E_4(z)$ and $E_6(z)$ are algebraically independent. His proof used crucially the basic properties of the ring $\widetilde{M}_*(\Gamma_1)^{\mathbb{Q}}$ discussed in §5, namely, that it is closed under differentiation and that each of its elements is a power series in $q = e^{2\pi i z}$ with rational coefficients of bounded denominator and polynomial growth. Specialized to $z = i$, Nesterenko's result implies that the three numbers π, e^π and $\Gamma(\frac{1}{4})$ are algebraically independent. The algebraic independence of π and e^π (even without $\Gamma(\frac{1}{4})$) had been a famous open problem.   ♡

♠ Hurwitz Numbers

Euler's famous result of 1734 that $\zeta(2r)/\pi^{2r}$ is rational for every $r \geq 1$ can be restated in the form

$$\sum_{\substack{n \in \mathbb{Z} \\ n \neq 0}} \frac{1}{n^k} = \text{(rational number)} \cdot \pi^k \qquad \text{for all } k \geq 2, \qquad (98)$$

where the "rational number" of course vanishes for k odd since then the contributions of n and $-n$ cancel. This result can be obtained, for instance, by looking at the Laurent expansion of $\cot x$ near the origin. Hurwitz asked the corresponding question if one replaces $\mathbb{Z}$ in (98) by the ring $\mathbb{Z}[i]$ of Gaussian integers and, by using an elliptic function instead of a trigonometric one, was able to prove the corresponding assertion

$$\sum_{\substack{\lambda \in \mathbb{Z}[i] \\ \lambda \neq 0}} \frac{1}{\lambda^k} = \frac{H_k}{k!}\, \omega^k \qquad \text{for all } k \geq 3, \tag{99}$$

for certain rational numbers $H_4 = \frac{1}{10}$, $H_8 = \frac{3}{10}$, $H_{12} = \frac{567}{130}$, ... (here $H_k = 0$ for $4 \nmid k$ since then the contributions of λ and $-\lambda$ or of λ and $i\lambda$ cancel), where

$$\omega = 4 \int_0^1 \frac{dx}{\sqrt{1 - x^4}} = \frac{\Gamma(\frac{1}{4})^2}{\sqrt{2\pi}} = 5.24411 \cdots.$$

Instead of using the theory of elliptic functions, we can see this result as a special case of Proposition 26, since the sum on the left of (99) is just the special value of the modular form $2G_k(z)$ defined in (10), which is related by (12) to the modular form $\mathbb{G}_k(z)$ with rational Fourier coefficients, at $z = i$ and ω is $2\pi\sqrt{2}$ times the Chowla–Selberg period $\Omega_{\mathbb{Q}(i)}$. Similar considerations, of course, apply to the sum $\sum \lambda^{-k}$ with λ running over the non-zero elements of the ring of integers (or of any other ideal) in any imaginary quadratic field, and more generally to the special values of L-series of Hecke "grossencharacters" which we will consider shortly. $\heartsuit$

We now turn to the second topic of this subsection: the Taylor expansions (as opposed to simply the values) of modular forms at CM points. As we already explained in the paragraph preceding equation (58), the "right" Taylor expansion for a modular form f at a point $z \in \mathfrak{H}$ is the one occurring on the left-hand side of that equation, rather than the straight Taylor expansion of f. The beautiful fact is that, if z is a CM point, then after a renormalization by dividing by suitable powers of the period Ω_z, each coefficient of this expansion is an algebraic number (and in many cases even rational). This follows from the following proposition, which was apparently first observed by Ramanujan.

Proposition 27. *The value of $E_2^*(\mathfrak{z})$ at a CM point $\mathfrak{z} \in K \cap \mathfrak{H}$ is an algebraic multiple of Ω_K^2.*

Corollary. *The value of $\partial^n f(\mathfrak{z})$, for any modular form f with algebraic Fourier coefficients, any integer $n \geq 0$ and any CM point $\mathfrak{z} \in K \cap \mathfrak{H}$, is an algebraic multiple of Ω_K^{k+2n}, where k is the weight of f.*

Proof. We will give only a sketch, since the proof is similar to that already given for $j(z)$. For $M = \begin{pmatrix} a & b \\ c & d \end{pmatrix} \in \mathcal{M}_m$ we define $(E_2^*|_2 M)(z) = m(cz + d)^{-2} E_2^*(Mz)$ (= the usual slash operator for the matrix $m^{-1/2}M \in \mathrm{SL}(2, \mathbb{R})$). From formula (1) and the fact that $E_2^*(z)$ is a linear combination of $1/y$ and a holomorphic function, we deduce immediately that the difference $E_2^* - E_2^*|_2 M$ is a holomorphic modular form of weight 2 (on the subgroup of finite index $M^{-1}\Gamma_1 M \cap \Gamma_1$ of Γ_1). It then follows by an argument similar to that in the proof of Proposition 23 that the function

$$P_m(z, X) := \prod_{M \in \Gamma_1 \backslash \mathcal{M}_m} \left(X + E_2^*(z) - (E_2^*|_2 M)(z)\right) \qquad (z \in \mathfrak{H},\, X \in \mathbb{C})$$

is a polynomial in X (of degree $\sigma_1(m)$) whose coefficients are modular forms of appropriate weights on Γ_1 with rational coefficients. (For example, $P_2(z, X) = X^3 - \frac{3}{4} E_4(z) X + \frac{1}{4} E_6(z)$.) The algebraicity of $E_2^*(\mathfrak{z})/\Omega_K^2$ for a CM point $\mathfrak{z}$ now follows from Proposition 26, since if $M \notin \mathbb{Z}\cdot \mathrm{Id}_2$ fixes $\mathfrak{z}$ then $E_2^*(\mathfrak{z}) - (E_2^*|_2 M)(\mathfrak{z})$ is a non-zero algebraic multiple of $E_2^*(\mathfrak{z})$. The corollary follows from (58) and the fact that each non-holomorphic derivative $\partial^n f(z)$ is a polynomial in $E_2^*(z)$ with coefficients that are holomorphic modular forms with algebraic Fourier coefficients, as one sees from equation (66).

Propositions 26 and 27 are illustrated for $\mathfrak{z} = \mathfrak{z}_D$ and $f = E_2^*$, E_4, E_6 and Δ in the following table, in which α in the penultimate row is the real root of $\alpha^3 - \alpha - 1 = 0$. Observe that the numbers in the final column of this table are all units; this is part of the general theory.

D	$\dfrac{\lvert D\rvert^{1/2} E_2^*(\mathfrak{z}_D)}{\Omega_D^2}$	$\dfrac{E_4(\mathfrak{z}_D)}{\Omega_D^4}$	$\dfrac{E_6(\mathfrak{z}_D)}{\lvert D\rvert^{1/2}\Omega_D^6}$	$\dfrac{\Delta(\mathfrak{z}_D)}{\Omega_D^{12}}$
-3	0	0	24	-1
-4	0	12	0	1
-7	3	15	27	-1
-8	4	20	28	1
-11	8	32	56	-1
-15	$6 + 3\sqrt{5}$	$15 + 12\sqrt{5}$	$42 + \frac{63}{\sqrt{5}}$	$\frac{3-\sqrt{5}}{2}$
-19	24	96	216	-1
-20	$12 + 4\sqrt{5}$	$40 + 12\sqrt{5}$	$72 + \frac{112}{\sqrt{5}}$	$\sqrt{5} - 2$
-23	$\frac{7 + 11\alpha + 12\alpha^2}{\alpha^{1/3}}$	$5\alpha^{1/3}(6 + 4\alpha + \alpha^2)$	$\frac{469 + 1176\alpha + 504\alpha^2}{23}$	$-\alpha^{-8}$
-24	$12 + 12\sqrt{2}$	$60 + 24\sqrt{2}$	$84 + 72\sqrt{2}$	$3 - 2\sqrt{2}$

In the paragraph preceding Proposition 17 we explained that the non-holomorphic derivatives $\partial^n f$ of a holomorphic modular form $f(z)$ are more natural and more fundamental than the holomorphic derivatives $D^n f$, because the Taylor series $\sum D^n f(z)\, t^n/n!$ represents f only in the disk $|z' - z| < \Im(z)$ whereas the series (58) represents f everywhere. The above corollary gives a second reason to prefer the $\partial^n f$: at a CM point $\mathfrak{z}$ they are monomials in the period $\Omega_{\mathfrak{z}}$ with algebraic coefficients, whereas the derivatives $D^n f(\mathfrak{z})$ are polynomials in $\Omega_{\mathfrak{z}}$ by equation (57) (in which there can be no cancellation since $\Omega_{\mathfrak{z}}$ is known to be transcendental). If we set

$$c_n = c_n(f, \mathfrak{z}, \Omega) = \frac{\partial^n f(\mathfrak{z})}{\Omega^{k+2n}} \qquad (n = 0, 1, 2, \dots), \qquad (100)$$

where Ω is a suitably chosen algebraic multiple of Ω_K, then the c_n (up to a possible common denominator which can be removed by multiplying f by a suitable integer) are even algebraic *integers*, and they still can be considered to be "Taylor coefficients of f at $\mathfrak{z}$" since by (58) the series $\sum_{n=0}^{\infty} c_n t^n/n!$ equals $(Ct + D)^{-k} f\left(\frac{At+B}{Ct+D}\right)$ where $\left(\begin{smallmatrix} A & B \\ C & D \end{smallmatrix}\right) = \left(\begin{smallmatrix} -\bar{z} & z \\ -1 & 1 \end{smallmatrix}\right) \left(\begin{smallmatrix} 1/4\pi y\Omega & 0 \\ 0 & \Omega \end{smallmatrix}\right) \in \mathrm{GL}(2, \mathbb{C})$. These normalized Taylor coefficients have several interesting number-theoretical applications, as we will see below, and from many points of view are actually better number-theoretic invariants than the more familiar coefficients a_n of the Fourier expansion $f = \sum a_n q^n$. Surprisingly enough, they are also much easier to calculate: unlike the a_n, which are mysterious numbers (think of the Ramanujan function $\tau(n)$) and have to be calculated anew for each modular form, the Taylor coefficients $\partial^n f$ or c_n are always given by a simple recursive procedure. We illustrate the procedure by calculating the numbers $\partial^n E_4(i)$, but the method is completely general and works the same way for all modular forms (even of half-integral weight, as we will see in §6.4).

Proposition 28. *We have* $\partial^n E_4(i) = p_n(0) E_4(i)^{1+n/2}$, *with* $p_n(t) \in \mathbb{Z}[\frac{1}{6}][t]$ *defined recursively by*

$$p_0(t) = 1, \quad p_{n+1}(t) = \frac{t^2 - 1}{2} p_n'(t) - \frac{n+2}{6} t p_n(t) - \frac{n(n+3)}{144} p_{n-1}(t) \quad (n \geq 0).$$

Proof. Since $E_2^*(i) = 0$, equation (66) implies that $\tilde{f}_{\partial}(i, X) = \tilde{f}_{\vartheta}(i, X)$ for any $f \in M_k(\Gamma_1)$, i.e., we have $\partial^n f(i) = \vartheta^{[n]} f(i)$ for all $n \geq 0$, where $\{\vartheta^{[n]} f\}_{n=0,1,2,\ldots}$ is defined by (64). We use Ramanujan's notations Q and R for the modular forms $E_4(z)$ and $E_6(z)$. By (54), the derivation ϑ sends Q and R to $-\frac{1}{3}R$ and $-\frac{1}{2}Q^2$, respectively, so ϑ acts on $M_*(\Gamma_1) = \mathbb{C}[Q, R]$ as $-\frac{R}{3}\frac{\partial}{\partial Q} - \frac{Q^2}{2}\frac{\partial}{\partial R}$. Hence $\vartheta^{[n]} Q = P_n(Q, R)$ for all n, where the polynomials $P_n(Q, R) \in \mathbb{Q}[Q, R]$ are given recursively by

$$P_0 = Q, \quad P_{n+1} = -\frac{R}{3}\frac{\partial P_n}{\partial Q} - \frac{Q^2}{2}\frac{\partial P_n}{\partial R} - n(n+3)\frac{Q P_{n-1}}{144}.$$

Since $P_n(Q, R)$ is weighted homogeneous of weight $2n + 4$, where Q and R have weight 4 and 6, we can write $P_n(Q, R)$ as $Q^{1+n/2} p_n(R/Q^{3/2})$ where p_n is a polynomial in one variable. The recursion for P_n then translates into the recursion for p_n given in the proposition.

The first few polynomials p_n are $p_1 = -\frac{1}{3}t$, $p_2 = \frac{5}{36}$, $p_3 = -\frac{5}{72}t$, $p_4 = \frac{5}{216}t^2 + \frac{5}{288}$, $\ldots$, giving $\partial^2 E_4(i) = \frac{5}{36} E_4(i)^2$, $\partial^4 E_4(i) = \frac{5}{288} E_4(i)^3$, etc. (The values for n odd vanish because i is a fixed point of the element $S \in \Gamma_1$ of order 2.) With the same method we find $\partial^n f(i) = q_n(0) E_4(i)^{(k+2n)/4}$ for any modular form $f \in M_k(\Gamma_1)$, with polynomials q_n satisfying the same recursion as p_n but with $n(n+3)$ replaced by $n(n+k-1)$ (and of course with a different initial value $q_0(t)$). If we consider a CM point $\mathfrak{z}$ other than i, then the method and result are similar but we have to use (68) instead of (66), where ϕ is a quasimodular form differing from E_2 by a (meromorphic) modular form of weight 2 and chosen so that $\phi^*(\mathfrak{z})$ vanishes. If we replace Γ_1 by some other group Γ, then the same method works in principle but we need an explicit description of the ring $M_*(\Gamma)$ to replace the description of $M_*(\Gamma_1)$ as $\mathbb{C}[Q, R]$ used above, and if the genus of the group is larger than 0 then the polynomials $p_n(t)$ have to be replaced by elements q_n of some fixed finite algebraic extension of $\mathbb{Q}(t)$, again satisfying a recursion of the form $q_{n+1} = Aq'_n + (nB+C)q_n + n(n+k-1)Dq_{n-1}$ with A, B, C and D independent of n. The general result says that the values of the non-holomorphic derivatives $\partial^n f(z_0)$ of any modular form at any point $z_0 \in \mathfrak{H}$ are given "quasi-recursively" as special values of a sequence of algebraic functions in one variable which satisfy a differential recursion.

♠ Generalized Hurwitz Numbers

In our last "application" we studied the numbers $G_k(i)$ whose values, up to a factor of 2, are given by the Hurwitz formula (99). We now discuss the meaning of the non-holomorphic derivatives $\partial^n G_k(i)$. From (55) we find

$$\partial_k\left(\frac{1}{(mz+n)^k}\right) = \frac{k}{2\pi i(z-\bar{z})} \cdot \frac{m\bar{z}+n}{(mz+n)^{k+1}}$$

and more generally

$$\partial_k^r\left(\frac{1}{(mz+n)^k}\right) = \frac{(k)_r}{(2\pi i(z-\bar{z}))^r} \cdot \frac{(m\bar{z}+n)^r}{(mz+n)^{k+r}}$$

for all $r \geq 0$, where $\partial_k^r = \partial_{k+2r-2} \circ \cdots \circ \partial_{k+2} \circ \partial_k$ and $(k)_r = k(k+1)\cdots(k+r-1)$ as in §5. Thus

$$\partial^n G_k(i) = \frac{(k)_n}{2(-4\pi)^n} \sum_{\substack{\lambda \in \mathbb{Z}[i] \\ \lambda \neq 0}} \frac{\bar{\lambda}^n}{\lambda^{k+n}} \qquad (n = 0, 1, 2, \dots)$$

(and similarly for $\partial^n G_k(\mathfrak{z})$ for any CM point $\mathfrak{z}$, with $\mathbb{Z}[i]$ replaced by $\mathbb{Z}\mathfrak{z} + \mathbb{Z}$ and -4π by $-4\pi\Im(\mathfrak{z})$). If we observe that the class number of the field $K = \mathbb{Q}(i)$ is 1 and that any integral ideal of K can be written as a principal ideal (λ) for exactly four numbers $\lambda \in \mathbb{Z}[i]$, then we can write

$$\sum_{\substack{\lambda \in \mathbb{Z}[i] \\ \lambda \neq 0}} \frac{\bar{\lambda}^n}{\lambda^{k+n}} = \sum_{\substack{\lambda \in \mathbb{Z}[i] \\ \lambda \neq 0}} \frac{\bar{\lambda}^{k+2n}}{(\lambda\bar{\lambda})^{k+n}} = 4\sum_{\mathfrak{a}} \frac{\psi_{k+2n}(\mathfrak{a})}{N(\mathfrak{a})^{k+n}}$$

where the sum runs over the integral ideals $\mathfrak{a}$ of $\mathbb{Z}[i]$ and $\psi_{k+2n}(\mathfrak{a})$ is defined as $\bar{\lambda}^{k+2n}$, where λ is any generator of $\mathfrak{a}$. (This is independent of λ if $k+2n$ is divisible by 4, and in the contrary case the sum vanishes.) The functions ψ_{k+2n} are called Hecke "grossencharacters" (the German original of this semi-anglicized word is "Größencharaktere", with five differences of spelling, and means literally "characters of size," referring to the fact that these characters, unlike the usual ideal class characters of finite order, depend on the size of the generator of a principal ideal) and their L-series $L_K(s, \psi_{k+2n}) = \sum_\mathfrak{a} \psi_{k+2n}(\mathfrak{a})N(\mathfrak{a})^{-s}$ are an important class of L-functions with known analytic continuation and functional equations. The above calculation shows that $\partial^n G_k(i)$ (or more generally the non-holomorphic derivatives of any holomorphic Eisenstein series at any CM point) are simple multiples of special values of these L-series at integral arguments, and Proposition 28 and its generalizations give us an algorithmic way to compute these values in closed form. $\heartsuit$

6.4 CM Elliptic Curves and CM Modular Forms

In the introduction to §6, we defined elliptic curves with complex multiplication as quotients $E = \mathbb{C}/\Lambda$ where $\lambda\Lambda \subseteq \Lambda$ for some non-real complex number λ. In that case, as we have seen, the lattice Λ is homothetic to $\mathbb{Z}\mathfrak{z} + \mathbb{Z}$ for some CM point $\mathfrak{z} \in \mathfrak{H}$ and the singular modulus $j(E) = j(\mathfrak{z})$ is algebraic, so E has a model over $\overline{\mathbb{Q}}$. The map from E to E induced by multiplication by λ is also algebraic and defined over $\overline{\mathbb{Q}}$. For simplicity, we concentrate on those $\mathfrak{z}$ whose discriminant is one of the 13 values $-3, -4, -7, \ldots, -163$ with class number 1, so that $j(\mathfrak{z}) \in \mathbb{Q}$ and E (but not the complex multiplication) can be defined over $\mathbb{Q}$. For instance, the three elliptic curves

$$y^2 = x^3 + x, \qquad y^2 = x^3 + 1, \qquad y^2 = x^3 - 35x - 98 \qquad (101)$$

have j-invariants 1728, 0 and -3375, corresponding to multiplication by the orders $\mathbb{Z}[i]$, $\mathbb{Z}\left[\frac{1+\sqrt{-3}}{2}\right]$, and $\mathbb{Z}\left[\frac{1+\sqrt{-7}}{2}\right]$, respectively. For the first curve (or more generally any curve of the form $y^2 = x^3 + Ax$ with $A \in \mathbb{Z}$) the multiplication by i corresponds to the obvious endomorphism $(x, y) \mapsto (-x, iy)$ of the curve, and similarly for the second curve (or any curve of the form $y^2 = x^3 + B$) we have the equally obvious endomorphism $(x, y) \mapsto (\omega x, y)$ of order 3, where ω is a non-trivial cube root of 1. For the third curve (or any of its "twists" $E : Cy^2 = x^3 - 35x - 98$) the existence of a non-trivial endomorphism is less obvious. One checks that the map

$$\phi : (x, y) \mapsto \left(\gamma^2\left(x + \frac{\beta^2}{x + \alpha}\right), \gamma^3 y\left(1 - \frac{\beta^2}{(x + \alpha)^2}\right)\right),$$

where $\alpha = (7 + \sqrt{-7})/2$, $\beta = (7 + 3\sqrt{-7})/2$, and $\gamma = (1 + \sqrt{-7})/4$, maps E to itself and satisfies $\phi(\phi(P)) - \phi(P) + 2P = 0$ for any point P on E, where the addition is with respect to the group law on the curve, so that we have a map from $\mathcal{O}_{-7}$ to the endomorphisms of E sending $\lambda = m\frac{1+\sqrt{-7}}{2} + n$ to the endomorphism $P \mapsto m\phi(P) + nP$.

The key point about elliptic curves with complex multiplication is that the number of their points over finite fields is given by a simple formula. For the three curves above this looks as follows. Recall that the number of points over $\mathbb{F}_p$ of an elliptic curve $E/\mathbb{Q}$ given by a Weierstrass equation

$y^2 = F(x) = x^3 + Ax + B$ equals $p + 1 - a_p$ where a_p (for p odd and not dividing the discriminant of F) is given by $-\sum_{x \;(\mathrm{mod}\; p)} \left(\frac{F(x)}{p} \right)$. For the three curves in (101) we have

$$\sum_{x \;(\mathrm{mod}\; p)} \left(\frac{x^3 + x}{p} \right) = \begin{cases} 0 & \text{if } p \equiv 3 \;(\mathrm{mod}\; 4), \\ -2a & \text{if } p = a^2 + 4b^2, \, a \equiv 1 \;(\mathrm{mod}\; 4) \end{cases} \qquad (102a)$$

$$\sum_{x \;(\mathrm{mod}\; p)} \left(\frac{x^3 + 1}{p} \right) = \begin{cases} 0 & \text{if } p \equiv 2 \;(\mathrm{mod}\; 3), \\ -2a & \text{if } p = a^2 + 3b^2, \, a \equiv 1 \;(\mathrm{mod}\; 3) \end{cases} \qquad (102b)$$

$$\sum_{x \;(\mathrm{mod}\; p)} \left(\frac{x^3 - 35x - 98}{p} \right) = \begin{cases} 0 & \text{if } (p/7) = -1, \\ -2a & \text{if } p = a^2 + 7b^2, \, (a/7) = 1. \end{cases} \qquad (102c)$$

(For other D with $h(D) = 1$ we would get a formula for $a_p(E)$ as $\pm A$ where $4p = A^2 + |D|B^2$.) The proofs of these assertions, and of the more general statements needed when $h(D) > 1$, will be omitted since they would take us too far afield, but we give the proof of the first (due to Gauss), since it is elementary and quite pretty. We prove a slightly more general but less precise statement.

Proposition 29. *Let p be an odd prime and A an integer not divisible by p. Then*

$$\sum_{x \;(\mathrm{mod}\; p)} \left(\frac{x^3 + Ax}{p} \right) = \begin{cases} 0 & \text{if } p \equiv 3 \;(\mathrm{mod}\; 4), \\ \pm 2a & \text{if } p \equiv 1 \;(\mathrm{mod}\; 4) \text{ and } (A/p) = 1, \\ \pm 4b & \text{if } p \equiv 1 \;(\mathrm{mod}\; 4) \text{ and } (A/p) = -1, \end{cases} \qquad (103)$$

where $|a|$ and $|b|$ in the second and third lines are defined by $p = a^2 + 4b^2$.

Proof. The first statement is trivial since if $p \equiv 3 \;(\mathrm{mod}\; 4)$ then $(-1/p) = -1$ and the terms for x and $-x$ in the sum cancel, so we can suppose that $p \equiv 1 \;(\mathrm{mod}\; 4)$. Denote the sum on the left-hand side of (103) by $s_p(A)$. Replacing x by rx with $r \not\equiv 0 \;(\mathrm{mod}\; p)$ shows that $s_p(r^2 A) = \left(\frac{r}{p} \right) s_p(A)$, so the number $s_p(A)$ takes on only four values, say $\pm 2\alpha$ for $A = g^{4i}$ or $A = g^{4i+2}$ and $\pm 2\beta$ for $A = g^{4i+1}$ or $A = g^{4i+3}$, where g is a primitive root modulo p. (That $s_p(A)$ is always even is obvious by replacing x by $-x$.) Now we take the sum of the squares of $s_p(A)$ as A ranges over all integers modulo p, noting that $s_p(0) = 0$. This gives

$$2(p-1)(\alpha^2 + \beta^2) = \sum_{A, x, y \in \mathbb{F}_p} \left(\frac{x^3 + Ax}{p} \right) \left(\frac{y^3 + Ay}{p} \right)$$

$$= \sum_{x, y \in \mathbb{F}_p} \left(\frac{xy}{p} \right) \sum_{A \in \mathbb{F}_p} \left(\frac{(x^2 + A)(y^2 + A)}{p} \right)$$

$$= \sum_{x, y \in \mathbb{F}_p} \left(\frac{xy}{p} \right) (-1 + p\, \delta_{x^2, y^2}) = 2p(p-1),$$

where in the last line we have used the easy fact that $\sum_{z \in \mathbb{F}_p} \left(\frac{z(z+r)}{p} \right)$ equals $p - 1$ for $r \equiv 0 \pmod{p}$ and -1 otherwise. (Proof: the first statement is obvious; the substitution $z \mapsto rz$ shows that the sum is independent of r if $r \not\equiv 0$; and the sum of the values for all integers r mod p clearly vanishes.) Hence $\alpha^2 + \beta^2 = p$. It is also obvious that α is odd (for $(A/p) = +1$ there are $\frac{p-1}{2} - 1$ values of x between 0 and $p/2$ with $\left(\frac{x^3 \pm Ax}{p} \right)$ non-zero and hence odd) and that β is even (for $(A/p) = -1$ all $\frac{p-1}{2}$ values of $\left(\frac{x^3 \pm Ax}{p} \right)$ with $0 < x < p/2$ are odd), so $\alpha = \pm a$, $\beta = \pm 2b$ as asserted.

An almost exactly similar proof works for equation (102b): here one defines $s_p(B) = \sum_x \left(\frac{x^3 + B}{p} \right)$ and observes that $s_p(r^3 B) = \left(\frac{r}{p} \right) s_p(B)$, so that $s_p(B)$ takes on six values $\pm \alpha$, $\pm \beta$ and $\pm \gamma$ depending on the class of $i \pmod 6$, where $p \equiv 1 \pmod 6$ (otherwise all $s_p(B)$ vanish) and $B = g^i$ with g a primitive root modulo p. Summing $s_p(B)$ over all quadratic residues $B \pmod p$ shows that $\alpha + \beta + \gamma = 0$, and summing $s_p(B)^2$ over all B gives $\alpha^2 + \beta^2 + \gamma^2 = 6p$. These two equations imply that $\alpha \equiv \beta \equiv \gamma \not\equiv 0 \pmod 3$ and that $4p = \alpha^2 + 3((\beta - \gamma)/3)^2$, which gives (102b) since it is easily seen that α is even and β and γ are odd. I do not know of any elementary proof of this sort for equation (102c) or for the similar identities corresponding to the other imaginary quadratic fields of class number 1. The reader is urged to try to find such a proof.

♠ Factorization, Primality Testing, and Cryptography

We mention briefly one "practical" application of complex multiplication theory. Many methods of modern cryptography depend on being able to identify very large prime numbers quickly or on being able to factor (or being relatively sure that no one else will be able to factor) very large composite numbers. Several methods involve the arithmetic of elliptic curves over finite fields, which yield finite groups in which certain operations are easily performed but not easily inverted. The difficulty of the calculations, and hence the security of the method, depends on the structure of the group of points of the curve over the finite fields, so one would like to be able to construct, say, examples of elliptic curves E over $\mathbb{Q}$ whose reduction modulo p for some very large prime p has an order which itself contains a very large prime factor. Since counting the points on $E(\mathbb{F}_p)$ directly is impractical when p is very large, it is essential here to know curves E for which the number a_p, and hence the cardinality of $E(\mathbb{F}_p)$, is known a priori, and the existence of closed formulas like the ones in (102) implies that the curves with complex multiplication are suitable. In practice one wants the complex multiplication to be by an order in a quadratic field which is not too big but also not too small. For this purpose one needs effective ways to construct the Hilbert class fields (which is where the needed singular moduli will lie) efficiently, and here again the methods mentioned in 6.1, and in particular the simplifications arising by replacing the modular function $j(z)$ by better modular functions, become relevant. For more information, see the bibliography.   ♡

We now turn from elliptic curves with complex multiplication to an important and related topic, the so-called CM modular forms. Formula (102a) says that the coefficient a_p of the L-series $L(E, s) = \sum a_n n^{-s}$ of the elliptic curve $E : y^2 = x^3 + x$ for a prime p is given by $a_p = \text{Tr}(\lambda) = \lambda + \bar{\lambda}$, where λ is one of the two numbers of norm p in $\mathbb{Z}[i]$ of the form $a + bi$ with $a \equiv 1 \pmod 4$, $b \equiv 0 \pmod 2$ (or is zero if there is no such λ). Now using the multiplicative properties of the a_n we find that the full L-series is given by

$$L(E, s) = L_K(s, \psi_1), \tag{104}$$

where $L_K(s, \psi_1) = \sum_{0 \neq \mathfrak{a} \subseteq \mathbb{Z}[i]} \psi_1(\mathfrak{a}) N(\mathfrak{a})^{-s}$ is the L-series attached as in §6.3 to the field $K = \mathbb{Q}(i)$ and the grossencharacter ψ_1 defined by

$$\psi_1(\mathfrak{a}) = \begin{cases} 0 & \text{if } 2 \mid N(\mathfrak{a}), \\ \lambda & \text{if } 2 \nmid N(\mathfrak{a}), \, \mathfrak{a} = (\lambda), \, \lambda \in 4\mathbb{Z} + 1 + 2\mathbb{Z}\,i. \end{cases} \tag{105}$$

In fact, the L-series of the elliptic curve E belongs to *three* important classes of L-series: it is an L-function coming from algebraic geometry (by definition), the L-series of a generalized character of an algebraic number field (by (104)), and also the L-series of a modular form, namely $L(E, s) = L(f_E, s)$ where

$$f_E(z) = \sum_{0 \neq \mathfrak{a} \subseteq \mathbb{Z}[i]} \psi_1(\mathfrak{a})\, q^{N(\mathfrak{a})} = \sum_{\substack{a \equiv 1 \,(\text{mod } 4) \\ b \equiv 0 \,(\text{mod } 2)}} a\, q^{a^2 + b^2} \tag{106}$$

which is a theta series of the type mentioned in the final paragraph of §3, associated to the binary quadratic form $Q(a, b) = a^2 + b^2$ and the spherical polynomial $P(a, b) = a$ (or $a + ib$) of degree 1. The same applies, with suitable modifications, for any elliptic curve with complex multiplication, so that the Taniyama-Weil conjecture, which says that the L-series of an elliptic curve over $\mathbb{Q}$ should coincide with the L-series of a modular form of weight 2, can be seen explicitly for this class of curves (and hence was known for them long before the general case was proved).

The modular form f_E defined in (106) can also be denoted θ_{ψ_1}, where ψ_1 is the grossencharacter (105). More generally, the L-series of the powers $\psi_d = \psi_1^d$ of ψ_1 are the L-series of modular forms, namely $L_K(\psi_d, s) = L(\theta_{\psi_d}, s)$ where

$$\theta_{\psi_d}(z) = \sum_{0 \neq \mathfrak{a} \subseteq \mathbb{Z}[i]} \psi_d(\mathfrak{a})\, q^{N(\mathfrak{a})} = \sum_{\substack{a \equiv 1 \,(\text{mod } 4) \\ b \equiv 0 \,(\text{mod } 2)}} (a + ib)^d\, q^{a^2 + b^2}, \tag{107}$$

which is a modular form of weight $d + 1$. Modular forms constructed in this way – i.e., theta series associated to a binary quadratic form $Q(a, b)$ and a spherical polynomial $P(a, b)$ of arbitrary degree d – are called *CM modular forms*, and have several remarkable properties. First of all, they are always linear combinations of the theta series $\theta_\psi(z) = \sum_\mathfrak{a} \psi(\alpha)\, q^{N(\mathfrak{a})}$ associated to some grossencharacter ψ of an imaginary quadratic field. (This is because any positive definite binary quadratic form over $\mathbb{Q}$ is equivalent to the norm form $\lambda \mapsto \lambda\bar{\lambda}$ on an ideal in an imaginary quadratic field, and since the Laplace operator associated to this form is $\frac{\partial^2}{\partial\lambda\,\partial\bar{\lambda}}$, the only spherical polynomials of degree d are linear combinations of λ^d and $\bar{\lambda}^d$.) Secondly, since the L-series $L(\theta_\psi, s) = L_K(\psi, s)$ has an Euler product, the modular forms θ_ψ attached to grossencharacters are always Hecke eigenforms, and this is the *only* infinite family of Hecke eigenforms, apart from Eisenstein series, which are known explicitly. (Other eigenforms do not appear to have any systematic rule of construction, and even the number fields in which their Fourier coefficients lie are totally mysterious, an example being the field $\mathbb{Q}(\sqrt{144169})$ of coefficients of the cusp form of level 1 and weight 24 mentioned in §4.1.) Thirdly, sometimes modular forms constructed by other methods turn out to be of CM type, leading to new identities. For instance, in four cases the modular form defined in (107) is a product of eta-functions: $\theta_{\psi_0}(z) = \eta(8z)^4/\eta(4z)^2$, $\theta_{\psi_1}(z) = \eta(8z)^8/\eta(4z)^2\eta(16z)^2$, $\theta_{\psi_2}(z) = \eta(4z)^6$, $\theta_{\psi_4}(z) = \eta(4z)^{14}/\eta(8z)^4$. More generally, we can ask which eta-products are of CM type. Here I do not know the answer, but for pure powers there is a complete result, due to Serre. The fact that both $\eta(z) = \sum\limits_{n \equiv 1 \,(\mathrm{mod}\ 6)} (-1)^{(n-1)/6} q^{n^2/24}$ and

$$\eta(z)^3 = \sum\limits_{n \equiv 1 \,(\mathrm{mod}\ 4)} n\, q^{n^2/8}$$ are unary theta series implies that each of the functions η^2, η^4 and η^6 is a binary theta series and hence a modular form of CM type (the function $\eta(z)^6$ equals $\theta_{\psi_2}(z/4)$, as we just saw, and $\eta(z)^4$ equals $f_E(z/6)$, where E is the second curve in (101)), but there are other, less obvious, examples, and these can be completely classified:

Theorem (Serre). *The function $\eta(z)^n$ (n even) is a CM modular form for $n = 2, 4, 6, 8, 10, 14$ or 26 and for no other value of n.*

Finally, the CM forms have another property called "lacunarity" which is not shared by any other modular forms. If we look at (106) or (107), then we see that the only exponents which occur are sums of two squares. By the theorem of Fermat proved in §3.1, only half of all primes (namely, those congruent to 1 modulo 4) have this property, and by a famous theorem of Landau, only $O(x/(\log x)^{1/2})$ of the integers $\leq x$ do. The same applies to any other CM form and shows that 50% of the coefficients a_p (p prime) vanish if the form is a Hecke eigenform and that 100% of the coefficients a_n ($n \in \mathbb{N}$) vanish for any CM form, eigenform or not. Another difficult theorem, again due to Serre, gives the converse statements:

Theorem (Serre). *Let $f = \sum a_n q^n$ be a modular form of integral weight ≥ 2. Then:*

1. *If f is a Hecke eigenform, then the density of primes p for which $a_p \neq 0$ is equal to $1/2$ if f corresponds to a grossencharacter of an imaginary quadratic field, and to 1 otherwise.*
2. *The number of integers $n \leq x$ for which $a_n \neq 0$ is $O(x/\sqrt{\log x})$ as $x \to \infty$ if f is of CM type and is larger than a positive multiple of x otherwise.*

♠ Central Values of Hecke L-Series

We just saw above that the Hecke L-series associated to grossencharacters of degree d are at the same time the Hecke L-series of CM modular forms of weight $d + 1$, and also, if $d = 1$, sometimes the L-series of elliptic curves over $\mathbb{Q}$. On the other hand, the Birch–Swinnerton-Dyer conjecture predicts that the value of $L(E, 1)$ for any elliptic curve E over $\mathbb{Q}$ (not just one with CM) is related as follows to the arithmetic of E: it vanishes if and only if E has a rational point of infinite order, and otherwise is (essentially) a certain period of E multiplied by the order of a mysterious group Ш, the Tate–Shafarevich group of E. Thanks to the work of Kolyvagin, one knows that this group is indeed finite when $L(E, 1) \neq 0$, and it is then a standard fact (because Ш admits a skew-symmetric non-degenerate pairing with values in $\mathbb{Q}/\mathbb{Z}$) that the order of Ш is a perfect square. In summary, the value of $L(E, 1)$, normalized by dividing by an appropriate period, is always a perfect square. This suggests looking at the central point ($=$ point of symmetry with respect to the functional equation) of other types of L-series, and in particular of L-series attached to grossencharacters of higher weights, since these can be normalized in a nice way using the Chowla–Selberg period (97), to see whether these numbers are perhaps also always squares.

An experiment to test this idea was carried out over 25 years ago by B. Gross and myself and confirmed this expectation. Let K be the field $\mathbb{Q}(\sqrt{-7})$ and for each $d \geq 1$ let $\psi_d = \psi_1^d$ be the grossencharacter of K which sends an ideal $\mathfrak{a} \subseteq \mathcal{O}_K$ to λ^d if $\mathfrak{a}$ is prime to $\mathfrak{p}_7 = (\sqrt{-7})$ and to 0 otherwise, where λ in the former case is the generator of $\mathfrak{a}$ which is congruent to a square modulo $\mathfrak{p}_7$. For $d = 1$, the L-series of ψ_d coincides with the L-series of the third elliptic curve in (101), while for general odd values of d it is the L-series of a modular form of weight $d + 1$ and trivial character on $\Gamma_0(49)$. The L-series $L(\psi_{2m-1}, s)$ has a functional equation sending s to $2m - s$, so the point of symmetry is $s = m$. The central value has the form

$$L(\psi_{2m-1}, m) = \frac{2A_m}{(m-1)!} \left(\frac{2\pi}{\sqrt{7}}\right)^m \Omega_K^{2m-1} \tag{108}$$

where

$$\Omega_K = \sqrt[4]{7} \left| \eta\left(\frac{1 + \sqrt{-7}}{2}\right) \right|^2 = \frac{\Gamma(\frac{1}{7})\Gamma(\frac{2}{7})\Gamma(\frac{4}{7})}{4\pi^2}$$

is the Chowla–Selberg period attached to K and (it turns out) $A_m \in \mathbb{Z}$ for all $m > 1$. (We have $A_1 = \frac{1}{4}$.) The numbers A_m vanish for m even because the functional equation of $L(\psi_{2m-1}, s)$ has a minus sign in that case, but the numerical computation suggested that the others were indeed all perfect squares: $A_3 = A_5 = 1^2$, $A_7 = 3^2$, $A_9 = 7^2$, $\ldots$, $A_{33} = 44762286327255^2$. Many years later, in a paper with Fernando Rodriguez Villegas, we were able to confirm this prediction:

Theorem. *The integer A_m is a square for all $m > 1$. More precisely, we have $A_{2n+1} = b_n(0)^2$ for all $n \geq 0$, where the polynomials $b_n(x) \in \mathbb{Q}[x]$ are defined recursively by $b_0(x) = \frac{1}{2}$, $b_1(x) = 1$ and $21\, b_{n+1}(x) = -(x-7)(64x-7)b_n'(x) + (32nx - 56n + 42)b_n(x) - 2n(2n-1)(11x+7)b_{n-1}(x)$ for $n \geq 1$.*

The proof is too complicated to give here, but we can indicate the main idea. The first point is that the numbers A_m themselves can be computed by the method explained in §6.3. There we saw (for the full modular group, but the method works for higher level) that the value at $s = k + n$ of the L-series of a grossencharacter of degree $d = k + 2n$ is essentially equal to the nth non-holomorphic derivative of an Eisenstein series of weight k at a CM point. Here we want $d = 2m - 1$ and $s = m$, so $k = 1$, $n = m - 1$. More precisely, one finds that $L(\psi_{2m-1}, m) = \frac{(2\pi/\sqrt{7})^m}{(m-1)!} \partial^{m-1} \mathbb{G}_{1,\varepsilon}(\mathfrak{z}_7)$, where $\mathbb{G}_{1,\varepsilon}$ is the Eisenstein series

$$\mathbb{G}_{1,\varepsilon} = \frac{1}{2} + \sum_{n=1}^{\infty} \left(\sum_{d|n} \varepsilon(n) \right) q^n = \frac{1}{2} + q + 2q^2 + 3q^4 + q^7 + \cdots$$

of weight 1 associated to the character $\varepsilon(n) = \left(\frac{n}{7} \right)$ and $\mathfrak{z}_7$ is the CM point $\frac{1}{2}\left(1 + \frac{i}{\sqrt{7}} \right)$. These coefficients can be obtained by a quasi-recursion like the one in Proposition 28 (though it is more complicated here because the analogues of the polynomials $p_n(t)$ are now elements in a quadratic extension of $\mathbb{Q}[t]$), and therefore are very easy to compute, but this does not explain why they are squares. To see this, we first observe that the Eisenstein series $\mathbb{G}_{1,\varepsilon}$ is one-half of the binary theta series $\Theta(z) = \sum_{m,n \in \mathbb{Z}} q^{m^2 + mn + 2n^2}$. In his thesis, Villegas proved a beautiful formula expressing certain linear combinations of values of binary theta series at CM points as the squares of linear combinations of values of unary theta series at other CM points. The same turns out to be true for the higher non-holomorphic derivatives, and in the case at hand we find the remarkable formula

$$\partial^{2n}\Theta(\mathfrak{z}_7) = 2^{2n}7^{2n+1/4} \left| \partial^n \theta_2(\mathfrak{z}_7^*) \right|^2 \qquad \text{for all } n \geq 0, \tag{109}$$

where $\theta_2(z) = \sum_{n \in \mathbb{Z}} q^{(n+1/2)^2/2}$ is the Jacobi theta-series defined in (32) and $\mathfrak{z}_7^* = \frac{1}{2}(1 + i\sqrt{7})$. Now the values of the non-holomorphic derivatives $\partial^n \theta_2(\mathfrak{z}_7^*)$ can be computed quasi-recursively by the method explained in 6.3. The result is the formula given in the theorem.

Remarks 1. By the same method, using that every Eisenstein series of weight 1 can be written as a linear combination of binary theta series, one can show that the correctly normalized central values of all L-series of grossencharacters of odd degree are perfect squares.

2. Identities like (109) seem very surprising. I do not know of any other case in mathematics where the Taylor coefficients of an analytic function at some point are in a non-trivial way the squares of the Taylor coefficients of another analytic function at another point.

3. Equation (109) is a special case of a yet more general identity, proved in the same paper, which expresses the non-holomorphic derivative $\partial^{2n}\theta_\psi(\mathfrak{z}_0)$ of the CM modular form associated to a grossencharacter ψ of degree $d = 2r$ as a simple multiple of $\partial^n\theta_2(\mathfrak{z}_1)\,\partial^{n+r}\theta_2(\mathfrak{z}_2)$, where $\mathfrak{z}_0$, $\mathfrak{z}_1$ and $\mathfrak{z}_2$ are CM points belonging to the quadratic field associated to ψ. $\heartsuit$

We end with an application of these ideas to a classical Diophantine equation.

♠ Which Primes are Sums of Two Cubes?

In §3 we gave a modular proof of Fermat's theorem that a prime number can be written as a sum of two squares if and only if it is congruent to 1 modulo 4. The corresponding question for cubes was studied by Sylvester in the 19th century. For squares the answer is the same whether one considers integer or rational squares (though in the former case the representation, when it exists, is unique and in the latter case there are infinitely many), but for cubes one only considers the problem over the rational numbers because there seems to be no rule for deciding which numbers have a decomposition into integral cubes. The question therefore is equivalent to asking whether the Mordell-Weil group $E_p(\mathbb{Q})$ of the elliptic curve $E_p : x^3 + y^3 = p$ is non-trivial. Except for $p = 2$, which has the unique decomposition $1^3 + 1^3$, the group $E_p(\mathbb{Q})$ is torsion-free, so that if there is even one rational solution there are infinitely many. An equivalent question is therefore: for which primes p is the rank r_p of $E_p(\mathbb{Q})$ greater than 0 ?

Sylvester's problem was already mentioned at the end of §6.1 in connection with Heegner points, which can be used to construct non-trivial solutions if $p \equiv 4$ or 7 mod 9. Here we consider instead an approach based on the Birch–Swinnerton-Dyer conjecture, according to which $r_p > 0$ if and only if the L-series of E_p vanishes at $s = 1$. By a famous theorem of Coates and Wiles, one direction of this conjecture is known: if $L(E_p, 1) \neq 0$ then $E_p(\mathbb{Q})$ has rank 0. The question we want to study is therefore: when does $L(E_p, 1)$ vanish? For five of the six possible congruence classes for p (mod 9) (we assume that $p > 3$, since $r_2 = r_3 = 0$) the answer is known. If $p \equiv 4$, 7 or 8 (mod 9), then the functional equation of $L(E_p, s)$ has a minus sign, so $L(E_p, 1) = 0$ and r_p is expected (and, in the first two cases, known) to be ≥ 1; it is also known by an "infinite descent" argument to be ≤ 1 in these cases. If $p \equiv 2$ or 5 (mod 9), then the functional equation has a plus sign and $L(E_p, 1)$ divided by a suitable period is $\equiv 1$ (mod 3) and hence $\neq 0$, so by the Coates-Wiles theorem these primes can never be sums of two cubes, a result which can also be proved in an elementary way by descent. In the remaining case $p \equiv 1$ (mod 9), however, the answer can vary: here the functional equation has a plus sign, so $\mathrm{ord}_{s=1}L(E_p, s)$ is even and r_p is also expected to be even, and descent gives $r_p \leq 2$, but both cases $r_p = 0$ and $r_p = 2$ occur. The following result, again proved jointly with F. Villegas, gives a criterion for these primes.

Theorem. *Define a sequence of numbers* $c_0 = 1$, $c_1 = 2$, $c_2 = -152$, $c_3 = 6848$, ... *by* $c_n = s_n(0)$ *where* $s_0(x) = 1$, $s_1(x) = 3x^2$ *and* $s_{n+1}(x) = (1 - 8x^3)s_n'(x) + (16n + 3)x^2 s_n(x) - 4n(2n - 1)x s_{n-1}(x)$ *for* $n \geq 1$. *If* $p = 9k + 1$ *is prime, then* $L(E_p, 1) = 0$ *if and only if* $p | c_k$.

For instance, the numbers $c_2 = -152$ and $c_4 = -8103296$ are divisible by $p = 19$ and $p = 37$, respectively, so $L(E_p, 1)$ vanishes for these two primes (and indeed $19 = 3^3 + (-2)^3$, $37 = 4^3 + (-3)^3$), whereas $c_8 = 532650564250569441280$ is not divisible by 73, which is therefore not a sum of two rational cubes. Note that the numbers c_n grow very quickly (roughly like n^{3n}), but to apply the criterion for a given prime number $p = 9k + 1$ one need only compute the polynomials $s_n(x)$ modulo p for $0 \leq n \leq k$, so that no large numbers are required.

We again do not give the proof of this theorem, but only indicate the main ingredients. The central value of $L(E_p, s)$ for $p \equiv 1 \pmod 9$ is given by

$$L(E_p, 1) = \frac{3\,\Omega}{\sqrt[3]{p}}\, S_p \text{ where } \Omega = \frac{\Gamma(\frac{1}{3})^3}{2\pi\sqrt{3}} \text{ and } S_p \text{ is an integer which is supposed}$$

to be a perfect square (namely the order of the Tate-Shafarevich group of E_p if $r_p = 0$ and 0 if $r_p > 0$). Using the methods from Villegas's thesis, one can show that this is true and that both S_p and its square root can be expressed as the traces of certain algebraic numbers defined as special values of modular functions at CM points: $S_p = \mathrm{Tr}(\alpha_p) = \mathrm{Tr}(\beta_p)^2$, where $\alpha_p = \dfrac{\sqrt[3]{p}}{54}\dfrac{\Theta(p\mathfrak{z}_0)}{\Theta(\mathfrak{z}_0)}$

and $\beta_p = \dfrac{\sqrt[6]{p}}{\sqrt{\pm 12}}\dfrac{\eta(p\mathfrak{z}_1)}{\eta(\mathfrak{z}_1/p)}$ with $\Theta(z) = \sum\limits_{m,n} q^{m^2 + mn + n^2}$, $\mathfrak{z}_0 = \dfrac{1}{2} + \dfrac{i}{6\sqrt{3}}$ and

$\mathfrak{z}_1 = \dfrac{r + \sqrt{-3}}{2}$ $(r \in \mathbb{Z}, r^2 \equiv -3 \pmod{4p})$. This gives an explicit formula for $L(E_p, 1)$, but it is not very easy to compute since the numbers α_p and β_p have large degree ($18k$ and $6k$, respectively if $p = 9k + 1$) and lie in a different number field for each prime p. To obtain a formula in which everything takes place over $\mathbb{Q}$, one observes that the L-series of E_p is the L-series of a cubic twist of a grossencharacter ψ_1 of $K = \mathbb{Q}(\sqrt{-3})$ which is independent of p. More precisely, $L(E_p, s) = L(\chi_p\psi_1, s)$ where ψ_1 is defined (just like the grossencharacters for $\mathbb{Q}(i)$ and $\mathbb{Q}(\sqrt{-7})$ defined in (105) and in the previous "application") by $\psi_1(\mathfrak{a}) = \lambda$ if $\mathfrak{a} = (\lambda)$ with $\lambda \equiv 1 \pmod 3$ and $\psi_1(\alpha) = 0$ if $3 | N(\mathfrak{a})$, and χ_p is the cubic character which sends $\mathfrak{a} = (\lambda)$ to the unique cube root of unity in K which is congruent to $(\bar{\lambda}/\lambda)^{(p-1)/3}$ modulo p. This means that formally the L-value $L(E_p, 1)$ for $p = 9k + 1$ is congruent modulo p to the central value $L(\psi^{12k+1}, 6k + 1)$. Of course both of these numbers are transcendental, but the theory of p-adic L-functions shows that their "algebraic parts" are in fact congruent modulo p: we have $L(\psi_1^{12k+1}, 6k + 1) = \dfrac{3^{9k-1}\,\Omega^{12k+1}}{(2\pi)^{6k}}\dfrac{C_k}{(6k)!}$

for some integer $C_k \in \mathbb{Z}$, and $S_p \equiv C_k \pmod p$. Now the calculation of C_k proceeds exactly like that of the number A_m defined in (108): C_k is, up to normalizing constants, equal to the value at $z = \mathfrak{z}_0$ of the $6k$-th non-holomorphic derivative of $\Theta(z)$, and can be computed quasi-recursively, and C_k is also equal to c_k^2 where c_k is, again up to normalizing constants, the value of the $3k$-th non-holomorphic derivative of $\eta(z)$ at $z = \frac{1}{2}(-1 + \sqrt{-3})$ and is given by the formula in the theorem. Finally, an estimate of the size of $L(E_p, 1)$ shows that $|S_p| < p$, so that S_p vanishes if and only if $c_k \equiv 0 \pmod{p}$, as claimed. The numbers c_k can also be described by a generating function rather than a quasi-recursion, using the relations between modular forms and differential equations discussed in §5, namely

$$\left(1 - x\right)^{1/24} F\left(\frac{1}{3}, \frac{1}{3}, \frac{2}{3}; x\right)^{1/2} = \sum_{n=0}^{\infty} \frac{c_n}{(3n)!} \left(\frac{x\, F(\frac{2}{3}, \frac{2}{3}; \frac{4}{3}; x)^3}{8\, F(\frac{1}{3}, \frac{1}{3}; \frac{2}{3}; x)^3}\right)^n,$$

where $F(a, b; c; x)$ denotes Gauss's hypergeometric function. $\heartsuit$

References and Further Reading

There are several other elementary texts on the theory of modular forms which the reader can consult for a more detailed introduction to the field. Four fairly short introductions are the classical book by Gunning (*Lectures on Modular Forms*, Annals of Math. Studies **48**, Princeton, 1962), the last chapter of Serre's *Cours d'Arithmétique* (Presses Universitaires de France, 1970; English translation *A Course in Arithmetic*, Graduate Texts in Mathematics **7**, Springer 1973), Ogg's book *Modular Forms and Dirichlet Series* (Benjamin, 1969; especially for the material covered in §§3–4 of these notes), and my own chapter in the book *From Number Theory to Physics* (Springer 1992). The chapter on elliptic curves by Henri Cohen in the last-named book is also a highly recommended and compact introduction to a field which is intimately related to modular forms and which is touched on many times in these notes. Here one can also recommend N. Koblitz's *Introduction to Elliptic Curves and Modular forms* (Springer Graduate Texts **97**, 1984). An excellent book-length Introduction to Modular Forms is Serge Lang's book of that title (Springer Grundlehren **222**, 1976). Three books of a more classical nature are B. Schoeneberg's *Elliptic Modular Functions* (Springer Grundlehren **203**, 1974), R.A. Rankin's *Modular Forms and Modular Functions* (Cambridge, 1977) and (in German) *Elliptische Funktionen und Modulformen* by M. Koecher and A. Krieg (Springer, 1998).

The point of view in these notes leans towards the analytic, with as many results as possible (like the algebraicity of $j(\mathfrak{z})$ when $\mathfrak{z}$ is a CM point) being derived purely in terms of the theory of modular forms over the complex numbers, an approach which was sufficient – and usually simpler – for the type of applications which I had in mind. The books listed above also belong to this category. But for many other applications, including the deepest ones in Diophantine equations and arithmetic algebraic geometry, a more arithmetic and more advanced approach is required. Here the basic reference is

Shimura's classic *Introduction to the Arithmetic Theory of Automorphic Functions* (Princeton 1971), while two later books that can also be recommended are Miyake's *Modular Forms* (Springer 1989) and the very recent book *A First Course in Modular Forms* by Diamond and Shurman (Springer 2005).

We now give, section by section, some references (not intended to be in any sense complete) for various of the specific topics and examples treated in these notes.

1–2. The material here is all standard and can be found in the books listed above, except for the statement in the final section of §1.2 that the class numbers of negative discriminants are the Fourier coefficients of some kind of modular form of weight $3/2$, which was proved in my paper in CRAS Paris (1975), 883–886.

3.1. Proposition 11 on the number of representations of integers as sums of four squares was proved by Jacobi in the *Fundamenta Nova Theoriae Ellipticorum*, 1829. We do not give references for the earlier theorems of Fermat and Lagrange. For more information about sums of squares, one can consult the book *Representations of Integers as Sums of Squares* (Springer, 1985) by E. Grosswald. The theory of Jacobi forms mentioned in connection with the two-variable theta functions $\theta_i(z, u)$ was developed in the book *The Theory of Jacobi Forms* (Birkhäuser, 1985) by M. Eichler and myself. Mersmann's theorem is proved in his Bonn Diplomarbeit, "Holomorphe η-Produkte und nichtverschwindende ganze Modulformen für $\Gamma_0(N)$" (Bonn, 1991), unfortunately never published in a journal. The theorem of Serre and Stark is given in their paper "Modular forms of weight $1/2$" in *Modular Forms of One Variable VI* (Springer Lecture Notes **627**, 1977, editors J-P. Serre and myself; this is the sixth volume of the proceedings of two big international conferences on modular forms, held in Antwerp in 1972 and in Bonn in 1976, which contain a wealth of further material on the theory). The conjecture of Kac and Wakimoto appeared in their article in *Lie Theory and Geometry in Honor of Bertram Kostant* (Birkhäuser, 1994) and the solutions by Milne and myself in the Ramanujan Journal and Mathematical Research Letters, respectively, in 2000.

3.2. The detailed proof and references for the theorem of Hecke and Schoenberg can be found in Ogg's book cited above. Niemeier's classification of unimodular lattices of rank 24 is given in his paper "Definite quadratische Formen der Dimension 24 und Diskriminante 1" (J. Number Theory **5**, 1973). Siegel's mass formula was presented in his paper "Über die analytische Theorie der quadratischen Formen" in the Annals of Mathematics, 1935 (No. 20 of his *Gesammelte Abhandlungen*, Springer, 1966). A recent paper by M. King (Math. Comp., 2003) improves by a factor of more than 14 the lower bound on the number of inequivalent unimodular even lattices of dimension 32. The paper of Mallows-Odlyzko-Sloane on extremal theta series appeared in J. Algebra in 1975. The standard general reference for the theory of lattices, which contains an immense amount of further material, is the book by Conway and

Sloane (*Sphere Packings, Lattices and Groups*, Springer 1998). Milnor's example of 16-dimensional tori as non-isometric isospectral manifolds was given in 1964, and 2-dimensional examples (using modular groups!) were found by Vignéras in 1980. Examples of pairs of truly drum-like manifolds – i.e., domains in the flat plane – with different spectra were finally constructed by Gordon, Webb and Wolpert in 1992. For references and more on the history of this problem, we refer to the survey paper in the book *What's Happening in the Mathematical Sciences* (AMS, 1993).

4.1–4.3. For a general introduction to Hecke theory we refer the reader to the books of Ogg and Lang mentioned above or, of course, to the beautifully written papers of Hecke himself (if you can read German). Van der Blij's example is given in his paper "Binary quadratic forms of discriminant -23" in Indagationes Math., 1952. A good exposition of the connection between Galois representations and modular forms of weight one can be found in Serre's article in *Algebraic Number Fields: L-Functions and Galois Properties* (Academic Press 1977).

4.4. Book-length expositions of the Taniyama–Weil conjecture and its proof using modular forms are given in *Modular Forms and Fermat's Last Theorem* (G. Cornell, G. Stevens and J. Silverman, eds., Springer 1997) and, at a much more elementary level, *Invitation to the Mathematics of Fermat-Wiles* (Y. Hellegouarch, Academic Press 2001), which the reader can consult for more details concerning the history of the problem and its solution and for further references. An excellent survey of the content and status of Serre's conjecture can be found in the book *Lectures on Serre's Conjectures* by Ribet and Stein (http://modular.fas.harvard.edu/papers/serre/ribet-stein.ps). For the final proof of the conjecture and references to all earlier work, see "Modularity of 2-adic Barsotti-Tate representations" by M. Kisin (http://www.math.uchicago.edu/~kisin/preprints.html). Livné's example appeared in his paper "Cubic exponential sums and Galois representations" (Contemp. Math. **67**, 1987). For an exposition of the conjectural and known examples of higher-dimensional varieties with modular zeta functions, in particular of those coming from mirror symmetry, we refer to the recent paper "Modularity of Calabi-Yau varieties" by Hulek, Kloosterman and Schütt in *Global Aspects of Complex Geometry* (Springer, 2006) and to the monograph *Modular Calabi-Yau Threefolds* by C. Meyer (Fields Institute, 2005). However, the proof of Serre's conjectures means that the modularity is now known in many more cases than indicated in these surveys.

5.1. Proposition 15, as mentioned in the text, is due to Ramanujan (eq. (30) in "On certain Arithmetical Functions," Trans. Cambridge Phil. Soc., 1916). For a good discussion of the Chazy equation and its relation to the "Painlevé property" and to SL$(2, \mathbb{C})$, see the article "Symmetry and the Chazy equation" by P. Clarkson and P. Olver (J. Diff. Eq. **124**, 1996). The result by Gallagher on means of periodic functions which we describe as our second application is described in his very nice paper "Arithmetic of means of squares and cubes" in Internat. Math. Res. Notices, 1994.

5.2. Rankin–Cohen brackets were defined in two stages: the general conditions needed for a polynomial in the derivatives of a modular form to itself be modular were described by R. Rankin in "The construction of automorphic forms from the derivatives of a modular form" (J. Indian Math. Soc., 1956), and then the specific bilinear operators $[\,\cdot\,,\,\cdot\,]_n$ satisfying these conditions were given by H. Cohen as a lemma in "Sums involving the values at negative integers of L functions of quadratic characters" (Math. Annalen, 1977). The Cohen–Kuznetsov series were defined in the latter paper and in the paper "A new class of identities for the Fourier coefficients of a modular form" (in Russian) by N.V. Kuznetsov in Acta Arith., 1975. The algebraic theory of "Rankin–Cohen algebras" was developed in my paper "Modular forms and differential operators" in Proc. Ind. Acad. Sciences, 1994, while the papers by Manin, P. Cohen and myself and by the Untenbergers discussed in the second application in this subsection appeared in the book *Algebraic Aspects of Integrable Systems: In Memory of Irene Dorfman* (Birkhäuser 1997) and in the J. Anal. Math., 1996, respectively.

5.3. The name and general definition of quasimodular forms were given in the paper "A generalized Jacobi theta function and quasimodular forms" by M. Kaneko and myself in the book *The Moduli Space of Curves* (Birkhäuser 1995), immediately following R. Dijkgraaf's article "Mirror symmetry and elliptic curves" in which the problem of counting ramified coverings of the torus is presented and solved.

5.4. The relation between modular forms and linear differential equations was at the center of research on automorphic forms at the turn of the (previous) century and is treated in detail in the classical works of Fricke, Klein and Poincaré and in Weber's *Lehrbuch der Algebra*. A discussion in a modern language can be found in §5 of P. Stiller's paper in the Memoirs of the AMS **299**, 1984. Beukers's modular proof of the Apéry identities implying the irrationality of $\zeta(2)$ and $\zeta(3)$ can be found in his article in Astérisque **147–148** (1987), which also contains references to Apéry's original paper and other related work. My paper with Kleban on a connection between percolation theory and modular forms appeared in J. Statist. Phys. **113** (2003).

6.1. There are several references for the theory of complex multiplication. A nice book giving an introduction to the theory at an accessible level is *Primes of the form $x^2 + ny^2$* by David Cox (Wiley, 1989), while a more advanced account is given in the Springer Lecture Notes Volume 21 by Borel, Chowla, Herz, Iwasawa and Serre. Shanks's approximation to π is given in a paper in J. Number Theory in 1982. Heegner's original paper attacking the class number one problem by complex multiplication methods appeared in Math. Zeitschrift in 1952. The result quoted about congruent numbers was proved by Paul Monsky in "Mock Heegner points and congruent numbers"

(Math. Zeitschrift 1990) and the result about Sylvester's problem was announced by Noam Elkies, in "Heegner point computations" (Springer Lecture Notes in Computing Science, 1994). See also the recent preprint "Some Diophantine applications of Heegner points" by S. Dasgupta and J. Voight.

6.2. The formula for the norms of differences of singular moduli was proved in my joint paper "Singular moduli" (J. reine Angew. Math. **355**, 1985) with B. Gross, while our more general result concerning heights of Heegner points appeared in "Heegner points and derivatives of L-series" (Invent. Math. **85**, 1986). The formula describing traces of singular moduli is proved in my paper of the same name in the book *Motives, Polylogarithms and Hodge Theory* (International Press, 2002). Borcherds's result on product expansions of automorphic forms was published in a celebrated paper in Invent. Math. in 1995.

6.3. The Chowla–Selberg formula is discussed, among many other places, in the last chapter of Weil's book *Elliptic Functions According to Eisenstein and Kronecker* (Springer Ergebnisse **88**, 1976), which contains much other beautiful historical and mathematical material. Chudnovsky's result about the transcendence of $\Gamma(\frac{1}{4})$ is given in his paper for the 1978 (Helskinki) International Congress of Mathematicians, while Nesterenko's generalization giving the algebraic independence of π and e^π is proved in his paper "Modular functions and transcendence questions" (in Russian) in Mat. Sbornik, 1996. A good summary of this work, with further references, can be found in the "featured review" of the latter paper in the 1997 Mathematical Reviews. The algorithmic way of computing Taylor expansions of modular forms at CM points is described in the first of the two joint papers with Villegas cited below, in connection with the calculation of central values of L-series.

6.4. A discussion of formulas like (102) can be found in any of the general references for the theory of complex multiplication listed above. The applications of such formulas to questions of primality testing, factorization and cryptography is treated in a number of papers. See for instance "Efficient construction of cryptographically strong elliptic curves" by H. Baier and J. Buchmann (Springer Lecture Notes in Computer Science **1977**, 2001) and its bibliography. Serre's results on powers of the eta-function and on lacunarity of modular forms are contained in his papers "Sur la lacunarité des puissances de η" (Glasgow Math. J., 1985) and "Quelques applications du théorème de densité de Chebotarev" (Publ. IHES, 1981), respectively. The numerical experiments concerning the numbers defined in (108) were given in a note by B. Gross and myself in the memoires of the French mathematical society (1980), and the two papers with F. Villegas on central values of Hecke L-series and their applications to Sylvester's problems appeared in the proceedings of the third and fourth conferences of the Canadian Number Theory Association in 1993 and 1995, respectively.

Hilbert Modular Forms and Their Applications

Jan Hendrik Bruinier

Fachbereich Mathematik, Technische Universität Darmstadt, Schloßgartenstraße 7, 64289 Darmstadt, Germany
E-mail: bruinier@mathematik.tu-darmstadt.de

Introduction

The present notes contain the material of the lectures given by the author at the summer school on "Modular Forms and their Applications" at the Sophus Lie Conference Center in the summer of 2004.

We give an introduction to the theory of Hilbert modular forms and some geometric and arithmetic applications. We tried to keep the informal style of the lectures. In particular, we often do not work in greatest possible generality, but rather consider a reasonable special case, in which the main ideas of the theory become clear.

For a more comprehensive account to Hilbert modular varieties, we refer to the books by Freitag [Fr], Garrett [Ga], van der Geer [Ge1], and Goren [Go]. We hope that the present text will be a useful addition to these references.

Hilbert modular surfaces can also be realized as modular varieties corresponding to the orthogonal group of a rational quadratic space of type $(2, 2)$. This viewpoint leads to several interesting features of these surfaces. For instance, they come with a natural family of divisors arising from embeddings of "smaller" orthogonal groups, the so-called Hirzebruch–Zagier divisors. Their study led to important discoveries and triggered generalizations in various directions. Moreover, the theta correspondence provides a source of automorphic forms related to the geometry of Hirzebruch–Zagier divisors.

A more recent development is the regularized theta lifting due to Borcherds, Harvey and Moore, which yields to automorphic products and automorphic Green functions. The focus of the present text is on these topics, highlighting the role of the orthogonal group. We added some background material on quadratic spaces and orthogonal groups, to make the connection explicit.

I thank G. van der Geer and D. Zagier for several interesting conversations during the summer school at the Sophus Lie Conference Center. Moreover, I thank J. Funke and T. Yang for their helpful comments on earlier versions of this manuscript.

1 Hilbert Modular Surfaces

In this section we give a brief introduction to Hilbert modular surfaces associated to real quadratic fields. For details we refer to [Fr], [Ga], [Ge1], [Go].

1.1 The Hilbert Modular Group

Let $d > 1$ be a squarefree integer. Then $F = \mathbb{Q}(\sqrt{d})$ is a real quadratic field, which we view as a subfield of $\mathbb{R}$. The discriminant of F is

$$D = \begin{cases} d, & \text{if } d \equiv 1 \pmod 4, \\ 4d, & \text{if } d \equiv 2, 3 \pmod 4. \end{cases} \tag{1.1}$$

We write $\mathcal{O}_F$ for the ring of integers in F, so

$$\mathcal{O}_F = \begin{cases} \mathbb{Z} + \frac{1+\sqrt{d}}{2}\mathbb{Z}, & \text{if } d \equiv 1 \pmod 4, \\ \mathbb{Z} + \sqrt{d}\mathbb{Z}, & \text{if } d \equiv 2, 3 \pmod 4. \end{cases} \tag{1.2}$$

The ring $\mathcal{O}_F$ is a Dedekind domain, that is, it is a noetherian integrally closed integral domain in which every non-zero prime ideal is maximal.

We denote by $\mathcal{O}_F^*$ the group of units in $\mathcal{O}_F$. By the Dirichlet unit theorem there is a unique unit $\varepsilon_0 > 1$ such that $\mathcal{O}_F^* = \{\pm 1\} \times \{\varepsilon_0^n; \, n \in \mathbb{Z}\}$. It is called the fundamental unit of F. We write $x \mapsto x'$ for the conjugation, $\mathrm{N}(x) = xx'$ for the norm in F, and $\mathrm{tr}(x) = x + x'$ for the trace in F. The different of F is denoted by $\mathfrak{d}_F$. Note that $\mathfrak{d}_F = (\sqrt{D})$.

Recall that an (integral) ideal of $\mathcal{O}_F$ is a $\mathcal{O}_F$-submodule of $\mathcal{O}_F$. A fractional ideal of F is a finitely generated $\mathcal{O}_F$-submodule of F. Fractional ideals form a group together with the ideal multiplication. The neutral element is $\mathcal{O}_F$ and the inverse of a fractional ideal $\mathfrak{a} \subset F$ is

$$\mathfrak{a}^{-1} = \{x \in F; \, x\mathfrak{a} \subset \mathcal{O}_F\}.$$

Since F is a quadratic extension of $\mathbb{Q}$, we have the useful formula $\mathfrak{a}^{-1} = \frac{1}{\mathrm{N}(\mathfrak{a})}\mathfrak{a}'$, where $\mathfrak{a}'$ is the conjugate of $\mathfrak{a}$. Two fractional ideals $\mathfrak{a}, \mathfrak{b}$ are called equivalent, if there is a $r \in F$ such that $\mathfrak{a} = r\mathfrak{b}$. The group of equivalence classes $\mathrm{Cl}(F)$ is called the ideal class group of F. It is a finite abelian group. Two fractional ideals $\mathfrak{a}, \mathfrak{b}$ are called equivalent in the narrow sense, if there is a totally positive $r \in F$ such that $\mathfrak{a} = r\mathfrak{b}$. The group of equivalence classes $\mathrm{Cl}^+(F)$ is called the narrow ideal class group of F. It is equal to $\mathrm{Cl}(F)$, if and only if ε_0 has norm -1. Otherwise it is an extension of degree 2 of $\mathrm{Cl}(F)$. The (narrow) class number of F is the order of the (narrow) ideal class group. It measures how far $\mathcal{O}_F$ is from being a principal ideal domain.

If the class number of F is greater than 1, there are ideals which cannot be generated by a single element. However, we have the following fact, which holds in any Dedekind ring.

Remark 1.1. If $\mathfrak{a} \subset F$ is a fractional ideal, then there exist $\alpha, \beta \in F$ such that $\mathfrak{a} = \alpha \mathcal{O}_F + \beta \mathcal{O}_F$. $\qquad\qquad\qquad\qquad\qquad\qquad\qquad\qquad\qquad\qquad\qquad\square$

The group $\mathrm{SL}_2(F)$ is embedded into $\mathrm{SL}_2(\mathbb{R}) \times \mathrm{SL}_2(\mathbb{R})$ by the two real embeddings of F. It acts on $\mathbb{H} \times \mathbb{H}$, where $\mathbb{H} = \{\tau \in \mathbb{C};\ \Im(\tau) > 0\}$ is the complex upper half plane, via fractional linear transformations,

$$\begin{pmatrix} a & b \\ c & d \end{pmatrix} z = \left(\frac{az_1 + b}{cz_1 + d}, \frac{a'z_2 + b'}{c'z_2 + d'} \right). \tag{1.3}$$

Here and throughout we use $z = (z_1, z_2)$ as a standard variable on $\mathbb{H}^2$. If $z \in \mathbb{H}^2$ and $\left(\begin{smallmatrix} a & b \\ c & d \end{smallmatrix} \right) \in \mathrm{SL}_2(F)$, we write

$$\mathrm{N}(cz + d) = (cz_1 + d)(c'z_2 + d'). \tag{1.4}$$

Lemma 1.2. *For $z \in \mathbb{H}^2$ and $\gamma = \left(\begin{smallmatrix} a & b \\ c & d \end{smallmatrix} \right) \in \mathrm{SL}_2(F)$ we have*

$$\Im(\gamma z) = \frac{\Im(z)}{|\mathrm{N}(cz + d)|^2}.$$

Proof. This follows immediately from the analogous assertion in the one-dimensional case. $\qquad\qquad\qquad\qquad\qquad\qquad\qquad\qquad\qquad\qquad\qquad\qquad\square$

If $\mathfrak{a}$ is a fractional ideal of F, we write

$$\Gamma(\mathcal{O}_F \oplus \mathfrak{a}) = \left\{ \begin{pmatrix} a & b \\ c & d \end{pmatrix} \in \mathrm{SL}_2(F);\quad a, d \in \mathcal{O}_F,\ b \in \mathfrak{a}^{-1},\ c \in \mathfrak{a} \right\} \tag{1.5}$$

for the *Hilbert modular group* corresponding to $\mathfrak{a}$. Moreover, we write

$$\Gamma_F = \Gamma(\mathcal{O}_F \oplus \mathcal{O}_F) = \mathrm{SL}_2(\mathcal{O}_F). \tag{1.6}$$

Let $\Gamma \subset \mathrm{SL}_2(F)$ be a subgroup which is commensurable with Γ_F, i.e., $\Gamma \cap \Gamma_F$ has finite index in both, Γ and Γ_F. Then Γ acts properly discontinuously on $\mathbb{H}^2$, i.e., if $W \subset \mathbb{H}^2$ is compact, then $\{\gamma \in \Gamma;\ \gamma W \cap W \neq \emptyset\}$ is finite (see Corollary 1.17). In particular, for any $a \in \mathbb{H}^2$, the stabilizer $\Gamma_a = \{\gamma \in \Gamma;\ \gamma a = a\}$ is a finite subgroup of Γ. Let $\bar{\Gamma}_a$ be the image of Γ_a in $\mathrm{PSL}_2(F) = \mathrm{SL}_2(F)/\{\pm 1\}$. If $\#\bar{\Gamma}_a > 1$ then a is called an *elliptic fixed point* for Γ and $\#\bar{\Gamma}_a$ is called the order of a. The order of a only depends of the Γ-class. Moreover, there are only finitely many Γ-classes of elliptic fixed points. It can be shown that Γ always has a finite index subgroup which has no elliptic fixed points.

The quotient

$$Y(\Gamma) = \Gamma \backslash \mathbb{H}^2 \tag{1.7}$$

is a normal complex surface. The singularities are given by the elliptic fixed points. They are finite quotient singularities.

The surface $Y(\Gamma)$ is non-compact. It can be compactified by adding a finite number of points, the *cusps* of Γ. They can be described as follows. The group $\mathrm{SL}_2(F)$ also acts on $\mathbb{P}^1(F) = F \cup \{\infty\}$ by

$$\begin{pmatrix} a & b \\ c & d \end{pmatrix} \frac{\alpha}{\beta} = \frac{a\frac{\alpha}{\beta}+b}{c\frac{\alpha}{\beta}+d} = \frac{a\alpha+b\beta}{c\alpha+d\beta}.$$

Notice that, since $\begin{pmatrix} a & b \\ c & d \end{pmatrix} \infty = \frac{a}{c}$, the action of $\mathrm{SL}_2(F)$ is transitive. The Γ-classes of $\mathbb{P}^1(F)$ are called the *cusps* of Γ.

Lemma 1.3. *The map*

$$\varphi : \Gamma_F \backslash \mathbb{P}^1(F) \longrightarrow \mathrm{Cl}(F),$$
$$(\alpha : \beta) \longmapsto \alpha \mathcal{O}_F + \beta \mathcal{O}_F,$$

is bijective.

Proof. We begin by showing that φ is well-defined: It is clear that $\varphi(\alpha : \beta) = \varphi(r\alpha : r\beta)$. Now let $\begin{pmatrix} a & b \\ c & d \end{pmatrix} \in \Gamma_F$, and let $\frac{\gamma}{\delta} = \begin{pmatrix} a & b \\ c & d \end{pmatrix} \frac{\alpha}{\beta}$. We need to show that $\varphi(\gamma : \delta) = \varphi(\alpha : \beta)$. We have

$$\varphi(\gamma : \delta) = \gamma\mathcal{O}_F + \delta\mathcal{O}_F$$
$$= (a\alpha+b\beta)\mathcal{O}_F + (c\alpha+d\beta)\mathcal{O}_F$$
$$\subset \varphi(\alpha : \beta).$$

Interchanging the roles of $(\gamma : \delta)$ and $(\alpha : \beta)$, we see

$$\varphi(\alpha : \beta) = (d\gamma - b\delta)\mathcal{O}_F + (-c\gamma + a\delta)\mathcal{O}_F$$
$$\subset \varphi(\gamma : \delta).$$

Consequently, $\varphi(\gamma : \delta) = \varphi(\alpha : \beta)$.

The surjectivity of φ follows from Remark 1.1.

Finally, we show that φ is injective. Let $\mathfrak{a} = \varphi(\alpha : \beta) = \varphi(\gamma : \delta)$. Then $1 \in \mathcal{O}_F = \mathfrak{a}\mathfrak{a}^{-1} = \alpha\mathfrak{a}^{-1} + \beta\mathfrak{a}^{-1}$. So there exist $\tilde{\alpha}, \tilde{\beta} \in \mathfrak{a}^{-1}$ such that $1 = \alpha\tilde{\beta} - \beta\tilde{\alpha}$. We find that

$$M_1 := \begin{pmatrix} \alpha & \tilde{\alpha} \\ \beta & \tilde{\beta} \end{pmatrix} \in \begin{pmatrix} \mathfrak{a} & \mathfrak{a}^{-1} \\ \mathfrak{a} & \mathfrak{a}^{-1} \end{pmatrix} \cap \mathrm{SL}_2(F),$$

and $M_1\infty = (\alpha : \beta)$. In the same way we find

$$M_2 := \begin{pmatrix} \gamma & \tilde{\gamma} \\ \delta & \tilde{\delta} \end{pmatrix} \in \begin{pmatrix} \mathfrak{a} & \mathfrak{a}^{-1} \\ \mathfrak{a} & \mathfrak{a}^{-1} \end{pmatrix} \cap \mathrm{SL}_2(F)$$

such that $M_2\infty = (\gamma : \delta)$. Therefore we have

$$M_2M_1^{-1} \in \begin{pmatrix} \mathfrak{a} & \mathfrak{a}^{-1} \\ \mathfrak{a} & \mathfrak{a}^{-1} \end{pmatrix} \begin{pmatrix} \mathfrak{a}^{-1} & \mathfrak{a}^{-1} \\ \mathfrak{a} & \mathfrak{a} \end{pmatrix} = \begin{pmatrix} \mathcal{O}_F & \mathcal{O}_F \\ \mathcal{O}_F & \mathcal{O}_F \end{pmatrix}.$$

Hence, $M_2M_1^{-1} \in \Gamma_F$ and $M_2M_1^{-1}(\alpha : \beta) = (\gamma : \delta)$. This concludes the proof of the Lemma. $\qquad\square$

Corollary 1.4. *The number of cusps of Γ_F is equal to the class number $h(F)$ of F. A subgroup $\Gamma \subset \mathrm{SL}_2(F)$ which is commensurable with Γ_F has finitely many cusps.*

Remark 1.5. Let $\Gamma_\infty \subset \Gamma$ be the stabilizer of ∞. Then there is a $\mathbb{Z}$-module $M \subset F$ of rank 2 and a finite index subgroup $V \subset \mathcal{O}_F^*$ acting on M such that the group

$$G(M, V) = \left\{ \begin{pmatrix} \varepsilon & \mu \\ 0 & \varepsilon^{-1} \end{pmatrix} ; \ \mu \in M \text{ and } \varepsilon \in V \right\} \tag{1.8}$$

is contained in Γ_∞ with finite index.

Example 1.6. If $\mathfrak{a} \subset F$ is a fractional ideal, then

$$\Gamma(\mathcal{O}_F \oplus \mathfrak{a})_\infty = \left\{ \begin{pmatrix} \varepsilon & \mu \\ 0 & \varepsilon^{-1} \end{pmatrix} ; \ \mu \in \mathfrak{a}^{-1}, \ \varepsilon \in \mathcal{O}_F^* \right\}.$$

1.2 The Baily–Borel Compactification

We embed $\mathbb{P}^1(F)$ into $\mathbb{P}^1(\mathbb{R}) \times \mathbb{P}^1(\mathbb{R})$ via the two real embeddings of F. Then we may view $\mathbb{P}^1(F)$ as the set of rational boundary points of $\mathbb{H}^2$ in the same way as $\mathbb{P}^1(\mathbb{Q})$ is viewed as the set of rational boundary points of $\mathbb{H}$. Here we consider

$$(\mathbb{H}^2)^* = \mathbb{H}^2 \cup \mathbb{P}^1(F). \tag{1.9}$$

By introducing a suitable topology on $(\mathbb{H}^2)^*$, the quotient $\Gamma \backslash (\mathbb{H}^2)^*$ can be made into a compact Hausdorff space. This leads to the Baily–Borel compactification of $Y(\Gamma)$.

Proposition 1.7. *On $(\mathbb{H}^2)^*$ there is a unique topology with the following properties:*

(i) The induced topology on $\mathbb{H}^2$ agrees with the usual one.
(ii) $\mathbb{H}^2$ is open and dense in $(\mathbb{H}^2)^*$.
(iii) The sets $U_C \cup \infty$, where

$$U_C = \left\{ (z_1, z_2) \in \mathbb{H}^2; \ \Im(z_1)\Im(z_2) > C \right\}$$

for $C > 0$, form a base of open neighborhoods of the cusp ∞.
(iv) If $\kappa \in \mathbb{P}^1(F)$ and $\rho \in \mathrm{SL}_2(F)$ with $\rho\infty = \kappa$, then the sets

$$\rho(U_C \cup \infty) \qquad (C > 0)$$

form a base of open neighborhoods of the cusp κ. $\qquad\square$

Remark 1.8. The system of open neighborhoods of κ defined by (iv) does not depend on the choice of ρ. The stabilizer Γ_∞ of ∞ acts on U_C. If $\gamma = \left(\begin{smallmatrix} \varepsilon & \mu \\ 0 & \varepsilon^{-1} \end{smallmatrix}\right) \in \Gamma_\infty$, then

$$\gamma z = \left(\varepsilon^2 z_1 + \varepsilon\mu, \varepsilon'^2 z_2 + \varepsilon'\mu'\right).$$

We consider the quotient

$$X(\Gamma) = \Gamma\backslash(\mathbb{H}^2)^*. \tag{1.10}$$

Theorem 1.9. *The quotient space $X(\Gamma)$, together with the quotient topology, is a compact Hausdorff space.* $\qquad\square$

Proposition 1.10. *For $C > 0$ sufficiently large, the canonical map*

$$\Gamma_\infty\backslash U_C \cup \infty \longrightarrow \Gamma\backslash(\mathbb{H}^2)^*$$

is an open embedding. $\qquad\square$

The group $\mathrm{SL}_2(F)$ acts by topological automorphisms on $(\mathbb{H}^2)^*$. Hence, for $\rho \in \mathrm{SL}_2(F)$, the natural map

$$X(\Gamma) \longrightarrow X(\rho^{-1}\Gamma\rho), \quad z \mapsto \rho^{-1}z$$

is topological. If $\rho\infty = \kappa$, it takes the cusp κ of Γ to the cusp ∞ of $\rho^{-1}\Gamma\rho$. In that way, local considerations near the cusps can often be reduced to considerations at the cusp ∞ (for a conjugate group), for which one can use Proposition 1.10.

We define a complex structure on $X(\Gamma)$ as follows. For an open subset $V \subset X(\Gamma)$ we let $U \subset (\mathbb{H}^2)^*$ be the inverse image under the canonical projection $\mathrm{pr} : (\mathbb{H}^2)^* \to X(\Gamma)$, and let U' be the inverse image in $\mathbb{H}^2$. We have the diagram

$$
\begin{array}{ccccc}
\mathbb{H}^2 & \longrightarrow & (\mathbb{H}^2)^* & \overset{\mathrm{pr}}{\longrightarrow} & X(\Gamma) \\
\uparrow & & \uparrow & & \uparrow \\
U' & \longrightarrow & U & \longrightarrow & V
\end{array}.
$$

We define $\mathcal{O}_{X(\Gamma)}(V)$ to be the ring of continuous functions $f : V \to \mathbb{C}$ such that $\mathrm{pr}^*(f)$ is holomorphic on U'. This defines a sheaf $\mathcal{O}_{X(\Gamma)}$ of rings on $X(\Gamma)$, and the pair $(X(\Gamma), \mathcal{O}_{X(\Gamma)})$ is a locally ringed space.

Theorem 1.11 (Baily–Borel). *The space $(X(\Gamma), \mathcal{O}_{X(\Gamma)})$ is a normal complex space.* $\qquad\square$

The proof is based on a criterion of Baily and Cartan for the continuation of complex structures, see [Fr] p. 112.

In contrast to the case of modular curves the resulting normal complex space $X(\Gamma)$ is not regular. There are finite quotient singularities at the elliptic fixed points, and more seriously, the cusps are highly singular points. By the theory of Hironaka the singularities can be resolved [Hi]. A weak form of Hironaka's result states that there exists a desingularization

$$\pi : \widetilde{X}(\Gamma) \longrightarrow X(\Gamma), \tag{1.11}$$

where $\widetilde{X}(\Gamma)$ is a non-singular connected projective variety such that π induces a biholomorphic map $\pi^{-1}(X(\Gamma)^{reg}) \to X(\Gamma)^{reg}$. Here $X(\Gamma)^{reg}$ is the regular locus of $X(\Gamma)$. One can further require that the complement of $\pi^{-1}(X(\Gamma)^{reg})$ is a divisor with normal crossings. The minimal resolution of singularities was constructed by Hirzebruch [Hz].

It can be shown that there is an ample line bundle on $X(\Gamma)$, the line bundle of modular forms (in sufficiently divisible weight). Consequently, the space $X(\Gamma)$ carries the structure of a projective algebraic variety over $\mathbb{C}$. The surface $Y(\Gamma)$ is a Zariski-open subvariety and therefore quasi-projective.

Remark 1.12. The Hilbert modular surfaces $Y(\Gamma)$ often have a moduli interpretation, analogously to the fact that $\mathrm{SL}_2(\mathbb{Z})\backslash\mathbb{H}$ parametrizes isomorphism classes of elliptic curves over $\mathbb{C}$. It can be used to construct integral models of Hilbert modular surfaces. For instance, $Y(\Gamma(\mathcal{O}_F \oplus \mathfrak{a}))$ is the coarse moduli space for isomorphism classes of triples $(A, \imath, m)$, where A is an abelian surface over $\mathbb{C}$, and $\imath : \mathcal{O}_F \to \mathrm{End}(A)$ is an embedding of rings (real multiplication), and m is an isomorphism from the polarization module of A to $(\mathfrak{a}\mathfrak{d}_F)^{-1}$ respecting the positivities, cf. [Go] Theorem 2.17. The variety $Y(\Gamma(\mathcal{O}_F \oplus \mathfrak{a}))$ can be interpreted as the complex points of a moduli stack over $\mathbb{Z}$. One can also construct toroidal compactifications and Baily–Borel compactifications over $\mathbb{Z}$, cf. [Rap], [DePa], [Ch].

Siegel Domains

Here we recall the properties of Siegel domains for Hilbert modular surfaces. They are nice substitutes for fundamental domains.

We write (x_1, x_2) for the real part and (y_1, y_2) for the imaginary part of (z_1, z_2). The top degree differential form

$$d\mu = \frac{dx_1 \, dy_1}{y_1^2} \frac{dx_2 \, dy_2}{y_2^2} \tag{1.12}$$

on $\mathbb{H}^2$ is invariant under the action of $\mathrm{SL}_2(\mathbb{R})^2$. It defines an invariant measure on $\mathbb{H}^2$, which is induced by the Haar measure on $\mathrm{SL}_2(\mathbb{R})$.

Definition 1.13. *A subset $S \subset \mathbb{H}^2$ is called a fundamental set for Γ, if*

$$\mathbb{H}^2 = \bigcup_{\gamma \in \Gamma} \gamma(S).$$

Definition 1.14. *A fundamental set S for Γ is called a fundamental domain for Γ, if*

(i) S is measurable.
(ii) There is a subset $N \subset S$ of measure zero, such that for all $z, w \in S \backslash N$ we have

$$z \sim_\Gamma w \;\Rightarrow\; z = w.$$

Remark 1.15. It can be shown that every measurable fundamental set contains a fundamental domain.

Nice fundamental sets for the action of Γ on $\mathbb{H}^2$ are given by *Siegel domains*: For a positive real number t we put

$$\mathcal{S}_t = \left\{ z \in \mathbb{H}^2; \;\; |x_i| < t \text{ and } y_i > \tfrac{1}{t} \text{ for } i = 1, 2 \right\}. \tag{1.13}$$

Proposition 1.16. *For any fixed $t \in \mathbb{R}_{>0}$ there exist only finitely many $\gamma \in \Gamma$ such that*

$$\gamma \mathcal{S}_t \cap \mathcal{S}_t \neq \emptyset. \tag{1.14}$$

Proof. It is clear that there are only finitely many $\gamma = \left(\begin{smallmatrix} a & b \\ c & d \end{smallmatrix} \right) \in \Gamma$ with $c = 0$ satisfying condition (1.14).

On the other hand, assume that $\gamma \in \Gamma$ with $c \neq 0$, and assume that there is a $z \in \gamma \mathcal{S}_t \cap \mathcal{S}_t$. Then we have

$$\frac{y_1}{|cz_1 + d|^2} > \frac{1}{t}, \tag{1.15}$$

$$\frac{y_2}{|c'z_2 + d'|^2} > \frac{1}{t}. \tag{1.16}$$

The first inequality implies that

$$y_1 > \frac{1}{t}\left((cx_1 + d)^2 + c^2 y_1^2 \right) \geq \frac{1}{t} c^2 y_1^2 > \frac{1}{t^2} c^2 y_1, \tag{1.17}$$

and therefore $|c| < t$. In the same way, inequality (1.16) implies that $|c'| < t$. Hence there are only finitely many possibilities for c. For these, by (1.17) and its analogue for y_2, the imaginary part (y_1, y_2) is bounded, and there are also just finitely many possibilities for d.

Moreover, replacing γ by γ^{-1} in the above argument, we find that there are only finitely many possibilities for a. But a, c, d determine γ. $\square$

Corollary 1.17. *The action of Γ on $\mathbb{H}^2$ is properly discontinuous, that is, if $W \subset \mathbb{H}^2$ is compact, then $\{\gamma \in \Gamma; \; \gamma W \cap W \neq \emptyset\}$ is finite.*

Proof. This follows from Proposition 1.16 and the fact that $\bigcup_{t \in \mathbb{R}_{>0}} \mathcal{S}_t = \mathbb{H}^2$. $\square$

Theorem 1.18. *Let $\kappa_1, \ldots, \kappa_r \in \mathbb{P}^1(F)$ be a set of representatives for the cusps of Γ, and let $\rho_1, \ldots, \rho_r \in \mathrm{SL}_2(F)$ such that $\rho_j \infty = \kappa_j$. There is a $t > 0$ such that*

$$\mathcal{S} = \bigcup_{j=1}^{r} \rho_j \mathcal{S}_t$$

is a measurable fundamental set for Γ.

Proof. See [Ga], Chapter 1.6.   □

1.3 Hilbert Modular Forms

Let $\Gamma \subset \mathrm{SL}_2(F)$ be a subgroup which is commensurable with Γ_F, and let $(k_1, k_2) \in \mathbb{Z}^2$.

Definition 1.19. *A meromorphic function $f : \mathbb{H}^2 \to \mathbb{C}$ is called a meromorphic Hilbert modular form of weight (k_1, k_2) for Γ, if*

$$f(\gamma z) = (cz_1 + d)^{k_1} (c'z_2 + d')^{k_2} f(z) \tag{1.18}$$

for all $\gamma = \left(\begin{smallmatrix} a & b \\ c & d \end{smallmatrix}\right) \in \Gamma$. If $k_1 = k_2 =: k$, then f is said to have parallel weight, and is simply called a meromorphic Hilbert modular form of weight k. If f is holomorphic on $\mathbb{H}^2$, then it is called a holomorphic Hilbert modular form. Finally, a Hilbert modular form f is called symmetric, if $f(z_1, z_2) = f(z_2, z_1)$.

For a function $f : \mathbb{H}^2 \to \mathbb{C}$ and $\gamma = \left(\begin{smallmatrix} a & b \\ c & d \end{smallmatrix}\right) \in \mathrm{SL}_2(F)$ we define the Petersson slash operator by

$$(f \mid_{k_1, k_2} \gamma)(z) = (cz_1 + d)^{-k_1} (c'z_2 + d')^{-k_2} f(\gamma z).$$

The assignment $f \mapsto f \mid_{k_1, k_2} \gamma$ defines a right action of $\mathrm{SL}_2(F)$ on complex valued functions on $\mathbb{H}^2$. Using it, we may rewrite condition (1.18) as

$$f \mid_{k_1, k_2} \gamma = f, \qquad \gamma \in \Gamma.$$

If $k_1 = k_2 =: k$, we simply write $f \mid_k \gamma$ instead of $f \mid_{k_1, k_2} \gamma$.

If f is a holomorphic Hilbert modular form for Γ, it has a Fourier expansion at the cusp ∞ of the following form. Let $M \subset F$ be a $\mathbb{Z}$-module of rank 2 and let $V \subset \mathcal{O}_F^*$ be a finite index subgroup acting on M such that the group

$$G(M, V) = \left\{ \begin{pmatrix} \varepsilon & \mu \\ 0 & \varepsilon^{-1} \end{pmatrix} ; \mu \in M \text{ and } \varepsilon \in V \right\} \tag{1.19}$$

is contained in Γ_∞ with finite index. The transformation law (1.18) for $\gamma \in G(M, V)$ implies that

$$f(z + \mu) = f(z)$$

for all $\mu \in M$. Therefore, f has a normally convergent Fourier expansion

$$f = \sum_{\nu \in M^\vee} a_\nu \, e(\mathrm{tr}(\nu z)), \qquad (1.20)$$

where $e(w) = e^{2\pi i w}$, $\mathrm{tr}(\nu z) = \nu z_1 + \nu' z_2$, and

$$M^\vee = \{\lambda \in F; \ \mathrm{tr}(\mu\lambda) \in \mathbb{Z} \text{ for all } \mu \in M\} \qquad (1.21)$$

is the dual lattice of M with respect to the trace form on F. The Fourier coefficients a_ν are given by

$$a_\nu = \frac{1}{\mathrm{vol}(\mathbb{R}^2/M)} \int_{\mathbb{R}^2/M} f(z) e(-\mathrm{tr}(\nu z)) \, dx_1 \, dx_2 \,. \qquad (1.22)$$

More generally, if $\kappa \in \mathbb{P}^1(F)$ is any cusp of Γ, we take $\rho \in \mathrm{SL}_2(F)$ such that $\rho\infty = \kappa$, and consider $f\,|_{k_1,k_2} \rho$ at ∞. The Fourier expansion of f at κ is the expansion of $f\,|_{k_1,k_2} \rho$ at ∞ (it depends on the choice of ρ).

It is a striking fact that, in contrast to the one-dimensional situation, a holomorphic Hilbert modular form is automatically holomorphic at the cusps by the Götzky–Koecher principle.

Theorem 1.20 (Götzky–Koecher principle). *Let $f : \mathbb{H}^2 \to \mathbb{C}$ be a holomorphic function satisfying $f\,|_{k_1,k_2} \gamma = f$ for all $\gamma \in G(M,V)$ as in (1.19). Then*

(i) $a_{\varepsilon^2\nu} = \varepsilon^{k_1}\varepsilon'^{k_2} a_\nu$ for all $\nu \in M^\vee$ and $\varepsilon \in V$,
(ii) $a_\nu \neq 0 \implies \nu = 0$ or $\nu \gg 0$.

Proof. (i) For $\varepsilon \in V$ we have $\left(\begin{smallmatrix} \varepsilon^{-1} & 0 \\ 0 & \varepsilon \end{smallmatrix}\right) \in G(M,V)$. The transformation law implies that

$$\varepsilon^{-k_1}\varepsilon'^{-k_2} \sum_{\nu \in M^\vee} a_\nu \, e(\mathrm{tr}(\nu\varepsilon^{-2}z)) = \sum_{\nu \in M^\vee} a_\nu \, e(\mathrm{tr}(\nu z)) \,.$$

Comparing Fourier coefficients, we obtain the first assertion.

(ii) Suppose that there is a $\nu \in M^\vee$ such that $a_\nu \neq 0$ and such that $\nu < 0$ or $\nu' < 0$. Without loss of generality we assume $\nu < 0$. There is an $\varepsilon \in V$ with $\varepsilon > 1$ and $0 < \varepsilon' < 1$ such that $\mathrm{tr}(\varepsilon\nu) < 0$. Then $\mathrm{tr}(\varepsilon^{2n}\nu)$ goes to $-\infty$ for $n \to \infty$.

The series

$$\sum_{n \geq 1} a_{\nu\varepsilon^{2n}} \, e(i\,\mathrm{tr}(\nu\varepsilon^{2n}))$$

is a subseries of the Fourier expansion of $f(z)$ at $z = (i,i)$ and therefore converges absolutely. But by (i) we have

$$\sum_{n \geq 1} |a_{\nu\varepsilon^{2n}} \, e(i\,\mathrm{tr}(\nu\varepsilon^{2n}))| = |a_\nu| \sum_{n \geq 1} \varepsilon^{k_1 n}\varepsilon'^{k_2 n} e^{-2\pi\,\mathrm{tr}(\nu\varepsilon^{2n})} \ \to \ \infty \,,$$

contradicting the convergence. $\square$

Corollary 1.21. *A holomorphic Hilbert modular form for the group Γ has a Fourier expansion at the cusp ∞ of the form*

$$f(z) = a_0 + \sum_{\substack{\nu \in M^\vee \\ \nu \gg 0}} a_\nu \, e(\operatorname{tr}(\nu z)) \,. \tag{1.23}$$

The constant term a_0 is called the value of f at ∞. We write $f(\infty) = a_0$. More generally, if $\kappa \in \mathbb{P}^1(F)$ is any cusp of Γ, we take $\rho \in \operatorname{SL}_2(F)$ such that $\rho\infty = \kappa$. We put $f(\kappa) = (f \mid_{k_1,k_2} \rho)(\infty)$. If $(k_1, k_2) \neq (0,0)$, the value $f(\kappa)$ of f at κ depends on the choice of ρ (by a non-zero factor).

Definition 1.22. *A holomorphic Hilbert modular form f is called a cusp form, if it vanishes at all cusps of Γ.*

Proposition 1.23. *Let f be a holomorphic modular form of weight (k_1, k_2) for Γ. If $k_1 \neq k_2$, then f is a cusp form.*

Proof. This follows from Theorem 1.20 (i), applied to the constant terms at the cusps. $\qquad\square$

Proposition 1.24. *Let f be a modular form of weight (k_1, k_2) for Γ. Then the function $h(z) = |f(z)y_1^{k_1/2}y_2^{k_2/2}|$ is Γ-invariant.*

Proof. This follows from Lemma 1.2. $\qquad\square$

Proposition 1.25. *Let f be a holomorphic modular form of weight (k_1, k_2) for Γ, and let $h(z) = |f(z)y_1^{k_1/2}y_2^{k_2/2}|$ be the Γ-invariant function of Proposition 1.24.*

(i) If f has parallel negative weight $k = k_1 = k_2$, then h attains a maximum on $\mathbb{H}^2$.

(ii) If f is a cusp form, then h vanishes at the cusps and attains a maximum on $\mathbb{H}^2$.

Proof. We only prove the first statement, the second is similar. By Proposition 1.24, it suffices to consider h on a fundamental set for Γ. In view of Theorem 1.18, it suffices to show that for any $\rho \in \operatorname{SL}_2(F)$ and any $t \in \mathbb{R}_{>0}$, the function $h(\rho z)$ is bounded and attains its maximum on the Siegel domain $\mathcal{S}_t$. Using the Fourier expansion of f at the cusp $\rho\infty$, we see that

$$h(\rho z) = (y_1 y_2)^{k/2} a_0 + (y_1 y_2)^{k/2} \sum_{\substack{\nu \in M^\vee \\ \nu \gg 0}} a_\nu \, e(\operatorname{tr}(\nu z))$$

for a suitable rank 2 lattice $M \subset F$. Since the weight k is negative, we find that $\lim_{y_1 y_2 \to \infty} h(\rho z) = 0$ on $\mathcal{S}_t$. Consequently, $h(\rho z)$ is bounded and attains a maximum on $\mathcal{S}_t$. $\qquad\square$

Proposition 1.26. *Let f be a holomorphic modular form of weight (k_1, k_2) for Γ. Then f vanishes identically unless k_1, k_2 are both positive or $k_1 = k_2 = 0$. In the latter case f is constant.*

Proof. Let us first consider the case that $k_1 = 0$ and $k_2 \neq 0$. By Proposition 1.23, f is a cusp form. The function $h(z) = y_2^{k_2/2} f(z)$ is holomorphic in z_1, and, by Proposition 1.25, $|h|$ attains a maximum on $\mathbb{H}^2$. According to the maximum principle, h must be constant as a function of z_1. Hence,

$$h(z_1, z_2) = h(z_1, \gamma' z_2)$$

for all $\gamma \in \Gamma$. Since $\{\gamma z_2;\ \gamma \in \Gamma\}$ is dense in $\mathbb{H}$, the function h must be constant on $\mathbb{H}^2$. Because h vanishes at the cusps, it must vanish identically. In the same way we see that $f = 0$, if $k_2 = 0$ and $k_1 \neq 0$.

Let us now consider the case that $k_1 = k_2 = 0$. If f is a cusp form, then Proposition 1.25 implies that $|f|$ attains a maximum on $\mathbb{H}^2$. Hence, by the maximum principle, f must be constant. Since f vanishes at ∞, we obtain that $f \equiv 0$. If f is holomorphic (but not necessarily cuspidal), we consider the cusp form

$$g(z) := \prod_{\kappa \in \Gamma \backslash \mathbb{P}^1(F)} (f(z) - f(\kappa)) .$$

We find that $g \equiv 0$, and therefore f is constant.

Finally, assume that some k_i, say k_1, is negative. If $k_1 \neq k_2$, then, by Proposition 1.23, f is a cusp form. If $k_1 = k_2$, then f has parallel negative weight. In both cases, Proposition 1.25 implies that $h(z) = |f(z) y_1^{k_1/2} y_2^{k_2/2}|$ is bounded by a constant $C > 0$ on $\mathbb{H}^2$. We consider the Fourier expansion of f at the cusp ∞ as in (1.20). The coefficients a_ν are given by (1.22). We find that

$$|a_\nu| \leq \frac{1}{\mathrm{vol}(\mathbb{R}^2/M)} \int_{\mathbb{R}^2/M} |f(z) e(-\mathrm{tr}(\nu z))| \, dx_1 \, dx_2$$

$$\leq C y_1^{-k_1/2} y_2^{-k_2/2} e^{-2\pi \, \mathrm{tr}(\nu y)} .$$

Taking the limit $y_1 \to 0$, we see that a_ν vanishes for all $\nu \in M^\vee$, and therefore $f \equiv 0$. $\qquad\square$

Corollary 1.27. *Let $\widetilde{X}(\Gamma) \to X(\Gamma)$ be a desingularization as in (1.11). Then any holomorphic 1-form on $\widetilde{X}(\Gamma)$ vanishes identically.*

Proof. Let ω be a holomorphic 1-form on $\widetilde{X}(\Gamma)$ and denote by η its pullback to the regular locus of $X(\Gamma)$. Viewing η as a Γ-invariant holomorphic 1-form on $\mathbb{H}^2$, we may write

$$\eta = f_1(z) \, dz_1 + f_2(z) \, dz_2 ,$$

where f_1 and f_2 are holomorphic Hilbert modular forms of weights $(2, 0)$ and $(0, 2)$, respectively. Hence, by Proposition 1.26, η vanishes identically. $\qquad\square$

In the same way one sees that any antiholomorphic 1-form on $\widetilde{X}(\Gamma)$ vanishes identically. Consequently,

$$H^1(\widetilde{X}(\Gamma), \mathcal{O}_{\widetilde{X}(\Gamma)}) = 0 \,,$$

that is, the surface $\widetilde{X}(\Gamma)$ is regular. Using Hodge theory we see that the first cohomolgy group $H^1(\widetilde{X}(\Gamma), \mathbb{C})$ vanishes. In particular, the interesting part of the cohomology of a Hilbert modular surface is in degree 2.

It can be shown that the Hilbert modular surfaces $\widetilde{X}(\Gamma(\mathcal{O}_F \oplus \mathfrak{a}))$ corresponding to the groups $\Gamma(\mathcal{O}_F \oplus \mathfrak{a})$ are simply connected. This also implies the vanishing of the first Betti number. However, there are also examples of Hilbert modular surfaces which are not simply connected. See [Ge1], Chapter IV.6. (Recall that the fundamental group of a complex surface is a birational invariant.)

For the rest of these notes we will only be considering Hilbert modular forms of parallel weight k.

Notation 1.28. *Let $k \in \mathbb{Z}$. We write $M_k(\Gamma)$ for the $\mathbb{C}$-vector space of holomorphic Hilbert modular forms of weight k for the group Γ, and denote by $S_k(\Gamma)$ the subspace of cusp forms.*

The codimension of $S_k(\Gamma)$ in $M_k(\Gamma)$ is clearly bounded by the number of cusps of Γ.

Proposition 1.29. *Let $\widetilde{X}(\Gamma) \to X(\Gamma)$ be a desingularization.*

 (i) Meromorphic Hilbert modular forms of weight 0 for Γ, can be identified with meromorphic functions on $\widetilde{X}(\Gamma)$.

 (ii) Meromorphic Hilbert modular forms of weight 2 for Γ, can be identified with meromorphic 2-forms on $\widetilde{X}(\Gamma)$.

(iii) Hilbert cusp forms of weight 2 for Γ, can be identified with holomorphic 2-forms on $\widetilde{X}(\Gamma)$.

Proof. The first two assertions are easy. For the third assertion we refer to [Fr], Chapter II.4.   □

In particular, the arithmetic genus of the surface $\widetilde{X}(\Gamma)$, that is, the Euler characteristic of the structure sheaf, is given by

$$\chi(\mathcal{O}_{\widetilde{X}(\Gamma)}) = \sum_{p=0}^{2}(-1)^p \dim H^p(\widetilde{X}(\Gamma), \mathcal{O}_{\widetilde{X}(\Gamma)}) = 1 + \dim(S_2(\Gamma)) \,.$$

Holomorphic Hilbert modular forms can be interpreted as sections of the *sheaf $\mathcal{M}_k(\Gamma)$ of modular forms*, which can be defined as follows: If we write $\mathrm{pr} : \mathbb{H}^2 \to Y(\Gamma)$ for the canonical projection, then the sections over an open subset $U \subset \Gamma \backslash \mathbb{H}^2$ are holomorphic functions on $\mathrm{pr}^{-1}(U)$, satisfying the transformation law (1.18). By the Koecher principle, this sheaf on $Y(\Gamma)$ extends to $X(\Gamma)$. It is a coherent $\mathcal{O}_{X(\Gamma)}$-module.

Let $n(\Gamma)$ denote the least common multiple of the orders of all elliptic fixed points for Γ. When $n(\Gamma) \mid k$, then $\mathcal{M}_k(\Gamma)$ is a line bundle. One can show that this line bundle is ample and thereby prove that $X(\Gamma)$ is algebraic. Notice that $\mathcal{M}_{nk}(\Gamma) = \mathcal{M}_k(\Gamma)^{\otimes n}$ for any positive integer n.

1.4 $M_k(\Gamma)$ is Finite Dimensional

In this section we show that $M_k(\Gamma)$ is finite dimensional. The argument is based on the comparison of two different norms on the space of cusp forms. It is a rather general principle and works in a much more general setting (cf. [Fr], Chapter I.6).

We begin by defining the Petersson scalar product on $M_k(\Gamma)$. The top degree differential form $d\mu = \frac{dx_1\,dy_1}{y_1^2}\,\frac{dx_2\,dy_2}{y_2^2}$ on $\mathbb{H}^2$ is invariant under the action of $\mathrm{SL}_2(\mathbb{R})^2$.

Definition 1.30. *Let $f, g \in M_k(\Gamma)$ such that the product fg is a cusp form. We define the Petersson scalar product of f and g by*

$$\langle f, g \rangle = \int_{\mathcal{F}} f(z)\overline{g(z)}(y_1 y_2)^k\,d\mu\,,$$

where $\mathcal{F}$ is a fundamental domain for Γ.

Lemma 1.31. *For f, g as above the Petersson scalar product converges absolutely and is independent of the choice of the fundamental domain.*

Proof. Arguing as in Proposition 1.25, we see that $f(z)\overline{g(z)}\,(y_1 y_2)^k$ is invariant under Γ and bounded on $\mathbb{H}^2$. Hence, that the integral does not depend on the choice of $\mathcal{F}$ follows from the absolute convergence using the theorem on dominated convergence for the Lebesgue integral. To prove the absolute convergence, it suffices to show that

$$\int_{\mathcal{F}} d\mu < \infty\,.$$

In view of Proposition 1.18, it suffices to show that

$$\int_{\mathcal{S}_t} d\mu < \infty$$

for all $t > 0$. This follows from the fact that $\int_{1/t}^{\infty} \frac{dy}{y^2} < \infty$. $\qquad\square$

In particular, the Petersson scalar product defines a hermitian scalar product on $S_k(\Gamma)$. We denote the corresponding L^2-norm on $S_k(\Gamma)$ by

$$\|f\|_2 := \sqrt{\langle f, f \rangle}\,. \tag{1.24}$$

On the other hand we have the maximum norm on $S_k(\Gamma)$ which is defined by

$$\|f\|_\infty = \max_{z \in \mathcal{F}} \left(|f(z)|(y_1 y_2)^{k/2} \right)\,. \tag{1.25}$$

Lemma 1.32. *There is a constant $A = A(\Gamma, k) > 0$ such that*

$$\|f\|_\infty \le A \cdot \|f\|_2$$

for all $f \in S_k(\Gamma)$.

Proof. The L^2-norm can be estimated by considering the Fourier expansions of f at the cusps of Γ and using Siegel domains (Proposition 1.18). See [Fr], Lemma 6.2 for details. $\qquad\square$

Theorem 1.33. *The vector space $M_k(\Gamma)$ is finite dimensional.*

Proof. It suffices to show that $\dim S_k(\Gamma) < \infty$. Let $f_1, \ldots, f_m$ be an orthonormal set with respect to the Petersson scalar product, that is, $\langle f_i, f_j \rangle = \delta_{ij}$. For an arbitrary linear combination

$$f = \sum_{j=1}^{m} c_j f_j$$

with coefficients $c_j \in \mathbb{C}$, we have $\|f\|_\infty \le A\|f\|_2$ by Lemma 1.32. Hence, for all $z \in \mathbb{H}^2$ we have

$$\left| \sum_{j=1}^{m} c_j f_j(z)(y_1 y_2)^{k/2} \right| \le A \left(\sum_{j=1}^{m} |c_j|^2 \right)^{1/2}.$$

We consider the inequality for $c_j = \overline{f_j(z)}$. We find

$$\sum_{j=1}^{m} |f_j(z)|^2 (y_1 y_2)^{k/2} \le A \left(\sum_{j=1}^{m} |f_j(z)|^2 \right)^{1/2}.$$

Dividing by the sum on the right hand side and taking the square we obtain

$$\sum_{j=1}^{m} |f_j(z)|^2 (y_1\, y_2)^{k} \le A^2.$$

Integrating over $\mathcal{F}$ we find that

$$m \le A^2 \operatorname{vol}(\Gamma \backslash \mathbb{H}^2).$$

This concludes the proof of the theorem. $\qquad\square$

1.5 Eisenstein Series

Here we define Eisenstein series for Hilbert modular groups. For simplicity we only consider the full Hilbert modular group $\Gamma_F = \mathrm{SL}_2(\mathcal{O}_F)$ of the real quadratic field F. We write $N(x)$ for the norm of $x \in F$, and $N(\mathfrak{a})$ for the norm of a fractional ideal $\mathfrak{a} \subset F$.

Let $B \in \mathrm{Cl}(F)$ be an ideal class of F. There is a zeta function associated with B which is defined by

$$\zeta_B(s) = \sum_{\substack{\mathfrak{c} \in B \\ \mathfrak{c} \subset \mathcal{O}_F}} N(\mathfrak{c})^{-s}. \tag{1.26}$$

Here s is a complex variable, and the sum runs through all integral ideals in the ideal class B. The series converges for $\Re(s) > 1$. It has a meromorphic continuation to the full complex plane and the completed zeta function

$$\Lambda_B(s) = D^{s/2} \pi^{-s} \Gamma(s/2)^2 \zeta_B(s) \tag{1.27}$$

satisfies the functional equation

$$\Lambda_B(s) = \Lambda_{\mathfrak{d} B^{-1}}(1 - s) \tag{1.28}$$

(see e.g. [Ne], Chapter VII.5). Here, $\mathfrak{d} = \mathfrak{d}_F = \sqrt{D} \mathcal{O}_F$ is the different of F. The Dedekind zeta function $\zeta_F(s)$ of F is given by

$$\zeta_F(s) = \sum_{B \in \mathrm{Cl}(F)} \zeta_B(s) = \sum_{\substack{\mathfrak{c} \subset \mathcal{O}_F \\ \text{integral ideal}}} N(\mathfrak{c})^{-s}. \tag{1.29}$$

Let $\mathfrak{b}$ be a fractional ideal in the ideal class B. The group of units $\mathcal{O}_F^*$ acts on $\mathfrak{b} \times \mathfrak{b}$ by $(c, d) \mapsto (\varepsilon c, \varepsilon d)$ for $\varepsilon \in \mathcal{O}_F^*$. Recall that for $\left(\begin{smallmatrix} a & b \\ c & d \end{smallmatrix}\right) \in \mathrm{SL}_2(F)$ and $z = (z_1, z_2) \in \mathbb{H}^2$, we write $N(cz + d) = (cz_1 + d)(c'z_2 + d')$.

Definition 1.34. *Let $k > 2$ be an even integer. We define the Eisenstein series of weight k associated to B by*[1]

$$G_{k,B}(z) = N(\mathfrak{b})^k \sum_{(c,d) \in \mathcal{O}_F^* \backslash \mathfrak{b} \times \mathfrak{b}}{}^{\prime} N(cz + d)^{-k}.$$

The Eisenstein series $G_{k,B}$ does not depend on the choice of the representative $\mathfrak{b}$ of the ideal class B and converges uniformly absolutely in every Siegel domain $\mathcal{S}_t$ ($t > 0$). Consequently, it defines an element of $M_k(\Gamma_F)$. The value at the cusp ∞ is given by

$$G_{k,B}(\infty) = \lim_{\substack{z \in \mathcal{S}_t \\ y_1 y_2 \to \infty}} G_{k,B}(z)$$

$$= N(\mathfrak{b})^k \sum_{d \in \mathcal{O}_F^* \backslash \mathfrak{b}}{}^{\prime} N(d)^{-k}$$

$$= \sum_{d \in \mathcal{O}_F^* \backslash \mathfrak{b}}{}^{\prime} N(d\mathfrak{b}^{-1})^{-k}$$

$$= \zeta_{B^{-1}}(k).$$

---

[1] The superscript at the summation sign means that the zero summand is omitted.

If $\kappa \in \mathbb{P}^1(F)$ is any cusp and $\rho = \begin{pmatrix} \alpha & \beta \\ \gamma & \delta \end{pmatrix} \in SL_2(F)$ with $\rho\infty = \kappa$, then

$$G_{k,B}(\kappa) = \lim_{\substack{z \in S_t \\ y_1 y_2 \to \infty}} (G_{k,B} |_k \rho)(z)$$

$$= N(\mathfrak{a})^k \zeta_{[\mathfrak{a}]B^{-1}}(k),$$

where $\mathfrak{a} = \mathcal{O}_F\alpha + \mathcal{O}_F\gamma$.

Theorem 1.35. *Let $k > 2$ be an even integer. The Eisenstein series $G_{k,B} \in M_k(\Gamma_F)$, where $B \in Cl(F)$, are linearly independent. The space $M_k(\Gamma_F)$ can be decomposed as a direct sum*

$$M_k(\Gamma_F) = S_k(\Gamma_F) \oplus \bigoplus_{B \in Cl(F)} \mathbb{C}G_{k,B}.$$

Proof. See [Ge1], p. 21. □

Remark 1.36. One can define $G_{k,B}$ for $k > 2$ odd in the same way. In this case it is easily seen that $G_{k,B} \equiv 0$ if $\mathcal{O}_F$ contains a unit of negative norm. Moreover, the theorem also holds for $k = 2$. In this case one can define $G_{2,B}$ by analytic continuation using the "Hecke-trick". In turns out that all Eisenstein series of weight 2 are holomorphic (in contrast to the case of elliptic modular forms, where the constant term is sometimes non-holomorphic).

The Fourier expansion of the Eisenstein series can be computed in the same way as in the case of elliptic modular forms.

Theorem 1.37. *Let $k \geq 2$ even. The Eisenstein series $G_{k,B}$ has the Fourier expansion*

$$G_{k,B}(z) = \zeta_{B^{-1}}(k) + \frac{(2\pi i)^{2k}}{(k-1)!^2} D^{1/2-k} \sum_{\substack{\nu \in \mathfrak{d}^{-1} \\ \nu \gg 0}} \sigma_{k-1,\mathfrak{d}B}(\mathfrak{d}\nu)\, e(\mathrm{tr}(\nu z)).$$

Here, for $A \in Cl(F)$ and an integral ideal $\mathfrak{l} \subset \mathcal{O}_F$, $\sigma_{s,A}(\mathfrak{l})$ denotes the divisor sum

$$\sigma_{s,A}(\mathfrak{l}) = \sum_{\substack{\mathfrak{c} \in A \ integral \\ \mathfrak{c}|\mathfrak{l}}} N(\mathfrak{c})^s.$$

□

Using the functional equation of $\zeta_B(s)$, we may write

$$G_{k,B}(z) = \frac{(2\pi i)^{2k}}{(k-1)!^2} D^{1/2-k} \left[\frac{1}{4}\zeta_{\mathfrak{d}B}(1-k) + \sum_{\substack{\nu \in \mathfrak{d}^{-1} \\ \nu \gg 0}} \sigma_{k-1,\mathfrak{d}B}(\mathfrak{d}\nu)\, e(\mathrm{tr}(\nu z)) \right].$$

$$(1.30)$$

Recall that a Hilbert modular form f of weight k is called symmetric, if $f(z_1, z_2) = f(z_2, z_1)$. It is easily seen that the Eisenstein series $G_{k,B}$ are symmetric.

Restriction to the Diagonal

If $f \in M_k(\Gamma_F)$ is a Hilbert modular form, we consider its restriction to the diagonal $g(\tau) = f(\tau, \tau)$. Since the elliptic modular group $SL_2(\mathbb{Z})$ is embedded diagonally into $\Gamma_F = SL_2(\mathcal{O}_F)$, the function g has the transformation behavior

$$g(\gamma\tau) = f(\gamma\tau, \gamma\tau) = (c\tau + d)^{2k} f(\tau, \tau)$$

for $\gamma = \left(\begin{smallmatrix} a & b \\ c & d \end{smallmatrix}\right) \in SL_2(\mathbb{Z})$. Therefore $g(\tau)$ is an elliptic modular form for $SL_2(\mathbb{Z})$ of weight $2k$. If f has the Fourier expansion

$$f(z) = a_0 + \sum_{\substack{\nu \in \mathfrak{d}^{-1} \\ \nu \gg 0}} a_\nu \, e(\mathrm{tr}(\nu z))$$

at the cusp ∞, then g has the expansion

$$g(\tau) = a_0 + \sum_{n \geq 1} \sum_{\substack{\nu \in \mathfrak{d}^{-1} \\ \nu \gg 0 \\ \mathrm{tr}(\nu) = n}} a_\nu \, e(n\tau). \tag{1.31}$$

The geometric interpretation is the following. The diagonal embedding $\mathbb{H} \to \mathbb{H}^2$, $\tau \mapsto (\tau, \tau)$ induces a morphism $\varphi : SL_2(\mathbb{Z}) \backslash \mathbb{H} \to Y(\Gamma_F)$, which is birational onto its image. If we view f as a section of the line bundle of modular forms of weight k over $Y(\Gamma_F)$, then $g = \varphi^*(f)$ is the pullback.

We now consider the restriction of the Eisenstein series $G_{k,B}$. Using the Fourier expansion (1.30), we see that

$$\frac{1}{4}\zeta_{\mathfrak{d}B}(1 - k) + \sum_{n \geq 1} \sum_{\substack{\nu \in \mathfrak{d}^{-1} \\ \nu \gg 0 \\ \mathrm{tr}(\nu) = n}} \sigma_{k-1, \mathfrak{d}B}(\mathfrak{d}\nu) \, e(n\tau) \tag{1.32}$$

is a modular form for $SL_2(\mathbb{Z})$ of weight $2k$. As a first consequence we see that the special values $\zeta_{\mathfrak{d}B}(1 - k)$ must be rational numbers. This follows from the fact that the divisor sums $\sigma_{k-1, \mathfrak{d}B}(\mathfrak{d}\nu)$ are rational (integers), and the fact that the spaces of elliptic modular forms for $SL_2(\mathbb{Z})$ have bases with rational Fourier coefficients. If $k = 2, 4$, then (1.32) must be a multiple of the elliptic Eisenstein series

$$E_{2k}(\tau) = -\frac{B_{2k}}{4k} + \sum_{n \geq 1} \sigma_{2k-1}(n) \, e(n\tau).$$

Here B_{2k} is the usual Bernoulli number and $\sigma_s(n) = \sum_{d | n} d^s$. Comparing the first Fourier coefficients we obtain a formula for $\zeta_{\mathfrak{d}B}(1 - k)$ due to Siegel.

Theorem 1.38 (Siegel). *If $k = 2, 4$, then*

$$\zeta_B(1 - k) = -\frac{B_{2k}}{k} \sum_{\substack{\nu \in \mathfrak{d}^{-1} \\ \nu \gg 0 \\ \mathrm{tr}(\nu) = 1}} \sigma_{k-1, B}(\mathfrak{d}\nu).$$

$\square$

By means of (1.29), and using $B_4 = B_8 = -1/30$, we obtain:

Corollary 1.39. *The special values of the Dedekind zeta function of F at the arguments $-1, -3$ are given by*

$$\zeta_F(-1) = \frac{1}{60} \sum_{\substack{x \in \mathbb{Z} \\ x^2 < D \\ x^2 \equiv D \,(4)}} \sigma_1 \left(\frac{D - x^2}{4}\right),$$

$$\zeta_F(-3) = \frac{1}{120} \sum_{\substack{x \in \mathbb{Z} \\ x^2 < D \\ x^2 \equiv D \,(4)}} \sigma_3 \left(\frac{D - x^2}{4}\right).$$

$\square$

We end this section with a table for these special values.

Table 1. Special values of $\zeta_F(s)$

D	5	8	12	13	17	21	24	28	29	33	37	40	41	44
$\zeta_F(-1)$	$\frac{1}{30}$	$\frac{1}{12}$	$\frac{1}{6}$	$\frac{1}{6}$	$\frac{1}{3}$	$\frac{1}{3}$	$\frac{1}{2}$	$\frac{2}{3}$	$\frac{1}{2}$	1	$\frac{5}{6}$	$\frac{7}{6}$	$\frac{4}{3}$	$\frac{7}{6}$
$\zeta_F(-3)$	$\frac{1}{60}$	$\frac{11}{120}$	$\frac{23}{60}$	$\frac{29}{60}$	$\frac{41}{30}$	$\frac{77}{30}$	$\frac{87}{20}$	$\frac{113}{15}$	$\frac{157}{20}$	$\frac{141}{10}$	$\frac{1129}{60}$	$\frac{1577}{60}$	$\frac{448}{15}$	$\frac{2153}{60}$

The Example $\mathbb{Q}(\sqrt{5})$

Eisenstein series and restriction to the diagonal can be used to determine the graded algebra of holomorphic Hilbert modular forms in some cases where the discriminant of F is small. Here we illustrate this for $F = \mathbb{Q}(\sqrt{5})$. The class number of F is 1, and the fundamental unit of $\mathcal{O}_F$ is given by $\varepsilon_0 = \frac{1+\sqrt{5}}{2}$. The graded algebra of modular forms for the group Γ_F was determined by Gundlach [Gu], see also [Mü]. For further examples see [Ge1], Chapter 8.

We denote by g_k the Eisenstein series for Γ_F of weight k normalized such that the constant term is 1 (so $g_k = \frac{1}{\zeta_F(k)} G_{k, \mathcal{O}_F}$).

Theorem 1.40 (Gundlach). *The graded algebra $M_{2*}^{sym}(\Gamma_F)$ of holomorphic symmetric Hilbert modular forms of even weight for Γ_F is the (weighted) polynomial ring $\mathbb{C}[g_2, g_6, g_{10}]$.*

Often it is more convenient to replace the generators g_6 and g_{10} by the cusp forms

$$s_6 = 67 \cdot (2^5 \cdot 3^3 \cdot 5^2)^{-1} \cdot (g_2^3 - g_6),$$
$$s_{10} = (2^{10} \cdot 3^5 \cdot 5^5 \cdot 7)^{-1} \cdot (2^2 \cdot 3 \cdot 7 \cdot 4231\, g_2^5 - 5 \cdot 67 \cdot 2293\, g_2^2 g_6 + 412751\, g_{10}).$$

Then Gundlach's result can be restated as

$$M_{2*}^{sym}(\Gamma_F) = \mathbb{C}[g_2, s_6, s_{10}]. \tag{1.33}$$

Notice that g_2, s_6, s_{10} all have rational integral and coprime Fourier coefficients.

The key idea for the proof is to show that there is a "square root" for s_{10}. Gundlach constructed a cusp form s_5 of weight 5 for Γ_F as a product of 10 theta constants. (Later, in Section 3.2, we will construct it as the Borcherds lift Ψ_1.) One can show that s_5 is anti-symmetric, that is, $s_5(z_1, z_2) = -s_5(z_2, z_1)$. Hence it has to vanish on the diagonal. It turns out that the divisor of s_5 is given by the image of the diagonal in $Y(\Gamma_F)$. Moreover, $s_5^2 = s_{10}$.

Proof of Theorem 1.40. It is clear that the restriction of g_2 to the diagonal is the Eisenstein series of weight 4 for $\mathrm{SL}_2(\mathbb{Z})$, normalized such that the constant term is 1. A quick computation shows that the restriction of s_6 is the delta function, the unique normalized cusp form of weight 12 for $\mathrm{SL}_2(\mathbb{Z})$. Consequently the restrictions of g_2 and s_6 generate the algebra of modular forms for $\mathrm{SL}_2(\mathbb{Z})$ of weight divisible by 4.

Suppose that f is a symmetric Hilbert modular form of even weight k for Γ_F. Then the restriction to the diagonal of f is a modular form for $\mathrm{SL}_2(\mathbb{Z})$ of weight divisible by 4 and therefore a polynomial in the restrictions of g_2 and s_6. Hence there is a polynomial $P \in \mathbb{C}[X, Y]$ such that

$$f_1 = f - P(g_2, s_6)$$

vanishes on the diagonal. Therefore f_1/s_5 is a holomorphic Hilbert modular form for Γ_F. It is anti-symmetric and therefore vanishes on the diagonal. Consequently, $f_1/s_5^2 \in M_{k-10}(\Gamma_F)$ is symmetric. Now the assertion follows by induction on the weight. $\qquad\square$

To get the full algebra $M_*(\Gamma_F)$ of Hilbert modular forms for Γ_F one needs in addition the existence of a symmetric Hilbert cusp form s_{15} of weight 15. Gundlach constructed it as a product of differences of Eisenstein series of weight 1 for a principal congruence subgroup of Γ_F. (In Section 3.2 we will construct it as the Borcherds lift Ψ_5.)

Theorem 1.41 (Gundlach). *The graded algebra $M_*(\Gamma_F)$ of Hilbert modular forms for Γ_F is generated by g_2, s_5, s_6, s_{15}. The anti-symmetric cusp form s_5 and the symmetric cusp form s_{15} satisfy the following relations over $M_{2*}^{sym}(\Gamma_F) = \mathbb{C}[g_2, s_6, s_{10}]$:*

$$s_5^2 = s_{10},$$

$$s_{15}^2 = 5^5 \cdot s_{10}^3 - 2^{-1} \cdot 5^3 \cdot g_2^2 s_6 s_{10}^2 + 2^{-4} \cdot g_2^5 s_{10}^2 + 2^{-1} \cdot 3^2 \cdot 5^2 \cdot g_2 s_6^3 s_{10}$$

$$- 2^{-3} \cdot g_2^4 s_6^2 s_{10} - 2 \cdot 3^3 \cdot s_6^5 + 2^{-4} \cdot g_2^3 s_6^4.$$

□

1.6 The L-function of a Hilbert Modular Form

In this section we briefly discuss how one can attach an L-function to a Hilbert modular form. First, one needs to know that the coefficients have polynomial growth.

Proposition 1.42 (Hecke estimate). *Let $f = \sum_\nu a_\nu \, e(\mathrm{tr}(\nu z)) \in M_k(\Gamma)$.*

(i) Then $a_\nu = O(\mathrm{N}(\nu)^k)$ for $\mathrm{N}(\nu) \to \infty$.
(ii) If f is a cusp form, we have the stronger estimate $a_\nu = O(\mathrm{N}(\nu)^{k/2})$ for $\mathrm{N}(\nu) \to \infty$.

Proof. Here we only carry out the proof for cusp forms. For non-cuspidal modular forms one has to slightly modify the argument. (For the group Γ_F one can also use Theorems 1.35 and 1.37.) According to (1.22) we have

$$a_\nu = \frac{1}{\mathrm{vol}(\mathbb{R}^2/M)} \int_{\mathbb{R}^2/M} f(z)e(-\mathrm{tr}(\nu z)) \, dx_1 \, dx_2.$$

By Proposition 1.25 we know that $|f(z)(y_1 y_2)^{k/2}|$ is bounded on $\mathbb{H}^2$. Hence there is a constant $C > 0$ such that

$$|a_\nu| \leq C \int_{\mathbb{R}^2/M} (y_1 y_2)^{-k/2} e^{-2\pi \, \mathrm{tr}(\nu y)} \, dx_1 \, dx_2$$

for all $y \in (\mathbb{R}_{>0})^2$. Choosing $y = (\frac{1}{\nu}, \frac{1}{\nu'})$, we see that

$$|a_\nu| \leq C \, \mathrm{vol}(\mathbb{R}^2/M) \, \mathrm{N}(\nu)^{k/2}.$$

This proves the proposition.

□

For the rest of this section we only consider the full Hilbert modular group Γ_F. Let $f \in M_k(\Gamma_F)$, and denote the Fourier expansion by

$$f = a_0 + \sum_{\substack{\nu \in \mathfrak{d}^{-1} \\ \nu \gg 0}} a_\nu\, e(\operatorname{tr}(\nu z))\,.$$

Let $\mathcal{O}_F^{*,+}$ be the group of totally positive units of $\mathcal{O}_F$, and let $U = \{\varepsilon^2; \varepsilon \in \mathcal{O}_F^{*,+}\}$. Then U has index 2 in the cyclic group $\mathcal{O}_F^{*,+}$. We have $a_\nu = a_{\varepsilon\nu}$ for all $\nu \in \mathfrak{d}^{-1}$ and all $\varepsilon \in U$.

Definition 1.43. *We define an L-series associated to f by*

$$L(f, s) = \sum_{\substack{\nu \in \mathfrak{d}^{-1}/U \\ \nu \gg 0}} a_\nu\, \operatorname{N}(\nu\mathfrak{d})^{-s}\,.$$

Example 1.44. For the Eisenstein series $G_{k,B}$ (see Definition 1.34) one easily checks that

$$L(G_{k,B}, s) = 2\frac{(2\pi i)^{2k}}{(k-1)!^2} D^{1/2-k}\zeta_{\mathfrak{d}^{-1}B^{-1}}(s)\,\zeta_{\mathfrak{d}B}(s+1-k)\,.$$

The functional equation (1.28) of the partial Dedekind zeta function $\zeta_B(s)$ implies that $L(G_{k,B}, s)$ has a meromorphic continuation and satisfies a functional equation relating s and $k-s$. Therefore it is reasonable to expect similar properties for the L-functions of cusp forms as well.

Theorem 1.45. *Let $f \in S_k(\Gamma_F)$. The completed L-function*

$$\Lambda(f, s) = D^s (2\pi)^{-2s} \Gamma(s)^2 L(f, s)$$

has a holomorphic continuation to $\mathbb{C}$, is entire and bounded in vertical strips, and satisfies the functional equation

$$\Lambda(f, s) = (-1)^k \Lambda(f, k-s)\,.$$

Proof. Using the Euler integral for the Gamma function, we see that

$$(2\pi)^{-2s} \Gamma(s)^2 \operatorname{N}(\nu)^{-s} = \int_0^\infty \int_0^\infty e^{-2\pi \operatorname{tr}(\nu y)} (y_1 y_2)^s \frac{dy_1}{y_1} \frac{dy_2}{y_2}\,.$$

Hence, by unfolding we find that

$$\Lambda(f, s) = \int_{(\mathbb{R}_{>0})^2/U} f(iy)(y_1 y_2)^s \frac{dy_1}{y_1} \frac{dy_2}{y_2}\,.$$

We split up the integral into an integral over $y_1 y_2 > 1$ and a second integral over $y_1 y_2 < 1$. The modularity of f implies that $f\left(i\left(\frac{1}{y_1}, \frac{1}{y_2}\right)\right) =$

$(-1)^k (y_1 y_2)^k f(iy)$. Hence the second integral can be transformed into an integral over $y_1 y_2 > 1$ as well. We find that

$$\Lambda(f,s) = \int\limits_{\substack{(\mathbb{R}_{>0})^2/U \\ y_1 y_2 > 1}} f(iy)(y_1 y_2)^s \frac{dy_1}{y_1} \frac{dy_2}{y_2} + (-1)^k \int\limits_{\substack{(\mathbb{R}_{>0})^2/U \\ y_1 y_2 > 1}} f(iy)(y_1 y_2)^{k-s} \frac{dy_1}{y_1} \frac{dy_2}{y_2} .$$

This integral representation converges for all $s \in \mathbb{C}$ and defines the holomorphic continuation of $\Lambda(f,s)$. Moreover, the functional equation is obvious now.
□

We now suppose that $\mathcal{O}_F$ contains a unit of negative norm. Then $M_k(\Gamma_F)$ contains no non-zero element for k odd. So we further suppose that k is even. Then the Fourier coefficients of $f \in M_k(\Gamma_F)$ satisfy $a_\nu = a_{\varepsilon\nu}$ for all $\nu \in \mathfrak{d}^{-1}$ and all $\varepsilon \in \mathcal{O}_F^{*,+}$. Thus, a_ν only depends on the ideal $(\nu\mathfrak{d}) \subset \mathcal{O}_F$, and we write $a((\nu\mathfrak{d})) = a_\nu$. Then we may rewrite the L-function of f in the form

$$L(f,s) = \sum_{\substack{\mathfrak{a} \subset \mathcal{O}_F \\ \text{principal ideal}}} a(\mathfrak{a}) \, \mathrm{N}(\mathfrak{a})^{-s} .$$

So this L-function is analogous to the zeta function $\zeta_B(s)$ associated to an ideal class B of F (here the unit class). It is natural to associate more general L-functions to f, for instance, an L-function where one sums over all integral ideals of F analogous to the full Dedekind zeta function of F. To this end it is more convenient to view Hilbert modular forms as automorphic functions on $(\mathrm{Res}_{F/\mathbb{Q}} \, \mathrm{SL}_2)(\mathbb{A})$, where $\mathrm{Res}_{F/\mathbb{Q}}$ denotes the Weil restriction of scalars, and $\mathbb{A}$ denotes the ring of adeles of $\mathbb{Q}$ (cf. [Ga]).

2 The Orthogonal Group $O(2,n)$

An important property of Hilbert modular surfaces is that they can also be regarded as modular varieties associated to the orthogonal group of a suitable rational quadratic space V of type $(2,2)$. There is an accidental isomorphism $\mathrm{Res}_{F/\mathbb{Q}} \, \mathrm{SL}_2 \cong \mathrm{Spin}_V$ of algebraic groups over $\mathbb{Q}$. Modular varieties for orthogonal groups $O(2,n)$ come with natural families of special algebraic cycles on them arising from embeddings of "smaller" orthogonal groups. They provide a rich source of extra structure and can be used to study geometric questions. In the $O(2,2)$-case of Hilbert modular surfaces these special cycles lead to Hirzebruch–Zagier divisors (codimension 1) and certain CM-points (codimension 2).

To put things in the right context, in this section we study quadratic spaces and modular forms for orthogonal groups in slightly greater generality than needed for the application to Hilbert modular surfaces. However, we hope that this will rather clarify things than complicate them. For a detailed account of the theory of quadratic forms and orthogonal groups we refer to [Ki], [Scha].

2.1 Quadratic Forms

Let R be a commutative ring with unity 1. We write R^* for the group of units in R. Let M be a finitely generated R-module. A *quadratic form* on M is a mapping $Q : M \to R$ such that

(i) $Q(rx) = r^2 Q(x)$ for all $r \in R$ and all $x \in M$,
(ii) $B(x,y) := Q(x+y) - Q(x) - Q(y)$ is a bilinear form.

The first condition follows from the second if 2 is invertible in R. Then we have $Q(x) = \frac{1}{2}B(x,x)$. The pair (M, Q) is called a quadratic module over R. If R is a field, we frequently say space instead of module. Two elements $x, y \in M$ are called orthogonal if $B(x,y) = 0$. If $A \subset M$ is a subset, we denote the orthogonal complement by

$$A^\perp = \{x \in M; \ B(x,y) = 0 \text{ for all } y \in A\}. \tag{2.1}$$

It is a submodule of M. For $x \in M$ we briefly write $x^\perp$ instead of $\{x\}^\perp$. The quadratic module M is called *non-degenerate*, if $M^\perp = \{0\}$. A non-zero vector $x \in M$ is called *isotropic* if $Q(x) = 0$, and *anisotropic*, if $Q(x) \neq 0$, respectively.

Let (M, Q) and (M', Q') be quadratic modules over R. An R-linear map $\sigma : M \to M'$ is called an *isometry*, if σ is injective and

$$Q'(\sigma(x)) = Q(x)$$

for all $x \in M$. If σ is also surjective then M and N are called isometric. The *orthogonal group* of M,

$$O_M = \{\sigma \in \mathrm{Aut}(M); \quad \sigma \text{ isometry}\}, \tag{2.2}$$

is the group of all isometries from M onto itself. The special orthogonal group is the subgroup

$$SO_M = \{\sigma \in O_M; \quad \det(\sigma) = 1\}. \tag{2.3}$$

Important examples of isometries are given by *reflections*. For an element $x \in M$ with $Q(x) \in R^*$ we define $\tau_x : M \to M$ by

$$\tau_x(y) = y - B(y,x)Q(x)^{-1}x, \qquad y \in M. \tag{2.4}$$

Then τ_x is an isometry and satisfies

$$\tau_x(x) = -x$$
$$\tau_x(y) = y, \quad \text{for } y \in x^\perp,$$
$$\tau_x^2 = \mathrm{id}.$$

So τ_x is the reflection in the hyperplane $x^\perp$.

Further examples of isometries are given by *Eichler elements*. Let $u \in M$ be isotropic and $v \in M$ with $B(u,v) = 0$. We define $E_{u,v} : M \to M$ by

$$E_{u,v}(y) = y + B(y,u)v - B(y,v)u - B(y,u)Q(v)u, \qquad y \in M. \qquad (2.5)$$

One easily checks that $E_{u,v}$ is an isometry and

$$E_{u,v}(u) = u,$$
$$E_{u,v}(v) = v - 2Q(v)u,$$
$$E_{u,v_1} E_{u,v_2} = E_{u,v_1+v_2} \quad \text{for } v_1, v_2 \in u^{\perp}.$$

If M is free, and $v_1, \ldots, v_n$ is a basis of M, we have the corresponding *Gram matrix* $S = (B(v_i, v_j))_{i,j}$. The class of $\det(S)$ in $R^*/(R^*)^2$ is independent of the choice of the basis. It is called the discriminant of M and is denoted by $d(M)$. Note that if $v_1, \ldots, v_n$ is an orthogonal basis of M, we have

$$d(M) = 2^n Q(v_1) \cdots Q(v_n). \qquad (2.6)$$

For the rest of this subsection, let M be a quadratic space of dimension n over a field k of characteristic $\neq 2$. The space is non-degenerate, if and only if $d(M) \neq 0$. If M is non-degenerate and $v_1 \in M$ is an anisotropic vector, there exist anisotropic vectors $v_2, \ldots, v_n \in M$ such that $v_1, \ldots, v_n$ is an orthogonal basis of M. Consequently, the reflection corresponding to v_1 satisfies $\det(\tau_{v_1}) = -1$.

Theorem 2.1. *Let M be a regular quadratic space over a field k of characteristic $\neq 2$. Then O_M is generated by reflections. Moreover, SO_M is the subgroup of elements of O_M which can be written as a product of an even number of reflections.* $\qquad \square$

Example 2.2. Let p, q be non-negative integers. We denote by $\mathbb{R}^{p,q}$ the quadratic space over $\mathbb{R}$ given by $\mathbb{R}^{p+q}$ with the quadratic form

$$Q(x) = x_1^2 + \cdots + x_p^2 - x_{p+1}^2 \cdots - x_{p+q}^2.$$

If V is a finite dimensional quadratic space over $\mathbb{R}$, then there exist non-negative integers p, q such that V is isometric to $\mathbb{R}^{p,q}$. The pair (p, q) is uniquely determined by V and is called the *type* of V. Moreover, $p - q$ is called the *signature* of V. The orthogonal group of $\mathbb{R}^{p,q}$ is also denoted by $O(p, q)$.

2.2 The Clifford Algebra

As before, let R be a commutative ring with unity 1, and let (V, Q) be a finitely generated quadratic module over R. If A is any R-algebra, we write $Z(A)$ for the center of A.

We consider the tensor algebra

$$T_V = \bigoplus_{m=0}^{\infty} V^{\otimes m} = R \oplus V \oplus (V \otimes_R V) \oplus \cdots$$

of V. Let $I_V \subset T_V$ be the two-sided ideal generated by $v \otimes v - Q(v)$ for $v \in V$. The *Clifford algebra* of V is defined by

$$C_V = T_V/I_V. \qquad (2.7)$$

Observe that R and V are embedded into C_V via the canonical maps. For simplicity, the element of C_V represented by $v_1 \otimes \cdots \otimes v_m$ (where $v_i \in V$) is denoted by $v_1 \cdots v_m$. By definition, we have for $u, v \in V \subset C_V$:

$$v^2 = Q(v),$$
$$uv + vu = B(u, v).$$

In particular, $uv = -vu$ if and only if u and v are orthogonal. The Clifford algebra has the following universal property.

Proposition 2.3. *Let* $f : V \to A$ *be an R-linear map to an R-algebra A with unity 1_A such that $f(v)^2 = Q(v)1_A$ for all $v \in V$. Then there exists a unique R-algebra homomorphism $C_V \to A$ such that the following diagram commutes:*

The universal property implies that an isometry $\varphi : V \to V'$ of quadratic spaces over R induces a unique R-algebra homomorphism $\tilde{\varphi} : C_V \to C_{V'}$ compatible with the natural inclusions. Therefore, the assignment $V \mapsto C_V$ defines a functor from the category of quadratic spaces over R with isometries as morphisms to the category of associative R-algebras with unity.

Moreover, if we fix a quadratic space (V, Q) over R, for any commutative R-algebra S with unity we can consider the extension of scalars $V(S) = V \otimes_R S$, which is a quadratic module over S in a natural way (the quadratic form being defined by $Q(v \otimes s) = s^2 Q(v)$). In the same way, we consider $C_V(S) = C_V \otimes_R S$. One easily checks that $C_V(S) = C_{V(S)}$. So taking the Clifford algebra commutes with extension of scalars.

Example 2.4. We denote by $C_{p,q}$ the Clifford algebra of the real quadratic space $\mathbb{R}^{p,q}$ of Example 2.2. For small p, q we have

$$C_{0,0} = \mathbb{R}, \qquad C_{1,0} = \mathbb{R} \oplus \mathbb{R}, \qquad C_{0,1} = \mathbb{C},$$
$$C_{2,0} = \mathrm{M}_2(\mathbb{R}), \qquad C_{1,1} = \mathrm{M}_2(\mathbb{R}), \qquad C_{0,2} = \mathbb{H}.$$

Here $\mathbb{H}$ denotes the Hamilton quaternion algebra. (This follows from Examples 2.7 and 2.8 below).

Now assume that V is free. If $v_1, \ldots, v_n$ is a basis of V, then these vectors generate C_V as an R-algebra. The elements

$$v_{i_1} \cdots v_{i_r} \qquad (1 \leq i_1 < \cdots < i_r \leq n \text{ and } 0 \leq r \leq n)$$

form a basis of C_V. In particular, C_V is a free R-module of rank 2^n. Observe that for the trivial quadratic form $Q \equiv 0$ the Clifford algebra C_V is simply the Grassmann algebra of V.

We write C_V^0 for the R-subalgebra of C_V generated by products of an even number of basis vectors of V, and C_V^1 for the R-submodule of C_V generated by products of an odd number of basis vectors of V. This definition is meaningful, since the defining relations of C_V only involve products of an even number of basis vectors. We obtain a decomposition

$$C_V = C_V^0 \oplus C_V^1,$$

which is a $\mathbb{Z}/2\mathbb{Z}$-grading on C_V. The subalgebra C_V^0 is called the *even Clifford algebra* of V (or the second Clifford algebra of V).

Multiplication by -1 defines an isometry of V. By Proposition 2.3 it induces an algebra automorphism

$$J : C_V \longrightarrow C_V, \tag{2.8}$$

called the *canonical automorphism*. If 2 is invertible in R, then the even Clifford algebra can be characterized by

$$C_V^0 = \{v \in C_V;\ J(v) = v\}.$$

There is a second involution on C_V, which is an anti-automorphism. It is called the *canonical involution* on C_V and is defined by

$$^t : C_V \longrightarrow C_V, \qquad (x_1 \otimes \cdots \otimes x_m)^t = x_m \otimes \cdots \otimes x_1. \tag{2.9}$$

It reduces to the identity on $R \oplus V$. It is used to define the *Clifford norm* on C_V by

$$\mathrm{N} : C_V \longrightarrow C_V, \qquad \mathrm{N}(x) = x^t x. \tag{2.10}$$

For $x \in V$ we have $\mathrm{N}(x) = Q(x)$. So the norm map extends the quadratic form on V. Note that the Clifford norm is in general *not* multiplicative on C_V.

For the rest of this subsection, let k be a field of characteristic $\neq 2$, and let (V, Q) be a non-degenerate quadratic space over k of dimension n. Moreover, let $v_1, \ldots, v_n$ be an orthogonal basis of V. We put

$$\delta = v_1 \cdots v_n \in C_V.$$

Remark 2.5. When n is even, we have

$$\delta v_i = -v_i \delta, \quad \delta^2 = (-1)^{n/2} 2^{-n} d(V) \in k^*/(k^*)^2 \,.$$

When n is odd, we have

$$\delta v_i = v_i \delta, \quad \delta^2 = (-1)^{(n-1)/2} 2^{-n} d(V) \in k^*/(k^*)^2 \,.$$

$\square$

Theorem 2.6. *The center of C_V is given by*

$$Z(C_V) = \begin{cases} k & \text{if } n \text{ is even}, \\ k + k\delta & \text{if } n \text{ is odd}. \end{cases}$$

The center of C_V^0 is given by

$$Z(C_V^0) = \begin{cases} k + k\delta & \text{if } n \text{ is even}, \\ k & \text{if } n \text{ is odd}. \end{cases}$$

$\square$

Let A be a ring with unity such that $k \subset Z(A)$. Recall that A is called a *quaternion algebra* over k, if it has a basis $\{1, x_1, x_2, x_3\}$ as a k-vector space such that

$$x_1^2 = \alpha, \quad x_2^2 = \beta, \quad x_3 = x_1 x_2 = -x_2 x_1$$

for some $\alpha, \beta \in k^*$. Then it is denoted by (α, β). The parameters α, β determine A up to k-algebra isomorphism. It is easily seen that $k = Z(A)$. The conjugation in A is defined by

$$x = a_0 + a_1 x_1 + a_2 x_2 + a_3 x_3 \mapsto \bar{x} = a_0 - a_1 x_1 - a_2 x_2 - a_3 x_3$$

for $x \in A$. The norm is defined by $\mathrm{N}(x) = x\bar{x} \in k$. A quaternion algebra over k is either isomorphic to $\mathrm{M}_2(k)$ or it is a division algebra. For more details we refer to [Ki] Chapter 1.5. We end this section by giving some examples of Clifford algebras associated to low dimensional quadratic spaces (see [Ki], p. 28).

Example 2.7. If $n = 1$ then $C_V = k + k\delta$ and $\delta^2 = d(V)/2$. As a k-algebra, we have

$$C_V \cong k[X]/(X^2 - d(V)/2) \,.$$

When $d(V)/2$ is not a square in k, this is a quadratic field extension of k. When $d(V)/2$ is a square in k, then $C_V \cong k \oplus k$.

Example 2.8. Suppose that $n = 2$ and that V has the orthogonal basis v_1, v_2 with $Q(v_i) = q_i \in k^*$. Then $C_V = k \oplus kv_1 \oplus kv_2 \oplus kv_1v_2$ is isomorphic to the quaternion algebra (q_1, q_2) over k. Moreover, $C_V^0 \cong k[X]/(X^2 + d(V))$.

Example 2.9. Suppose that $n = 3$ and that V has the orthogonal basis v_1, v_2, v_3 with $Q(v_i) = q_i \in k^*$. Then $C_V^0 = k \oplus kv_1v_2 \oplus kv_2v_3 \oplus kv_1v_3$ is isomorphic to the quaternion algebra $(-q_1q_2, -q_2q_3)$ over k. The conjugation in the quaternion algebra is identified with the main involution of the Clifford algebra, and the norm with the Clifford norm.

Example 2.10. Suppose that $n = 4$ and that the space V has the orthogonal basis v_1, v_2, v_3, v_4 with $Q(v_i) = q_i \in k^*$. Then the center Z of the even Clifford algebra C_V^0 is $k + k\delta$, and we have

$$C_V^0 = Z + Zv_1v_2 + Zv_2v_3 + Zv_1v_3 \,.$$

Since $(v_1v_2)^2 = -q_1q_2$, $(v_2v_3)^2 = -q_2q_3$, and $(v_1v_2)(v_2v_3) = q_2(v_1v_3)$, the algebra C_V^0 is isomorphic to the quaternion algebra $(-q_1q_2, -q_2q_3)$ over Z. The conjugation in the quaternion algebra is identified with the main involution of the Clifford algebra, and the norm with the Clifford norm.

2.3 The Spin Group

As before, let R be a commutative ring with unity 1, and let (V, Q) be a finitely generated quadratic module over R. The *Clifford group* CG_V of V is defined by

$$\mathrm{CG}_V = \{x \in C_V; \quad x \text{ invertible and } xVJ(x)^{-1} = V\} \,. \tag{2.11}$$

By definition, every $x \in \mathrm{CG}_V$ defines an automorphism α_x of V by

$$\alpha_x(v) = xvJ(x)^{-1} \,.$$

We obtain a linear representation $\alpha : \mathrm{CG}_V \to \mathrm{Aut}_R(V)$, $x \mapsto \alpha_x$, called the *vector representation*. It is easily seen that the involution $x \mapsto x^t$ takes CG_V to itself. Consequently, if $x \in \mathrm{CG}_V$, then the Clifford norm $\mathrm{N}(x)$ belongs to CG_V as well.

Lemma 2.11. *Assume that $R = k$ is a field of characteristic $\neq 2$. The kernel of the vector representation $\alpha : \mathrm{CG}_V \to \mathrm{Aut}_k(V)$ is equal to k^*. The Clifford norm induces a homomorphism $\mathrm{CG}_V \to k^*$.*

Proof. It is clear that k^* is contained in $\ker(\alpha)$. We now show that $\ker(\alpha) \subset k^*$. Let $x \in \ker(\alpha) \subset \mathrm{CG}_V$. We write $x = x_0 + x_1$ with $x_0 \in C_V^0$ and $x_1 \in C_V^1$. Then we have $xvJ(x)^{-1} = v$ for all $v \in V$. Hence

$$x_0v = vx_0 \,,$$
$$x_1v = -vx_1 \,.$$

Since V generates C_V as an algebra, we see that $x_0 \in Z(C_V)^* \cap C_V^0 = k^*$. Moreover, one can check that the second condition implies that $x_1 = 0$.

We know that $N(x)$ acts trivially on V via α for $x \in CG_V$. Let $v \in V$. Since $w := \alpha_x(v) \in V$, we have $w = -J(w)^t$. This implies

$$xvJ(x)^{-1} = (x^t)^{-1}vJ(x^t) \,,$$

and therefore $N(x)vJ(N(x))^{-1} = v$. Hence $N(x) \in k^*$. The multiplicativity of the norm is a direct consequence.

Lemma 2.12. *For $x \in CG_V$ the automorphism $\alpha_x \in \mathrm{Aut}_R(V)$ is an isometry.*

Proof. Let $v \in V$. Since $w = \alpha_x(v) \in V$, we have

$$
\begin{aligned}
Q(w) &= N(w) \\
&= J(x^{-1})^t v^t x^t x v J(x^{-1}) \\
&= Q(v) \,.
\end{aligned}
$$

This shows that α_x is an isometry. $\qquad\square$

Consequently, the vector representation defines a homomorphism

$$\alpha : CG_V \to O_V \,. \tag{2.12}$$

Moreover, if $x \in CG_V \cap V$, then $Q(x) \in R^*$ and α_x is equal to the reflection τ_x in the hyperplane $x^\perp$.

Definition 2.13. *We define the general Spin group GSpin_V and the Spin group Spin_V of V by*

$$
\begin{aligned}
\mathrm{GSpin}_V &= CG_V \cap C_V^0, \\
\mathrm{Spin}_V &= \{x \in \mathrm{GSpin}_V; \ N(x) = 1\} \,.
\end{aligned}
$$

For the rest of this section we assume that $R = k$ is a field of characteristic $\neq 2$. We briefly discuss the structure of the Clifford and the Spin group.

In this case, by Theorem 2.1, the vector representation (2.12) is surjective onto O_V. Moreover, the kernel is given by k^* (see Lemma 2.11). Hence CG_V and GSpin_V are central extensions of O_V and SO_V, respectively,

$$1 \longrightarrow k^* \longrightarrow CG_V \longrightarrow O_V \longrightarrow 1 \,,$$

$$1 \longrightarrow k^* \longrightarrow \mathrm{GSpin}_V \longrightarrow SO_V \longrightarrow 1 \,.$$

According to Lemma 2.11 and Theorem 2.6, the Clifford norm defines a homomorphism $CG_V \to k^*$. It induces a homomorphism

$$\theta : O_V \longrightarrow k^*/(k^*)^2 \,, \tag{2.13}$$

called the *spinor norm*. It is characterized by the property that for the reflection τ_v corresponding to an anisotropic vector $v \in V$ we have

$$\theta(\tau_v) = Q(v) \in k^*/(k^*)^2 .$$

We obtain the exact sequence

$$1 \longrightarrow \{\pm 1\} \longrightarrow \mathrm{Spin}_V \overset{\alpha}{\longrightarrow} \mathrm{SO}_V \overset{\theta}{\longrightarrow} k^*/(k^*)^2 .$$

The groups CG_V, GSpin, and Spin_V can be viewed as the groups of k-valued points of an affine algebraic group over k. If A is a commutative k-algebra with unity, then the group of A-valued points of CG_V is $\mathrm{CG}_V(A) = \mathrm{CG}_{V(A)}$, and analogously for the other groups.

The following lemma will be useful in Section 2.7.

Lemma 2.14. *Assume that* $\dim(V) \leq 4$. *Then*

$$\mathrm{GSpin}_V = \{x \in C_V^0; \quad \mathrm{N}(x) \in k^*\},$$
$$\mathrm{Spin}_V = \{x \in C_V^0; \quad \mathrm{N}(x) = 1\}.$$

Proof. It is clear that the left hand sides are contained in the right hand sides.

Conversely, let $x \in C_V^0$ with $\mathrm{N}(x) \in k^*$. Then x is invertible, because $y = x^t \, \mathrm{N}(x)^{-1} \in C_V^0$ is inverse to x. Hence, it suffices to show that $xVx^{-1} \subset V$.

Let $v \in V$. It is clear that $w := xvx^{-1} \in C_V^1$. The assumption $\dim(V) \leq 4$ implies that

$$V = \{g \in C_V^1; \quad g^t = g\}.$$

Therefore it suffices to show that $w = w^t$. Since $\mathrm{N}(x) \in k^*$, we have $\mathrm{N}(x)v\,\mathrm{N}(x)^{-1} = v$. This implies that

$$xvx^{-1} = (x^t)^{-1}vx^t$$

and therefore $w = w^t$. $\qquad\square$

Quadratic Spaces in Dimension Four

We now consider the special cases that (V, Q) is a rational quadratic space of dimension 4 over the field k. Let v_1, v_2, v_3, v_4 be an orthogonal basis of V and put $q_i = Q(v_i) \in k^*$. By means of Example 2.10 and Lemma 2.14, we see that Spin_V is the group of norm 1 elements in the quaternion algebra $(-q_1q_2, -q_2q_3)$ over $Z := Z(C_V^0) = k + k\delta$, where $\delta = v_1v_2v_3v_4$.

We would like to describe the vector representation of Spin_V intrinsically in terms of C_V^0. This can be done by identifying V with an isometric copy $\tilde{V}$ in C_V^0. (Note that by definition $V \not\subset C_V^0$.) The vector representation on V translates to a "twisted" vector representation on $\tilde{V}$. We partly follow [KR] §0.

Lemma 2.15. *Let* $v_0 \in V$ *with* $q_0 = Q(v_0) \neq 0$, *and denote by* $\sigma = \mathrm{Ad}(v_0)$ *the adjoint automorphism of* C_V^0 *associated to* v_0, *i.e.,* $x^\sigma = v_0xv_0^{-1}$ *for* $x \in C_V^0$. *Then* $\delta^\sigma = -\delta$ *and the fixed algebra of* σ *in* C_V^0 *is a quaternion algebra* B_0 *over* k *such that* $C_V^0 = B_0 \otimes_k Z$.

Proof. See [KR] Lemma 0.2. □

In particular, on the center Z of C_V^0, the automorphism σ agrees with the conjugation in Z/k. Let

$$\tilde{V} = \{x \in C_V^0; \quad x^t = x^\sigma\}. \tag{2.14}$$

This is a quadratic space over k together with the quadratic form

$$\tilde{Q}(x) = q_0 \cdot x^\sigma x = q_0 \cdot N(x). \tag{2.15}$$

The group Spin_V acts on $\tilde{V}$ by

$$x \mapsto \tilde{\alpha}_g(x) := gxg^{-\sigma}, \tag{2.16}$$

for $x \in \tilde{V}$ and $g \in \mathrm{Spin}_V$. The quadratic form $\tilde{Q}$ is preserved under this action:

$$\tilde{Q}(gxg^{-\sigma}) = q_0 \cdot (gxg^{-\sigma})^t(gxg^{-\sigma}) = q_0 \cdot x^t x = \tilde{Q}(x). \tag{2.17}$$

Lemma 2.16. *The assignment $x \mapsto x \cdot v_0$ defines an isometry of quadratic spaces*

$$(\tilde{V}, \tilde{Q}) \longrightarrow (V, Q),$$

which is compatible with the actions of Spin_V.

Proof. See [KR] Lemma 0.3. □

2.4 Rational Quadratic Spaces of Type $(2, n)$

Let (V, Q) be a non-degenerate quadratic space over $\mathbb{Q}$. Then $V(\mathbb{R}) = V \otimes_{\mathbb{Q}} \mathbb{R}$ is isometric to $\mathbb{R}^{p,q}$ for a unique pair of non-negative integers (p, q), called the type of V. If $K \subset O_V(\mathbb{R})$ is a maximal compact subgroup, then $O_V(\mathbb{R})/K$ is a symmetric space. It is hermitian, i.e., has a complex structure, if and only if $p = 2$ or $q = 2$. Since this is the case of interest to us, throughout this subsection we assume that V has type $(2, n)$. We discuss several realizations of the corresponding hermitian symmetric domain. We frequently write $(\cdot, \cdot)$ for the bilinear form $B(\cdot, \cdot)$.

The Grassmannian Model

We consider the Grassmannian of 2-dimensional subspaces of $V(\mathbb{R})$ on which the quadratic form is positive definite,

$$\mathrm{Gr}(V) = \{v \subset V(\mathbb{R}); \quad \dim v = 2 \text{ and } Q|_v > 0\}.$$

By Witt's theorem, $O_V(\mathbb{R})$ acts transitively on $\mathrm{Gr}(V)$. If $v_0 \in \mathrm{Gr}(V)$ is fixed, then the stabilizer K of v_0 is a maximal compact subgroup of $O_V(\mathbb{R})$, and $K \cong O(2) \times O(n)$. Thus $\mathrm{Gr}(V) \cong O_V(\mathbb{R})/K$ is a realization of the hermitian symmetric space. The Grassmannian model has the advantage that it provides an easy description of $O_V(\mathbb{R})/K$, but unfortunately we do not see the complex structure.

The Projective Model

We consider the complexification $V(\mathbb{C})$ of V and the corresponding projective space

$$P(V(\mathbb{C})) = (V(\mathbb{C})\backslash\{0\})/\mathbb{C}^*. \tag{2.18}$$

The zero quadric

$$\mathcal{N} = \{[Z] \in P(V(\mathbb{C})); \quad (Z, Z) = 0\} \tag{2.19}$$

is a closed algebraic subvariety. The subset

$$\mathcal{K} = \{[Z] \in P(V(\mathbb{C})); \quad (Z, Z) = 0, (Z, \bar{Z}) > 0\} \tag{2.20}$$

of the zero quadric is a complex manifold of dimension n consisting of two connected components. The orthogonal group $O_V(\mathbb{R})$ acts transitively on $\mathcal{K}$. The subgroup $O_V^+(\mathbb{R})$ of elements whose spinor norm equals the determinant preserves the components of $\mathcal{K}$, whereas $O_V(\mathbb{R})\backslash O_V^+(\mathbb{R})$ interchanges them. We choose one fixed component of $\mathcal{K}$ and denote it by $\mathcal{K}^+$. If $Z \in V(\mathbb{C})$ we write $Z = X + iY$ with $X, Y \in V(\mathbb{R})$ for the decomposition into real and imaginary part.

Lemma 2.17. *The assignment* $[Z] \mapsto v(Z) := \mathbb{R}X + \mathbb{R}Y$ *defines a real analytic isomorphism* $\mathcal{K}^+ \to \mathrm{Gr}(V)$.

Proof. If $Z \in V(\mathbb{C})$, then the condition $[Z] \in \mathcal{K}$ is equivalent to

$$X \perp Y, \quad \text{and} \quad (X, X) = (Y, Y) > 0. \tag{2.21}$$

But this means that X and Y span a two dimensional positive definite subspace of $V(\mathbb{R})$ and thereby define an element of $\mathrm{Gr}(V)$. Conversely for a given $v \in \mathrm{Gr}(V)$ we may choose a (suitably oriented) orthogonal basis X, Y as in (2.21) and obtain a unique $[Z] = [X + iY] \in \mathcal{K}^+$. (Then $[\bar{Z}] \in \mathcal{K}$ corresponds to the same $v \in \mathrm{Gr}(V)$.) We get a real analytic isomorphism between $\mathcal{K}^+$ and $\mathrm{Gr}(V)$. $\square$

The advantage of the projective model is that it comes with a natural complex structure. However, it is not the direct analogue of the upper half plane, the standard model for the hermitian symmetric space for $\mathrm{SL}_2(\mathbb{R})$.

The Tube Domain Model

We may realize $\mathcal{K}^+$ as a tube domain in the following way. Suppose that $e_1 \in V$ is a non-zero isotropic vector and $e_2 \in V$ with $(e_1, e_2) = 1$. We define a rational quadratic subspace $W \subset V$ by $W = V \cap e_1^{\perp} \cap e_2^{\perp}$. Then W is Lorentzian, that is, has type $(1, n-1)$ and

$$V = W \oplus \mathbb{Q}e_2 \oplus \mathbb{Q}e_1 .$$

If $Z \in V(\mathbb{C})$ and $Z = z + ae_2 + be_1$ with $z \in W(\mathbb{C})$ and $a, b \in \mathbb{C}$, we briefly write $Z = (z, a, b)$. We consider the tube domain

$$\mathcal{H} = \{z \in W(\mathbb{C}); \quad Q(\Im(z)) > 0\} . \tag{2.22}$$

Lemma 2.18. *The assignment*

$$z \mapsto \psi(z) := [(z, 1, -Q(z) - Q(e_2))] \tag{2.23}$$

defines a biholomorphic map $\psi : \mathcal{H} \to \mathcal{K}$.

Proof. One easily checks that if $z \in \mathcal{H}$ then $\psi(z) \in \mathcal{K}$. Conversely assume that $[Z] \in \mathcal{K}$ with $Z = X + iY$. From the fact that X, Y span a two dimensional positive definite subspace of $V(\mathbb{R})$ it follows that $(Z, e_1) \neq 0$. Thus $[Z]$ has a unique representative of the form $(z, 1, b)$. The condition $Q(Z) = 0$ implies that $b = -Q(z) - Q(e_2)$, and thereby $[Z] = [(z, 1, -Q(z) - Q(e_2))]$. Moreover, from $(Z, \bar{Z}) > 0$ one easily deduces $Q(\Im(z)) > 0$. We may infer that the map ψ is biholomorphic. $\qquad\square$

The domain $\mathcal{H} \subset W(\mathbb{C}) \cong \mathbb{C}^n$ has two components corresponding to the two cones of positive norm vectors in the Lorentzian space $W(\mathbb{R})$. We denote by $\mathcal{H}^+$ the component which is mapped to $\mathcal{K}^+$ under the above isomorphism. It can be viewed as a generalized upper half plane. The group $O_V^+(\mathbb{R})$ acts transitively on it. In the $O(2, 1)$ case the domain $\mathcal{H}^+$ can be identified with the usual upper half plane $\mathbb{H}$. For $O(2, 2)$ it can be identified with $\mathbb{H}^2$ as we shall see below. However, a disadvantage of the tube domain model is that the action of $O_V^+(\mathbb{R})$ is not linear anymore.

Lattices

As before, let (V, Q) be a non-degenerate quadratic space over $\mathbb{Q}$ of type $(2, n)$.

Definition 2.19. *A lattice in V is a $\mathbb{Z}$-module $L \subset V$ such that $V = L \otimes_{\mathbb{Z}} \mathbb{Q}$.*

A lattice $L \subset V$ is called *integral* if the bilinear form is integral valued on L, that is, $(x, y) \in \mathbb{Z}$ for all $x, y \in L$. A lattice is called *even* if the quadratic form is integral valued on L, that is, $Q(x) \in \mathbb{Z}$ for all $x \in L$. So an even lattice is a free quadratic module over $\mathbb{Z}$ of finite rank. Clearly every even lattice is integral.

The dual lattice $L^{\vee}$ is defined by

$$L^{\vee} = \{x \in V; \ (x, y) \in \mathbb{Z} \text{ for all } y \in L\} .$$

The lattice L is integral if and only if $L \subset L^{\vee}$. In this case the quotient $L^{\vee}/L$ is a finite abelian group, called the *discriminant group*. If S is the Gram matrix corresponding to a lattice basis of L, we have

$$|L^{\vee}/L| = |\det(S)|.$$

For the rest of this section we assume that $L \subset V$ is an even lattice. The orthogonal group O_L is a discrete subgroup of $O_V(\mathbb{R}) \cong O(2,n)$. Let

$$\Gamma \subset O_L \cap O_V^+(\mathbb{R}) \tag{2.24}$$

be a subgroup of finite index. Then Γ acts properly discontinuously on $\mathrm{Gr}(V)$, $\mathcal{K}^+$, and $\mathcal{H}^+$. We consider the quotient

$$Y(\Gamma) = \Gamma \backslash \mathcal{H}^+ \tag{2.25}$$

similarly as in the construction of Hilbert modular surfaces in Section 1.1. It is a normal complex space, which is compact if and only if V is anisotropic.

If $Y(\Gamma)$ is non-compact, it can be compactified by adding rational boundary components (see e.g. [BrFr]). These boundary components are most easily described in the projective model $\mathcal{K}^+$. The boundary points of $\mathcal{K}^+$ in the zero quadric $\mathcal{N}$ correspond to non-trivial isotropic subspaces of $V(\mathbb{R})$.

Let $F \subset V(\mathbb{R})$ be an isotropic line. Then F represents a boundary point of $\mathcal{K}^+$. A boundary point of this type is called *special*, otherwise *generic*. A set consisting of one special boundary point is called a *zero-dimensional boundary component*.

Let $F \subset V(\mathbb{R})$ be a two-dimensional isotropic subspace. The set of all generic boundary points of $\mathcal{K}^+$ which can be represented by an element of $F(\mathbb{C})$ is called the *one-dimensional boundary component attached to F*.

By a boundary component we understand a one- or two-dimensional boundary component. One can show (see [BrFr], Section 2):

Lemma 2.20. *There is a bijective correspondence between boundary components of $\mathcal{K}^+$ in the zero quadric $\mathcal{N}$ and non-zero isotropic subspaces $F \subset V(\mathbb{R})$ of the corresponding dimension. The boundary of $\mathcal{K}^+$ is the disjoint union of the boundary components.* $\qquad\square$

A boundary component is called rational if the corresponding isotropic space F is defined over $\mathbb{Q}$. The union of $\mathcal{K}^+$ with all rational boundary components is denoted by $(\mathcal{K}^+)^*$. The rational orthogonal group $O_V(\mathbb{Q}) \cap O_V^+(\mathbb{R})$ acts on $(\mathcal{K}^+)^*$. By the theory of Baily–Borel, the quotient

$$X(\Gamma) = (\mathcal{K}^+)^*/\Gamma$$

together with the Baily–Borel topology is a compact Hausdorff space. There is a natural complex structure on $X(\Gamma)$ as a normal complex space. Moreover, using modular forms, one can construct an ample line bundle on $X(\Gamma)$. Therefore, $X(\Gamma)$ is projective algebraic. It is called the modular variety associated to Γ. Using the theory of canonical models, one can show that $X(\Gamma)$ is actually defined over a number field (see [Ku2]).

Heegner Divisors

Let Γ be as above, see (2.24). In order to understand the geometry of $X(\Gamma)$, we study special divisors on this variety, obtained from embeddings of modular varieties corresponding to quadratic subspaces of V.

Let $\lambda \in L^\vee$ with $Q(\lambda) < 0$. Then the orthogonal complement $V_\lambda = \lambda^\perp \subset V$ is a rational quadratic space of type $(2, n-1)$. Moreover, the orthogonal complement of λ in $\mathcal{K}^+$,

$$H_\lambda = \{[Z] \in \mathcal{K}^+; \ (Z, \lambda) = 0\},$$

is an analytic divisor. It is the hermitian symmetric domain corresponding to $(V_\lambda, Q|_{V_\lambda})$. Let us briefly look at the description of H_λ in the tube domain model $\mathcal{H}^+$ using the above notation. We write $\lambda = \lambda_W + ae_2 + be_1$ with $\lambda_W \in W$ and $a, b \in \mathbb{Q}$. Then

$$H_\lambda \cong \{z \in \mathcal{H}^+; \ aQ(z) - (z, \lambda_W) - aq(e_2) - b = 0\}$$

is given by a quadratic equation in the coordinates of $\mathcal{H}^+$. (Therefore it is sometimes called a rational quadratic divisor.)

If $\beta \in L^\vee/L$ is fixed and m is a fixed negative rational number, then

$$H(\beta, m) = \sum_{\substack{\lambda \in \beta+L \\ \mathbb{Q}(\lambda)=m}} H_\lambda \tag{2.26}$$

defines an analytic divisor on $\mathcal{K}^+$ called the *Heegner divisor* of discriminant (β, m). If Γ acts trivially on $L^\vee/L$, then, by Chow's lemma, this divisor descends to an algebraic divisor on $Y(\Gamma)$ (denoted in the same way). By [Ku2], it is defined over a number field. Here we mainly consider the composite Heegner divisor

$$H(m) = \frac{1}{2} \sum_{\beta \in L^\vee/L} H(\beta, m) = \sum_{\substack{\lambda \in L^\vee/\{\pm 1\} \\ \mathbb{Q}(\lambda)=m}} H_\lambda. \tag{2.27}$$

It is Γ-invariant and descends to an algebraic divisor on $Y(\Gamma)$. Hence, $Y(\Gamma)$ comes with a natural family of algebraic divisors indexed by negative rational numbers (with denominators bounded by the level of L). The existence of such a family is special for orthogonal and unitary groups.

2.5 Modular Forms for O(2, n)

Let V, L, Γ be as above. We write

$$\widetilde{\mathcal{K}}^+ = \{Z \in V(\mathbb{C})\backslash\{0\}; \ [Z] \in \mathcal{K}^+\}$$

for the cone over $\mathcal{K}^+$.

Definition 2.21. *Let $k \in \mathbb{Z}$, and let χ be character of Γ. A meromorphic function F on $\widetilde{\mathcal{K}}^+$ is called a meromorphic modular form of weight k and character χ for the group Γ, if*

(i) F is homogeneous of degree $-k$, i.e., $F(cZ) = c^{-k}F(Z)$ for any $c \in \mathbb{C}\backslash\{0\}$;
(ii) F is invariant under Γ, i.e., $F(gZ) = \chi(g)F(Z)$ for any $g \in \Gamma$;
(iii) F is meromorphic at the boundary.

If f is actually holomorphic on $\widetilde{\mathcal{K}}^+$ and at the boundary, it is called a holomorphic modular form.

By the Koecher principle the boundary condition is automatically fulfilled if the Witt rank of V, that is, the dimension of a maximal isotropic subspace, is smaller than n. (Note that because of the signature the Witt rank of L is always ≤ 2.)

2.6 The Siegel Theta Function

Examples of modular forms on orthogonal groups can be constructed using Eisenstein series similarly as in Section 1.5. However, we do not discuss this. Here we consider a rather different source of modular forms, the so called theta lifting. The groups $\mathrm{SL}_2(\mathbb{R})$ and $\mathrm{O}(2, n)$ form a dual reductive pair in the sense of Howe [Ho]. Hence, Howe duality gives rise to a correspondence between automorphic representations for the two groups. Often one can realize this correspondence as a lifting from automorphic forms on one group to the other, by integrating against certain kernel functions given by theta functions.

Let V, L, Γ be as above and assume that n is even so that $\dim(V)$ is even. Let

$$N = \min\{a \in \mathbb{Z}_{>0}; \quad aQ(\lambda) \in \mathbb{Z} \text{ for all } \lambda \in L^\vee\}$$

be the *level* of L. We modify the discriminant of L by a sign and consider

$$\Delta = (-1)^{\frac{n+2}{2}} \det S,$$

where S is the Gram matrix for a lattice basis of L. One can show that $\Delta \equiv 1, 0$ (mod 4). Therefore $\chi_\Delta = \left(\frac{\Delta}{\cdot}\right)$ is a quadratic Dirichlet character modulo N.

For $\lambda \in V(\mathbb{R})$ and $v \in \mathrm{Gr}(V)$ we have a unique decomposition $\lambda = \lambda_v + \lambda_{v^\perp}$, where λ_v and $\lambda_{v^\perp}$ are the orthogonal projections of λ to v and $v^\perp$, respectively. The positive definite quadratic form

$$Q_v(\lambda) = Q(\lambda_v) - Q(\lambda_{v^\perp})$$

on V is called the majorant associated to v. If $Z \in \widetilde{\mathcal{K}}^+$, we briefly write λ_Z and Q_Z instead of $\lambda_{v(Z)}$ and $Q_{v(Z)}$, where $v(Z)$ is the positive definite plane corresponding to Z via Lemma 2.17.

Definition 2.22. *Let $r \in \mathbb{Z}_{\geq 0}$. The Siegel theta function of weight r of the lattice L is defined by*

$$\Theta_r(\tau, Z) = v^{n/2} \sum_{\lambda \in L^\vee} \frac{(\lambda, Z)^r}{(Z, \bar{Z})^r} e\big(Q(\lambda_Z)N\tau + Q(\lambda_{Z\perp})N\bar{\tau}\big)$$

$$= v^{n/2} \sum_{\lambda \in L^\vee} \frac{(\lambda, Z)^r}{(Z, \bar{Z})^r} e\big(Q(\lambda)Nu + Q_Z(\lambda)Niv\big),$$

for $\tau = u + iv \in \mathbb{H}$ and $Z \in \widetilde{\mathcal{K}}^+$ (see e.g. [Bo4], [Od1], [RS]). Here we denote $e(w) = e^{2\pi i w}$ as usual.

Because of the rapid decay of the exponential term $e(Q_Z(\lambda)Niv)$, the series converges normally on $\mathbb{H} \times \widetilde{\mathcal{K}}^+$. It defines a real analytic function, which is non-holomorphic in both variables, τ and Z. Using the Poisson summation formula, or the theory of the Weil representation, one can show that as a function in τ, the Siegel theta function satisfies

$$\Theta_r(\gamma\tau, Z) = \chi_\Delta(d)(c\tau + d)^{r + \frac{2-n}{2}} \Theta_r(\tau, Z) \tag{2.28}$$

for all $\gamma = \left(\begin{smallmatrix} a & b \\ c & d \end{smallmatrix}\right) \in \Gamma_0(N)$, where

$$\Gamma_0(N) = \left\{ \begin{pmatrix} a & b \\ c & d \end{pmatrix} \in \mathrm{SL}_2(\mathbb{Z}); \quad c \equiv 0 \pmod{N} \right\}. \tag{2.29}$$

Moreover, in the variable Z, the function $\overline{\Theta_r(\tau, Z)}$ transforms as a modular form of weight r for Γ. This follows by direct inspection.

We may use the Siegel theta function as an integral kernel to lift elliptic modular forms for $\Gamma_0(N)$ to modular forms on the orthogonal group. More precisely, let $f \in S_k(\Gamma_0(N), \chi_\Delta)$ be a cusp form for $\Gamma_0(N)$ with character χ_Δ of weight $k = r + \frac{2-n}{2}$. We define the theta lift $\Phi_0(Z, f)$ of f by the integral

$$\Phi_0(Z, f) = \int_{\mathcal{F}} f(\tau)\overline{\Theta_r(\tau, Z)} v^k \frac{du\,dv}{v^2}, \tag{2.30}$$

where $\mathcal{F}$ denotes a fundamental domain for $\Gamma_0(N)$.

Theorem 2.23. *The theta lift $\Phi_0(Z, f)$ of f is a holomorphic modular form of weight $r = k - \frac{2-n}{2}$ for the orthogonal group Γ.*

Proof. The transformation properties of the Siegel theta function immediately imply that $\Phi_0(Z, f)$ transforms as a modular form of weight r for the group Γ. However, it is not clear at all, that $\Phi_0(Z, f)$ is holomorphic. This can be proved by computing the Fourier expansion. For details we refer to e.g. [Bo4] Theorem 14.3, [Od1] Section 5, Theorem 2, or [RS]. □

Remark 2.24. The linear map $f \mapsto \Phi_0(Z, f)$ often has a non-trivial kernel. The question when it vanishes is related to the vanishing of a special value of the standard L-function of f [Ral]. Therefore it can be rather difficult. However, in many cases it is also possible to obtain non-vanishing results by computing the Fourier expansion of the lift.

2.7 The Hilbert Modular Group as an Orthogonal Group

In this section we discuss the accidental isomorphism relating the Hilbert modular group to an orthogonal group of type $(2,2)$ in detail. The Heegner divisors of the previous section give rise to certain algebraic divisors on Hilbert modular surfaces, known as Hirzebruch–Zagier divisors [HZ].

Let $d \in \mathbb{Q}^*$ be not a square, and put $F = \mathbb{Q}(\sqrt{d})$. We consider the four dimensional $\mathbb{Q}$-vector space

$$V = \mathbb{Q} \oplus \mathbb{Q} \oplus F$$

together with the quadratic form $Q(a, b, \nu) = \nu\nu' - ab$, where $\nu \mapsto \nu'$ denotes the conjugation in F. So (V, Q) is a rational quadratic space of type $(2, 2)$ if $d > 0$ and of type $(3, 1)$ if $d < 0$. We consider the orthogonal basis

$$v_1 = (1, 1, 0), \qquad\qquad v_3 = (0, 0, 1),$$
$$v_2 = (1, -1, 0), \qquad\qquad v_4 = (0, 0, \sqrt{d}).$$

Then $\delta = v_1 v_2 v_3 v_4$ satisfies $\delta^2 = d$. According to Remark 2.5 and Theorem 2.6, the center $Z(C_V^0)$ of the even Clifford algebra of V is given by $Z(C_V^0) = \mathbb{Q} + \mathbb{Q}\delta \cong F$. Moreover, in view of Example 2.10,

$$C_V^0 = Z + Z v_1 v_2 + Z v_2 v_3 + Z v_1 v_3$$

is isomorphic to the split quaternion algebra $M_2(F)$ over F. This isomorphism can be realized by the assignment

$$1 \mapsto \begin{pmatrix} 1 & 0 \\ 0 & 1 \end{pmatrix}, \qquad\qquad v_2 v_3 \mapsto \begin{pmatrix} 0 & 1 \\ -1 & 0 \end{pmatrix},$$

$$v_1 v_2 \mapsto \begin{pmatrix} 1 & 0 \\ 0 & -1 \end{pmatrix}, \qquad\qquad v_1 v_3 \mapsto \begin{pmatrix} 0 & 1 \\ 1 & 0 \end{pmatrix}.$$

The canonical involution on C_V^0 corresponds to the conjugation

$$\begin{pmatrix} a & b \\ c & d \end{pmatrix}^* = \begin{pmatrix} d & -b \\ -c & a \end{pmatrix}$$

in $M_2(F)$. The Clifford norm corresponds to the determinant. Hence, by Lemma 2.14, Spin_V can be identified with $SL_2(F)$. As algebraic groups over $\mathbb{Q}$ we have $\mathrm{Spin}_V \cong \mathrm{Res}_{F/\mathbb{Q}} SL_2$. Consequently, the group $\Gamma_F = SL_2(\mathcal{O}_F)$ and commensurable groups can be viewed as arithmetic subgroups of Spin_V. For instance, using (2.32) below, it is easily seen that $\Gamma_F = \mathrm{Spin}_L$, where L denotes the lattice $\mathbb{Z} \oplus \mathbb{Z} \oplus \mathcal{O}_F \subset V$.

We now describe the vector representation explicitly using Lemmas 2.15 and 2.16. Let $\sigma = \mathrm{Ad}(v_1)$ be the adjoint automorphism of C_V^0 associated to the basis vector v_1, i.e., $x^\sigma = v_1 x v_1^{-1}$ for $x \in C_V^0$. Then $\delta^\sigma = -\delta$, and on the center F of C_V^0, the automorphism σ agrees with the conjugation in $F/\mathbb{Q}$. On $M_2(F)$ the action of σ is given by

$$\begin{pmatrix} a & b \\ c & d \end{pmatrix} \mapsto \begin{pmatrix} a & b \\ c & d \end{pmatrix}^\sigma = \begin{pmatrix} d' & -c' \\ -b' & a' \end{pmatrix}.$$

As in (2.14) let

$$\begin{aligned} \tilde{V} &= \{X \in M_2(F); \quad X^* = X^\sigma\} \\ &= \{X \in M_2(F); \quad X^t = X'\} \\ &= \left\{ \begin{pmatrix} a & \nu' \\ \nu & b \end{pmatrix}; \quad a, b \in \mathbb{Q} \text{ and } \nu \in F \right\}. \end{aligned}$$

This is a rational quadratic space together with the quadratic form

$$\tilde{Q}(X) = -X^\sigma \cdot X = -\det(X).$$

The corresponding bilinear form is

$$\tilde{B}(X_1, X_2) = -\operatorname{tr}(X_1 \cdot X_2^*),$$

for $X_1, X_2 \in \tilde{V}$. The group $\mathrm{SL}_2(F) \cong \mathrm{Spin}_V$ acts isometrically on $\tilde{V}$ by

$$x \mapsto g.X := gXg^{-\sigma} = gXg'^t, \tag{2.31}$$

for $X \in \tilde{V}$ and $g \in \mathrm{SL}_2(F)$. A computation shows that in the present case the isometry of quadratic spaces $\tilde{V} \to V$, $X \mapsto X \cdot v_1$, of Lemma 2.16 is given by

$$\begin{pmatrix} a & \nu' \\ \nu & b \end{pmatrix} \mapsto (a, b, \nu). \tag{2.32}$$

Throughout the rest of this section we work with $\tilde{V}$ and the twisted vector representation (2.31). We assume that d is positive so that F is real quadratic. We now describe the hermitian symmetric space corresponding to $\mathrm{O}_{\tilde{V}}$ as in Section 2.4.

The two real embeddings $x \mapsto (x, x') \in \mathbb{R}^2$ induce an embedding $\tilde{V} \to M_2(\mathbb{R})$. Hence we have $\tilde{V}(\mathbb{C}) = M_2(\mathbb{C})$ and

$$\mathcal{K} = \{[Z] \in P(M_2(\mathbb{C})); \quad \det(Z) = 0, -\operatorname{tr}(Z\bar{Z}^*) > 0\}.$$

We consider the isotropic vectors $e_1 = \begin{pmatrix} -1 & 0 \\ 0 & 0 \end{pmatrix}$ and $e_2 = \begin{pmatrix} 0 & 0 \\ 0 & 1 \end{pmatrix}$ in $\tilde{V}$, and the orthogonal complement $W = \tilde{V} \cap e_1^\perp \cap e_2^\perp$. For $z = (z_1, z_2) \in \mathbb{C}^2 \cong W(\mathbb{C})$ we put

$$M(z) = \begin{pmatrix} z_1 z_2 & z_1 \\ z_2 & 1 \end{pmatrix} \in M_2(\mathbb{C}).$$

Then $[M(z)]$ lies in the zero quadric in $P(\mathrm{M}_2(\mathbb{C}))$, and $[M(z)] \in \mathcal{K}$ if and only if $\mathfrak{I}(z_1)\mathfrak{I}(z_2) > 0$. Consequently, we may identify $\mathbb{H}^2$ with $\mathcal{H}^+$. If we denote by $\mathcal{K}^+$ the corresponding component of $\mathcal{K}$, we obtain a biholomorphic map

$$\mathbb{H}^2 \longrightarrow \mathcal{K}^+, \quad z \mapsto [M(z)]. \tag{2.33}$$

It commutes with the actions of $\mathrm{SL}_2(F)$, where the action on $\mathcal{K}^+$ is given by (2.31). More precisely, in the cone $\widetilde{\mathcal{K}}^+$ we have

$$\gamma.M(z) = \mathrm{N}(cz + d)M(\gamma z) \tag{2.34}$$

for $\gamma = \left(\begin{smallmatrix} a & b \\ c & d \end{smallmatrix}\right) \in \mathrm{SL}_2(F)$. This implies that modular forms of weight k in the sense of Definition 2.21 can be identified with Hilbert modular forms of parallel weight k in the sense of Definition 1.19.

We consider in V the lattice

$$L = \mathbb{Z} \oplus \mathbb{Z} \oplus \mathcal{O}_F \cong \left\{ \begin{pmatrix} a & \nu' \\ \nu & b \end{pmatrix} \in \tilde{V}; \quad a, b \in \mathbb{Z} \text{ and } \nu \in \mathcal{O}_F \right\}. \tag{2.35}$$

The dual lattice of L is

$$L^\vee = \mathbb{Z} \oplus \mathbb{Z} \oplus \mathfrak{d}_F^{-1} \cong \left\{ \begin{pmatrix} a & \nu' \\ \nu & b \end{pmatrix} \in \tilde{V}; \quad a, b \in \mathbb{Z} \text{ and } \nu \in \mathfrak{d}_F^{-1} \right\}. \tag{2.36}$$

The discriminant group is given by $L^\vee/L \cong \mathcal{O}_F/\mathfrak{d}_F$.

Proposition 2.25. *Under the isomorphism* $\mathrm{Spin}_V \cong \mathrm{SL}_2(F)$, *the subgroup* Spin_L *is identified with* Γ_F. $\qquad\square$

The map (2.33) induces an isomorphism of modular varieties $Y(\Gamma_F) \to Y(\mathrm{Spin}_L)$.

Remark 2.26. More generally, let $\mathfrak{a}$ be a fractional ideal of F and put $A = \mathrm{N}(\mathfrak{a})$. We may consider the lattices

$$L(\mathfrak{a}) = \left\{ \begin{pmatrix} a & \nu' \\ \nu & Ab \end{pmatrix} \in \tilde{V}; \quad a, b \in \mathbb{Z} \text{ and } \nu \in \mathfrak{a} \right\},$$

$$L^\vee(\mathfrak{a}) = \left\{ \begin{pmatrix} a & \nu' \\ \nu & Ab \end{pmatrix} \in \tilde{V}; \quad a, b \in \mathbb{Z} \text{ and } \nu \in \mathfrak{a}\mathfrak{d}_F^{-1} \right\}.$$

Observe that $L(\mathfrak{a})$ is A-integral (that is, the bilinear form has values in $A\mathbb{Z}$), and $L^\vee(\mathfrak{a})$ is the $A\mathbb{Z}$-dual of $L(\mathfrak{a})$. The group $\Gamma(\mathcal{O}_F \oplus \mathfrak{a}) \subset \mathrm{SL}_2(F)$ defined in (1.5) preserves these lattices.

Hirzebruch–Zagier Divisors

In view of the above discussion, the construction of Heegner divisors provides a natural family of algebraic divisors on a Hilbert modular surface, in this case known as *Hirzebruch–Zagier divisors* [HZ].

If $A = \left(\begin{smallmatrix} a & \lambda' \\ \lambda & b \end{smallmatrix} \right) \in \tilde{V}$ and $z = (z_1, z_2) \in \mathbb{H}^2$, then

$$(M(z), A) = -\operatorname{tr}(M(z) \cdot A^*) = -bz_1z_2 + \lambda z_1 + \lambda' z_2 - a.$$

The zero locus of the right hand side defines an analytic divisor on $\mathbb{H}^2$.

Definition 2.27. *Let m be a positive integer. The Hirzebruch–Zagier divisor T_m of discriminant m is defined as the Heegner divisor $H(-m/D)$ for the lattice $L \subset V$, i.e.,*

$$T_m = \sum_{\substack{(a,b,\lambda) \in L^\vee / \{\pm 1\} \\ ab - \lambda\lambda' = m/D}} \left\{ (z_1, z_2) \in \mathbb{H}^2; \quad az_1z_2 + \lambda z_1 + \lambda' z_2 + b = 0 \right\}.$$

It defines an algebraic divisor on the Hilbert modular surface $Y(\Gamma_F)$. Here the multiplicities of all irreducible components are 1. (There is no ramification in codimension 1.) By taking the closure, we also obtain a divisor on $X(\Gamma_F)$. We will denote these divisors by T_m as well, since it will be clear from the context where they are considered. It is well known that T_m is defined over $\mathbb{Q}$.

Remark 2.28. When m is not a square modulo D, then $T_m = \emptyset$.

Example 2.29. The divisor T_1 on $X(\Gamma_F)$ can be identified with the image of the modular curve $X(1) = \overline{\mathrm{SL}_2(\mathbb{Z}) \backslash \mathbb{H}}$ under the diagonal embedding considered in Section 1.5.

3 Additive and Multiplicative Liftings

Let $F \subset \mathbb{R}$ be the real quadratic field of discriminant D. Let (V, Q) be the corresponding rational quadratic space of type $(2, 2)$ as in Section 2.7, and let $L \subset V$ be the even lattice (2.35). The corresponding Siegel theta function $\Theta_k(\tau, z)$ in weight k is modular in both variables τ and z: As a function of τ, $\Theta_k(\tau, z)$ is a non-holomorphic modular form of weight k for the group $\Gamma_0(D)$ with character $\chi_D = \left(\frac{D}{\cdot} \right)$. As a function in z, $\overline{\Theta_k(\tau, z)}$ is a non-holomorphic modular form of weight k for the Hilbert modular group Γ_F. For a cusp form $f \in S_k(D, \chi_D)$ of weight k for $\Gamma_0(D)$ with character χ_D, we may consider the theta integral $\Phi_0(z, f)$ as in (2.30). By means of Theorem 2.23 we find that $\Phi_0(z, f)$ defines a Hilbert cusp form of weight k for the group Γ_F (which may vanish identically). Similar constructions can be done for the Hilbert modular groups $\Gamma(\mathcal{O}_F \oplus \mathfrak{a})$ and for their congruence subgroups.

3.1 The Doi–Naganuma Lift

In the following we discuss the theta lift in more detail. To keep the exposition simple, we assume that $D = p$ is a prime and $F = \mathbb{Q}(\sqrt{p})$. We consider the full Hilbert modular group Γ_F.

For explicit computations it is convenient to modify the theta lifting a bit. Let $M_k(p, \chi_p)$ denote the space of holomorphic modular forms of weight k for $\Gamma_0(p)$ and χ_p. Since this space is trivial when k is odd, we assume that k is even. A function $f \in M_k(p, \chi_p)$ has a Fourier expansion

$$f(\tau) = \sum_{n \geq 0} c(n) q^n \,,$$

where $q = e^{2\pi i \tau}$ as usual. We define the "plus" and "minus" subspace of $M_k(p, \chi_p)$ by

$$M_k^{\pm}(p, \chi_p) = \{f \in M_k(p, \chi_p); \quad \chi_p(n) = \mp 1 \Rightarrow c(n) = 0\} \,, \tag{3.1}$$

and write $S_k^{\pm}(p, \chi_p)$ for the subspace of cusp forms.

Examples of modular forms in $M_k^{\pm}(p, \chi_p)$ can be constructed by means of Eisenstein series. Recall that there are the two Eisenstein series

$$G_k(\tau) = 1 + \frac{2}{L(1-k, \chi_p)} \sum_{n=1}^{\infty} \sum_{d|n} d^{k-1} \chi_p(d) q^n, \tag{3.2}$$

$$H_k(\tau) = \sum_{n=1}^{\infty} \sum_{d|n} d^{k-1} \chi_p(n/d) q^n \tag{3.3}$$

in $M_k(p, \chi_p)$ (cf. [He] Werke p. 818), the former corresponding to the cusp ∞, the latter corresponding to the cusp 0. The linear combination

$$E_k^{\pm} = 1 + \sum_{n \geq 1} B_k^{\pm}(n) q^n = 1 + \frac{2}{L(1-k, \chi_p)} \sum_{n \geq 1} \sum_{d|n} d^{k-1} \left(\chi_p(d) \pm \chi_p(n/d)\right) q^n \tag{3.4}$$

belongs to $M_k^{\pm}(p, \chi_p)$.

Proposition 3.30 (Hecke). *The space $M_k(p, \chi_p)$ decomposes into the direct sum*

$$M_k(p, \chi_p) = M_k^{+}(p, \chi_p) \oplus M_k^{-}(p, \chi_p) \,.$$

Moreover,

$$M_k^{\pm}(p, \chi_p) = \mathbb{C} E_k^{\pm} \oplus S_k^{\pm}(p, \chi_p) \,.$$

$\square$

Modular forms in the plus space behave in many ways similarly as modular forms on the full elliptic modular group $SL_2(\mathbb{Z})$. In fact, Theorem 5 of [BB] states that $M_k^{\pm}(p, \chi_p)$ is isomorphic to the space of vector-valued modular forms of weight k for $SL_2(\mathbb{Z})$ transforming with the Weil representation of $L^{\vee}/L$.

Notation 3.31. *For a formal Laurent series* $\sum c(n)q^n \in \mathbb{C}((q))$ *we put*

$$\tilde{c}(n) = \begin{cases} c(n), & \text{if } p \nmid n, \\ 2c(n), & \text{if } p \mid n. \end{cases} \tag{3.5}$$

Proposition 3.32. *Let* $f = \sum c(n)q^n \in M_k^{\pm}(p, \chi_p)$ *and* $g = \sum b(n)q^n \in M_{k'}^{\pm}(p, \chi_p)$. *Then*

$$\langle f, g \rangle = \sum_{n \in \mathbb{Z}} \sum_{m \in \mathbb{Z}} \tilde{c}(m)b(pn - m)q^n$$

is a modular form of weight $k + k'$ *for* $SL_2(\mathbb{Z})$. *The assignment* $(f, g) \mapsto \langle f, g \rangle$ *defines a bilinear pairing.*

Proof. This can be proved by interpreting modular forms in the plus space as vector valued modular forms for $SL_2(\mathbb{Z})$, see [BB]. □

Remark 3.33. Proposition 3.32 implies some amusing identities of divisor sums arising from the equalities $\langle E_k^+, E_k^+ \rangle = E_{2k}$ for $k = 2, 4$. Here E_{2k} denotes the Eisenstein series of weight $2k$ for $SL_2(\mathbb{Z})$ normalized such that the constant term is 1.

Note that the statement of Proposition 3.32 does not depend on the holomorphicity of f. An analogous result holds for non-holomorphic modular forms. For instance, the complex conjugate of the Siegel theta function $\overline{\Theta_k(\tau, z)}$ of the lattice L satisfies the plus space condition. This follows from Definition 2.22, since for $(a, b, \lambda) \in L^\vee$ we have

$$-pQ(a, b, \lambda) = p(ab - \lambda\lambda') \equiv \square \pmod{p}.$$

Definition 3.34. *For* $f \in M_k^+(p, \chi_p)$ *we define the (modified) theta lift by the integral*

$$\Phi(z, f) = \int_{SL_2(\mathbb{Z}) \backslash \mathbb{H}} \langle f(\tau), \overline{\Theta_k(\tau, z)} \rangle v^k \frac{du\, dv}{v^2}.$$

The integral converges absolutely if f is a cusp form. If f is not cuspidal, the integral has to be regularized (see [Bo4]). By computing the Fourier expansion of $\Phi(z, f)$, the following theorem can be proved (cf. [Za1], [Bo4] Theorem 14.3).

Theorem 3.35. *Let* $f = \sum_n c(n)q^n \in M_k^+(p, \chi_p)$. *The theta lift* $\Phi(z, f)$ *has the following properties.*

(i) $\Phi(z, f)$ *is a Hilbert modular form of weight* k *for* Γ_F.

(ii) It has the Fourier expansion

$$\Phi(z, f) = -\frac{B_k}{2k}\tilde{c}(0) + \sum_{\substack{\nu \in \partial_F^{-1} \\ \nu \gg 0}} \sum_{d|\nu} d^{k-1}\tilde{c}\left(\frac{p\nu\nu'}{d^2}\right) q_1^\nu q_2^{\nu'} ,$$

where B_k denotes the k-th Bernoulli number, and $q_j = e^{2\pi i z_j}$.
(iii) The lift takes cusp forms to cusp forms.

$\square$

If we define in addition $\Phi(z, f)$ to be identically zero on $M_k^-(p, \chi_p)$, we obtain the *Doi–Naganuma lift* (see [DN], [Na]),

$$\mathrm{DN} : M_k(p, \chi_p) \longrightarrow M_k(\Gamma_F).$$

It is a fundamental fact that the Doi–Naganuma lift (and theta lifts in general) behave well with respect to the actions of the Hecke algebras.

Theorem 3.36 (Doi–Naganuma, Zagier).   *The Doi–Naganuma lifting takes Hecke eigenforms to Hecke eigenforms. For a normalized Hecke eigenform $f = \sum_n c(n)q^n \in M_k(p, \chi_p)$ we have*

$$L(\mathrm{DN}(f), s) = L(f, s) \cdot L(f^\rho, s),$$

where $L(f, s)$ denotes the Hecke L-function of f and $f^\rho = \sum \overline{c(n)}q^n$. $\square$

Let $\Lambda(f, s) = p^{s/2}(2\pi)^{-s}\Gamma(s)L(f, s)$ be the completed Hecke L-function of the eigenform f. It has the functional equation

$$\Lambda(f, s) = C \cdot \Lambda(f^\rho, k - s)$$

with a non-zero constant $C \in \mathbb{C}$. Therefore

$$R(s) = p^s (2\pi)^{-2s} \Gamma(s)^2 L(f, s) L(f^\rho, s)$$

has the functional equation

$$R(s) = R(k - s),$$

which looks as the functional equation of the L-function of a Hilbert modular form of weight k, see Theorem 1.45 in Section 1.6. Moreover, all further analytic properties of $R(s)$ agree with those of L-functions of Hilbert modular forms. Hence, using a converse theorem (similar to Hecke's converse theorem), one can infer that $R(s)$ really comes from a Hilbert modular form.

Originally, this argument led Doi and Naganuma to the discovery of the lifting. Using the converse theorem argument they were able to prove the existence of the lifting in the few cases where $\mathcal{O}_F$ is euclidian. Employing a later result of Vaserstein (see [Gel] Chapter IV.6) on generators of Hilbert modular groups, the proof can be generalized.

The theta lifting approach came up later, and was suggested by Eichler and Shintani and worked out by Kudla, Oda, Vignerás, and others.

3.2 Borcherds Products

Here we consider the Borcherds lift for Hilbert modular surfaces. It can be viewed as a multiplicative analogue of the Doi–Naganuma lift. It takes certain weakly holomorphic elliptic modular forms of weight 0 to meromorphic Hilbert modular forms which have an infinite product expansion resembling the Dedekind eta function. The zeros and poles of such *Borcherds products* are supported on Hirzebruch–Zagier divisors.

Local Borcherds Products

As a warm up, we study a local analogue of Borcherds products at the cusps of Hilbert modular surfaces. This is a special case of the more general results for $O(2, n)$ of [BrFr].

We return to the setup of Section 1.1. In particular, $F \subset \mathbb{R}$ is a real quadratic field of discriminant D and $\Gamma_F = \mathrm{SL}_2(\mathcal{O}_F)$ denotes the Hilbert modular group. We ask whether the Hirzebruch–Zagier divisors T_m on $X(\Gamma_F)$ are $\mathbb{Q}$-Cartier. Since the non-compact Hilbert modular surface $Y(\Gamma_F)$ is non-singular except for the finite quotient singularities corresponding to the elliptic fixed points, it is clear that T_m is $\mathbb{Q}$-Cartier on $Y(\Gamma_F)$. We only have to investigate the behavior at the cusps.

Lemma 3.37. *Let $A = (a, b, \lambda) \in L^\vee$ with $ab - \lambda\lambda' > 0$. The closure of the image in Y_F of*

$$\{(z_1, z_2) \in \mathbb{H}^2; \quad az_1z_2 + \lambda z_1 + \lambda' z_2 + b = 0\}$$

goes through the cusp ∞ if and only if $a = 0$.

Proof. This is an immediate consequence of Proposition 1.7 (iii). □

Let m be a positive integer. We define the local Hirzebruch–Zagier divisor at ∞ of discriminant m by

$$T_m^\infty = \sum_{\substack{\lambda \in \mathfrak{d}_F^{-1}/\{\pm 1\} \\ -\lambda\lambda' = m/D \\ b \in \mathbb{Z}}} \{(z_1, z_2) \in \mathbb{H}^2; \quad \lambda z_1 + \lambda' z_2 + b = 0\} \subset \mathbb{H}^2.$$

This divisor is invariant under the stabilizer $\Gamma_{F,\infty}$ of ∞.

Theorem 3.38. *The Hirzebruch–Zagier divisor T_m on $X(\Gamma_F)$ is $\mathbb{Q}$-Cartier.*

Proof. We have to investigate the behavior at the cusps. Here we only consider the cusp ∞, the other cusps can be treated in the same way.

We have to show that there is a small open neighborhood $U \subset X(\Gamma_F)$ of ∞ and a holomorphic function f on U such that

$$\mathrm{div}(f) = r \cdot T_m|_U \in \mathrm{Div}(U)$$

for some positive integer r. Here $T_m|_U$ denotes the restriction of T_m to U. In view of Proposition 1.10 and Lemma 3.37 it suffices to show that there exists a $\Gamma_{F,\infty}$-invariant holomorphic function $\tilde{f} : \mathbb{H}^2 \to \mathbb{C}$ such that $\mathrm{div}(\tilde{f}) = r \cdot \tilde{T}_m^\infty$. This follows from Proposition 3.40 below. □

Remark 3.39. The statement of Theorem 3.38 does *not* generalize to Heegner divisors on $O(2,n)$. For instance, for $n > 3$ there are obstructions to the $\mathbb{Q}$-Cartier property at generic boundary points, which are related to theta series of even definite lattices of rank $n - 2$ with harmonic polynomials of degree 2. (See [BrFr], [Lo].)

The local Hirzebruch–Zagier divisor T_m^∞ decomposes as a sum

$$T_m^\infty = \sum_{\substack{\lambda \in \mathfrak{d}_F^{-1}/\mathcal{O}_F^{*,2} \\ -\lambda\lambda' = m/D \\ \lambda > 0}} T_\lambda^\infty, \tag{3.6}$$

where $\mathcal{O}_F^{*,2}$ denotes the subgroup of squares in the unit group $\mathcal{O}_F^*$, and

$$T_\lambda^\infty = \sum_{\substack{u \in \mathcal{O}_F^{*,2} \\ b \in \mathbb{Z}}} \{(z_1, z_2) \in \mathbb{H}^2; \quad \lambda u z_1 + \lambda' u' z_2 + b = 0\}. \tag{3.7}$$

The divisor T_λ is invariant under $\Gamma_{F,\infty}$. In the following, we construct a holomorphic function on $\mathbb{H}^2/\Gamma_{F,\infty}$ whose divisor is T_λ^∞, using local Borcherds products [BrFr]. We start by introducing some notation.

The subset

$$S(m) = \bigcup_{\substack{\lambda \in \mathfrak{d}_F^{-1} \\ -\lambda\lambda' = m/D}} \{y \in (\mathbb{R}_{>0})^2; \quad \lambda y_1 + \lambda' y_2 = 0\} \tag{3.8}$$

of $(\mathbb{R}_{>0})^2$ is a union of hyperplanes. It is invariant under $\Gamma_{F,\infty}$. The complement $(\mathbb{R}_{>0})^2 \setminus S(m)$ is not connected. The connected components are called the *Weyl chambers* (of $\mathfrak{d}_F^{-1}$) of index m.

Let W be a subset of a Weyl chamber of index m and $\lambda \in \mathfrak{d}_F^{-1}$ with $-\lambda\lambda' = m/D$. Then λ is called *positive* with respect to W, if $\mathrm{tr}(\lambda w) > 0$ for all $w \in W$ (which is equivalent to requiring $\mathrm{tr}(\lambda w_0) > 0$ for some $w_0 \in W$). In this case we write

$$(\lambda, W) > 0.$$

Moreover, λ is called *reduced* with respect to W, if

$$(u\lambda, W) < 0, \quad \text{and} \quad (\lambda, W) > 0,$$

for all $u \in \mathcal{O}_F^{*,2}$ with $u < 1$. This condition is equivalent to

$$(\varepsilon_0^{-2}\lambda, W) < 0, \quad \text{and} \quad (\lambda, W) > 0.$$

It implies that $\lambda > 0$. We denote by $R(m, W)$ the set of all $\lambda \in \mathfrak{d}_F^{-1}$ with $-\lambda\lambda' = m/D$ which are reduced with respect to W. (Note that this definition slightly differs from the one in [BB].) It is a finite set and

$$\{\lambda \in \mathfrak{d}_F^{-1}; \quad -\lambda\lambda' = m/D\} = \{\pm\lambda u; \quad \lambda \in R(m, W) \text{ and } u \in \mathcal{O}_F^{*,2}\}. \quad (3.9)$$

Let W be a subset of a Weyl chamber of index m and $\lambda \in \mathfrak{d}_F^{-1}$ with $-\lambda\lambda' = m/D$. We define a holomorphic function $\psi_\lambda^\infty : \mathbb{H}^2 \to \mathbb{C}$ by

$$\psi_\lambda^\infty(z) = \prod_{u \in \mathcal{O}_F^{*,2}} \left[1 - e(\sigma_u \operatorname{tr}(u\lambda z))\right],$$

where

$$\sigma_u = \begin{cases} +1, & \text{if } (u\lambda, W) > 0, \\ -1, & \text{if } (u\lambda, W) < 0. \end{cases}$$

The sign σ_u has to be inserted to obtain a *convergent* infinite product. By construction we have $\psi_\lambda^\infty = \psi_{-\lambda}^\infty$ and

$$\operatorname{div}(\psi_\lambda^\infty) = T_\lambda^\infty.$$

Moreover, the product is invariant under translations $\left(\begin{smallmatrix} 1 & \mu \\ 0 & 1 \end{smallmatrix}\right) \in \Gamma_{F,\infty}$. However, ψ_λ^∞ is *not* invariant under the full stabilizer of ∞. It defines an automorphy factor

$$J(\gamma, z) = \psi_\lambda^\infty(\gamma z)/\psi_\lambda^\infty(z) \quad (3.10)$$

of $\Gamma_{F,\infty}$ acting on $\mathbb{H}^2$, that is, an element of $H^1(\Gamma_{F,\infty}, \mathcal{O}(\mathbb{H}^2)^*)$. We need to show that this automorphy factor is trivial up to torsion. It suffices to consider what happens under the generator $\left(\begin{smallmatrix} \varepsilon_0 & 0 \\ 0 & \varepsilon_0^{-1} \end{smallmatrix}\right)$ of the subgroup of diagonal matrices in $\Gamma_{F,\infty}$. We have

$$\frac{\psi_\lambda^\infty(\varepsilon_0^2 z)}{\psi_\lambda^\infty(z)} = \prod_{u \in \mathcal{O}_F^{*,2}} \frac{1 - e(\sigma_{u/\varepsilon_0^2} \operatorname{tr}(u\lambda z))}{1 - e(\sigma_u \operatorname{tr}(u\lambda z))}.$$

In this product only one factor is not equal to 1. If we assume that λ is reduced with respect to W, we obtain

$$\frac{\psi_\lambda^\infty(\varepsilon_0^2 z)}{\psi_\lambda^\infty(z)} = \frac{1 - e(-\operatorname{tr}(\lambda z))}{1 - e(\operatorname{tr}(\lambda z))}$$
$$= e(1/2 - \operatorname{tr}(\lambda z)).$$

On the other hand, we consider the invertible holomorphic function

$$I_\lambda(z) = e\left(\operatorname{tr}\left(\frac{\lambda}{\varepsilon_0^2 - 1} z\right)\right)$$

on $\mathbb{H}^2$. It satisfies

$$\frac{I_\lambda(\varepsilon_0^2 z)}{I_\lambda(z)} = e(\mathrm{tr}(\lambda z)).$$

Moreover, $I_\lambda(z + \mu) = I_\lambda(z)$ for all $\mu \in (\varepsilon_0^2 - 1)\mathcal{O}_F$. Therefore, up to torsion, the automorphy factor $J(\gamma, z)$ in (3.10) can be trivialized with $I_\lambda(z)$. The function

$$\Psi_\lambda^\infty(z) = I_\lambda(z) \cdot \psi_\lambda^\infty(z) = e\left(\mathrm{tr}\left(\frac{\lambda}{\varepsilon_0^2 - 1} z\right)\right) \prod_{u \in \mathcal{O}_F^{*,2}} [1 - e(\sigma_u \, \mathrm{tr}(u\lambda z))]$$

(3.11)

is holomorphic on $\mathbb{H}^2$, has divisor T_λ^∞, and a power of it is invariant under $\Gamma_{F,\infty}$. Observe that Ψ_λ^∞ does not depend on the choice of the Weyl chamber W, although the factors I_λ and ψ_λ^∞ do.

Now it is easy to construct an analogous function for T_m^∞. We define the Weyl vector of index m for the Weyl chamber W by

$$\rho_{m,W} = \sum_{\lambda \in R(m,W)} \frac{\lambda}{\varepsilon_0^2 - 1}.$$

(3.12)

Moreover, we define the local Borcherds product for T_m^∞ by

$$\Psi_m^\infty(z) = \prod_{\substack{\lambda \in \mathfrak{d}_F^{-1}/\mathcal{O}_F^{*,2} \\ -\lambda\lambda' = m/D \\ \lambda > 0}} \Psi_\lambda^\infty(z) = e\big(\mathrm{tr}(\rho_{m,W} z)\big) \prod_{\substack{\lambda \in \mathfrak{d}_F^{-1} \\ -\lambda\lambda' = m/D \\ (\lambda, W) > 0}} [1 - e(\mathrm{tr}(\lambda z))].$$

(3.13)

Proposition 3.40. *The divisor of Ψ_m^∞ is equal to T_m^∞. A power of Ψ_m^∞ is invariant under $\Gamma_{F,\infty}$.* ☐

Example 3.41. We compute Ψ_1^∞ more explicitly. The point $(1, \varepsilon_0) \in (\mathbb{R}_{>0})^2$ does not belong to $S(1)$. Hence it lies in a unique Weyl chamber W of index 1. The set of $\lambda \in \mathfrak{d}_F^{-1}$ with $-\lambda\lambda' = 1/D$ which are reduced with respect to W is given by

$$R(1, W) = \begin{cases} \{\varepsilon_0^2/\sqrt{D}\}, & \text{if } \varepsilon_0\varepsilon_0' = -1, \\ \{\varepsilon_0/\sqrt{D}, \varepsilon_0^2/\sqrt{D}\}, & \text{if } \varepsilon_0\varepsilon_0' = +1. \end{cases}$$

The corresponding Weyl vector is equal to

$$\rho_{1,W} = \begin{cases} \dfrac{\varepsilon_0}{\sqrt{D}} \dfrac{1}{\mathrm{tr}(\varepsilon_0)}, & \text{if } \varepsilon\varepsilon_0' = -1, \\[2ex] \dfrac{1 + \varepsilon_0}{\mathrm{tr}(\sqrt{D}\varepsilon_0)}, & \text{if } \varepsilon\varepsilon_0' = +1. \end{cases}$$

In the case $\varepsilon\varepsilon_0' = -1$, the point $(\varepsilon_0^{-1}, \varepsilon_0)$ lies in the same Weyl chamber W. It is often more convenient to work with this base point. If $\varepsilon\varepsilon_0' = 1$, then $(\varepsilon_0^{-1}, \varepsilon_0) \in S(1)$.

The Borcherds Lift

For the material of the next two sections we also refer to [Br3]. The Doi–Naganuma lift of the Section 3.1 only defines a non-trivial map when $k > 0$. (For $k = 0$ we have $M_k(D, \chi_D) = 0$.) It is natural to ask if one can also do something meaningful in the border case $k = 0$ where the Siegel theta function (2.22) reduces to the theta function $\Theta_0(\tau, z)$ associated to the standard Gaussian on $V(\mathbb{R})$. To get a feeling for this question, one can pretend that there is a non-trivial element $f = \sum_n c(n)q^n \in M_0^+(p, \chi_p)$ and formally write down its lifting according to Theorem 3.35. We find that it has the Fourier expansion

$$\Phi(z, f) = -\frac{B_0}{2k}\tilde{c}(0) + \sum_{\substack{\nu \in \partial_F^{-1} \\ \nu \gg 0}} \sum_{d|\nu} \frac{1}{d}\tilde{c}\left(\frac{p\nu\nu'}{d^2}\right) q_1^\nu q_2^{\nu'} .$$

Reordering the summation, this can be written as

$$\Phi(z, f) = -\frac{B_0}{2k}\tilde{c}(0) - \sum_{\substack{\nu \in \partial_F^{-1} \\ \nu \gg 0}} \log\left(1 - q_1^\nu q_2^{\nu'}\right)^{\tilde{c}(p\nu\nu')} .$$

Hence, the lifting looks as the logarithm of a "modular" infinite product, resembling the Dedekind eta function. The idea of Borcherds, Harvey and Moore was to drop the assumption on f being holomorphic and to replace it by something weaker [Bo1], [Bo2], [Bo4], [HM]. They consider a regularized theta lift for weakly holomorphic modular forms. It leads to meromorphic modular forms with infinite product expansions (roughly of the above type).

This construction works in greater generality for $O(2, n)$. It yields a lift from weakly holomorphic modular forms of weight $1 - n/2$ to meromorphic modular forms on $O(2, n)$ with zeros and poles supported on Heegner divisors. Here we only consider the $O(2, 2)$-case of Hilbert modular surfaces. Moreover, to simplify the exposition, we assume that the real quadratic field F has prime discriminant p.

Let Γ be a subgroup of $SL_2(\mathbb{Q})$ which is commensurable with $SL_2(\mathbb{Z})$. Recall that a meromorphic modular form of weight k with respect to Γ is called *weakly holomorphic* if it is holomorphic outside the cusps. At the cusp ∞ such a modular form f has a Fourier expansion of the form

$$f(\tau) = \sum_{\substack{n \in \mathbb{Z} \\ n \geq N}} c(n)q^{n/h} ,$$

where $N \in \mathbb{Z}$, and $h \in \mathbb{Z}_{>0}$ is the width of the cusp ∞. By an elementary argument it can be proved that the Fourier coefficients of f are bounded by

$$c(n) = O\left(e^{C\sqrt{n}}\right), \qquad n \to \infty, \tag{3.14}$$

for some positive constant $C > 0$ depending on the order of the poles at the various cusps of Γ (see [BrFu1] Section 3). This estimate is also a consequence of the (much more precise) Hardy–Rademacher–Ramanujan asymptotic for the coefficients of weakly holomorphic modular forms.

Let $W_k(p, \chi_p)$ be the space of weakly holomorphic modular forms of weight k for the group $\Gamma_0(p)$ with character χ_p. Any modular form f in this space has a Fourier expansion of the form $f = \sum_{n \gg -\infty} c(n)q^n$. Similarly as in (3.1) we denote by $W_k^+(p, \chi_p)$ the subspace of those $f \in W_k(p, \chi_p)$, whose coefficients $c(n)$ satisfy the plus space condition, that is, $c(n) = 0$ whenever $\chi_p(n) = -1$.

Lemma 3.42. *Let $k \leq 0$. A weakly holomorphic modular form $f = \sum_n c(n) \cdot q^n \in W_k^+(p, \chi_p)$ is uniquely determined by its principal part*

$$\sum_{n<0} c(n)q^n \in \mathbb{C}[q^{-1}].$$

Proof. The difference of two elements of $W_k^+(p, \chi_p)$ with the same principal part is holomorphic at the cups ∞. Using the plus space condition (Lemma 3 of [BB]), one infers that the difference is also holomorphic at the cusp 0. Hence, it is a holomorphic modular form of weight $k \leq 0$ with Nebentypus, and therefore vanishes identically. □

Corollary 3.43. *Let $k \leq 0$. Assume that $f \in W_k^+(p, \chi_p)$ has principal part in $\mathbb{Q}[q^{-1}]$. Then all Fourier coefficients of f are rational with bounded denominators.*

Proof. This follows from Lemma 3.42 and the properties of the Galois action on $W_k(p, \chi_p)$. □

Let $f = \sum_n c(n)q^n \in W_k^+(p, \chi_p)$. Then

$$(\mathbb{R}_{>0})^2 \setminus \bigcup_{\substack{m>0 \\ c(-m)\neq 0}} S(m)$$

is not connected. The connected components are called the *Weyl chambers* associated to f. If $W \subset (\mathbb{R}_{>0})^2$ is such a Weyl chamber, then the *Weyl vector* corresponding to f and W is defined by

$$\rho_{f,W} = \sum_{m>0} \tilde{c}(-m)\rho_{m,W} \in F. \tag{3.15}$$

Here $\rho_{m,W}$ is given by (3.12) and we have used the notation (3.5).

We are now ready to state Borcherds' theorem in a formulation that fits nicely our setting (see [Bo4] Theorem 13.3 and [BB] Theorem 9).

Theorem 3.44 (Borcherds). *Let $f = \sum_{n \gg -\infty} c(n)q^n$ be a weakly holomorphic modular form in $W_0^+(p, \chi_p)$ and assume that $\tilde{c}(n) \in \mathbb{Z}$ for all $n < 0$. Then there exists a meromorphic Hilbert modular form $\Psi(z, f)$ for Γ_F (with some multiplier system of finite order) such that:*

(i) The weight of Ψ is equal to the constant term $c(0)$ of f.

(ii) The divisor $Z(f)$ of Ψ is determined by the principal part of f at the cusp ∞. It equals

$$Z(f) = \sum_{n<0} \tilde{c}(n) T_{-n} .$$

(iii) Let W be a Weyl chamber associated to f and put $N = \min\{n;\ c(n) \neq 0\}$. The function Ψ has the Borcherds product expansion

$$\Psi(z, f) = q_1^\rho q_2^{\rho'} \prod_{\substack{\nu \in \mathfrak{d}_F^{-1} \\ (\nu, W) > 0}} \left(1 - q_1^\nu q_2^{\nu'}\right)^{\tilde{c}(p\nu\nu')} ,$$

which converges normally for all z with $y_1 y_2 > |N|/p$ outside the set of poles. Here $\rho = \rho_{f,W}$ is the Weyl vector corresponding to f and W, and $q_j^\nu = e^{2\pi i \nu z_j}$ for $\nu \in F$.

Proof. We indicate the idea of the proof. We consider the theta lift (Section 2.6) for the lattice L in the quadratic space $V = \mathbb{Q} \oplus \mathbb{Q} \oplus F$ (Section 2.7) and use the accidental isomorphism $\Gamma_F \cong \mathrm{Spin}_V$. The corresponding Siegel theta function $\overline{\Theta_0(\tau, z)}$ in weight 0 transforms as an element of $M_0^+(p, \chi_p)$ in the variable τ. As a function of z it is invariant under Γ_F. The pairing $\langle f(\tau), \overline{\Theta_0(\tau, z)} \rangle$ (see Proposition 3.32) is a $\mathrm{SL}_2(\mathbb{Z})$-invariant function in τ.

We consider the theta integral

$$\int_{\mathcal{F}} \langle f(\tau), \overline{\Theta_0(\tau, z)} \rangle \, \frac{du\, dv}{v^2} , \qquad (3.16)$$

where $\mathcal{F} = \{\tau \in \mathbb{H};\ |\tau| \geq 1,\ |u| \leq 1/2\}$ denotes the standard fundamental domain for $\mathrm{SL}_2(\mathbb{Z})$. Formally it defines a Γ_F-invariant function on $\mathbb{H}^2$. Unfortunately, because of the exponential growth of f at the cusps, the integral diverges. However, Harvey and Moore discovered that it can be regularized as follows [HM], [Bo4], [Kon]: If the constant term $c(0)$ of f vanishes, one can regularize (3.16) by taking

$$\lim_{t \to \infty} \int_{\mathcal{F}_t} \langle f(\tau), \overline{\Theta_0(\tau, z)} \rangle \, \frac{du\, dv}{v^2} , \qquad (3.17)$$

where $\mathcal{F}_t = \{\tau \in \mathcal{F};\ |v| \leq t\}$ denotes the truncated standard fundamental domain. So the regularization consists in prescribing the order of integration. We first integrate over u and then over v. If the constant term of f does not vanish, the limit in (3.17) still diverges. It can be regularized by considering

$$\Phi(z, f, s) = \lim_{t \to \infty} \int_{\mathcal{F}_t} \langle f(\tau), \overline{\Theta_0(\tau, z)} \rangle v^{-s} \frac{du\, dv}{v^2} \qquad (3.18)$$

for $s \in \mathbb{C}$. The limit exists for $\Re(s)$ large enough and has a meromorphic continuation to the whole complex plane. We define the regularized theta integral $\Phi(z, f)$ to be the constant term in the Laurent expansion of $\Phi(z, f, s)$ at $s = 0$.

One can show that $\Phi(z, f)$ defines a Γ_F-invariant real analytic function on $\mathbb{H}^2 \setminus Z(f)$ with a logarithmic singularity[2] along the divisor $-4Z(f)$ ([Bo4] §6). The Fourier expansion of $\Phi(z, f)$ can be computed explicitly by applying some partial Poisson summation on the theta kernel. It turns out that

$$\Phi(z, f) = -4 \log \left| \Psi(z, f)(y_1 y_2)^{c(0)/2} \right| - 2c(0) \left(\log(2\pi) + \Gamma'(1) \right),$$

in the domain of convergence of the infinite product for $\Psi(z, f)$. Using this identity and the properties of $\Phi(z, f)$, one can prove that the infinite product has a meromorphic continuation to $\mathbb{H}^2$ satisfying the hypotheses of the theorem. $\square$

Remark 3.45. The fact that $\Psi(f, z)$ only converges in a sufficiently small neighborhood of the cusp ∞ is due to the rapid growth of the Fourier coefficients of weakly holomorphic modular forms, see (3.14).

Meromorphic Hilbert modular forms that arise as liftings of weakly holomorphic modular forms by Theorem 3.44 are called *Borcherds products*.

The following two propositions highlight the arithmetic nature of Borcherds products. Via the q-expansion principle (see [Rap], [Ch]) they imply that a suitable power of a Borcherds product defines a rational section of the line bundle of Hilbert modular forms over $\mathbb{Z}$.

Proposition 3.46. *Any meromorphic Borcherds product is the quotient of two holomorphic Borcherds products.*

Proof. See [BBK] Proposition 4.5. $\square$

Proposition 3.47. *For any holomorphic Borcherds product Ψ there exists a positive integer n such that:*

(i) Ψ^n is a Hilbert modular form for Γ_F with trivial multiplier system.
(ii) All Fourier coefficients of Ψ^n are contained in $\mathbb{Z}$.
(iii) The greatest common divisor of the Fourier coefficients of Ψ^n equals 1.

Proof. The first assertion is clear. The second and the third follow by Corollary 3.43 from the infinite product expansion given in Theorem 3.44(iii). $\square$

---

[2] If X is a normal complex space, $D \subset X$ a Cartier divisor, and f a smooth function on $X \setminus \mathrm{supp}(D)$, then f has a logarithmic singularity along D, if for any local equation g for D on an open subset $U \subset X$, the function $f - \log|g|$ is smooth on U.

Obstructions

The Borcherds lift provides explicit relations among Hirzebruch–Zagier divisors on a Hilbert modular surface. It is natural to seek for a precise description of those linear combinations of Hirzebruch–Zagier divisors, which are divisors of Borcherds products. Since the divisor of a Borcherds product $\Psi(z, f)$ is determined by the principal part of the weakly holomorphic modular form f, it suffices to understand which Fourier polynomials $\sum_{n<0} c(n)q^n \in \mathbb{C}[q^{-1}]$ can occur as principal parts of elements of $W_0^+(p, \chi_p)$.

A necessary condition is easily obtained. If $f \in W_k^+(p, \chi_p)$ with Fourier coefficients $c(n)$, and $g \in M_{2-k}^+(p, \chi_p)$ with Fourier coefficients $b(n)$, then the pairing $\langle f, g \rangle$ is a weakly holomorphic modular form of weight 2 for $SL_2(\mathbb{Z})$. Thus

$$\langle f, g \rangle d\tau$$

is a meromorphic differential on the Riemann sphere whose only pole is at the cusp ∞. By the residue theorem its residue has to vanish. But the residue is just the constant term in the Fourier expansion of $\langle f, g \rangle$. We find that

$$\sum_{n \leq 0} \tilde{c}(n)b(-n) = 0. \tag{3.19}$$

Applying this condition to the Eisenstein series $E_{2-k}^+(\tau)$, see (3.4), one gets a formula for the constant term of f.

Proposition 3.48. *Let k be a non-positive integer. Let $f = \sum_n c(n)q^n \in W_k^+(p, \chi_p)$. Then*

$$c(0) = -\frac{1}{2} \sum_{n<0} \tilde{c}(n)B_{2-k}^+(-n).$$

$\square$

Using Serre duality for vector bundles on Riemann surfaces, Borcherds showed that the necessary condition is also sufficient (see [Bo6] and [BB] Theorem 6).

Theorem 3.49. *There exists an $f \in W_k^+(p, \chi_p)$ with prescribed principal part $\sum_{n<0} c(n)q^n$ (where $c(n) = 0$ if $\chi_p(n) = -1$), if and only if*

$$\sum_{n<0} \tilde{c}(n)b(-n) = 0$$

for every cusp form $g = \sum_{m>0} b(m)q^m$ in $S_{2-k}^+(p, \chi_p)$. $\square$

Corollary 3.50. *A formal power series $\sum_{m\geq 0} b(m)q^m \in \mathbb{C}[[q]]^+$ is the Fourier expansion of a modular form in $M_{2-k}^+(p, \chi_p)$, if and only if*

$$\sum_{n\leq 0} \tilde{c}(n)b(-n) = 0$$

for every $f = \sum_n c(n)q^n$ in $W_k^+(p, \chi_p)$.

Proof. This follows immediately from Theorem 3.49, see [Br3], Corollary 4.2.
□

If X is a regular projective algebraic variety, we write $\text{Div}(X)$ for the group of divisors of X, and $\text{Rat}(X)$ for the subgroup of divisors of rational functions on X. The first Chow group of X is the quotient

$$\text{CH}^1(X) = \text{Div}(X)/\text{Rat}(X).$$

Furthermore, we put $\text{CH}^1(X)_{\mathbb{Q}} = \text{CH}^1(X) \otimes_{\mathbb{Z}} \mathbb{Q}$. Recall that $\text{CH}^1(X)$ is isomorphic to the Picard group of X, the group of isomorphism classes of algebraic line bundles on X. The isomorphism is given by mapping a line bundle $\mathcal{L}$ to the class $c_1(\mathcal{L})$ of the divisor of a rational section of $\mathcal{L}$. The Chow group $\text{CH}^1(X)$ is an important invariant of X. It is finitely generated.

Let $\pi : \widetilde{X} \to X(\Gamma_F)$ be a desingularization. If k is a positive integer divisible by the order of all elliptic fixed points of Γ_F, then $\mathcal{M}_k := \pi^* \mathcal{M}_k(\Gamma_F)$, the pullback of the line bundle of modular forms of weight k, defines an element of $\text{Pic}(\widetilde{X})$. We consider its class in $\text{CH}^1(\widetilde{X})$. More generally, if k is any rational number, we chose an integer n such that nk is a positive integer divisible by $n(\Gamma_F)$ and put $c_1(\mathcal{M}_k) = \frac{1}{n} c_1(\mathcal{M}_{nk}) \in \text{CH}^1(\widetilde{X})_{\mathbb{Q}}$.

The Hirzebruch–Zagier divisors are $\mathbb{Q}$-Cartier on $X(\Gamma_F)$. Their pullbacks define elements in $\text{CH}^1(\widetilde{X})_{\mathbb{Q}}$. We want to describe their positions in this Chow group. To this end we consider the generating series

$$A(\tau) = c_1(\mathcal{M}_{-1/2}) + \sum_{m>0} \pi^*(T_m) q^m \in \mathbb{Q}[[q]] \otimes_{\mathbb{Q}} \text{CH}^1(\widetilde{X})_{\mathbb{Q}}. \qquad (3.20)$$

Combining Theorem 3.44 and Corollary 3.50 one obtains the following striking application.

Theorem 3.51. *The divisors $\pi^*(T_m)$ generate a subspace of $\text{CH}^1(\widetilde{X})_{\mathbb{Q}}$ of dimension $\leq \dim(M_2^+(p, \chi_p))$. The generating series $A(\tau)$ is a modular form in $M_2^+(p, \chi_p)$ with values in $\text{CH}^1(\widetilde{X})_{\mathbb{Q}}$, i.e., an element of $M_2^+(p, \chi_p) \otimes_{\mathbb{Q}} \text{CH}^1(\widetilde{X})_{\mathbb{Q}}$.*

In other words, if λ is a linear functional on $\text{CH}^1(\widetilde{X})_{\mathbb{Q}}$, then

$$\lambda\left(c_1(\mathcal{M}_{-1/2})\right) + \sum_{m>0} \lambda(\pi^* T_m) q^m \in M_2^+(p, \chi_p).$$

A typical linear functional, one can take for λ, is given by the intersection pairing with a fixed divisor on $\widetilde{X}$. Theorem 3.51 was first proved by Hirzebruch and Zagier by computing intersection numbers of Hirzebruch–Zagier divisors with other such divisors and with the exceptional divisors coming from Hirzebruch's resolution of the cusp singularities [HZ]. Their discovery triggered important investigations by several people, showing that more generally periods of certain special cycles in arithmetic quotients of orthogonal or unitary type can be viewed as the coefficients of Siegel modular forms. For instance, Oda considered cycles on quotients of $O(2, n)$ given by embedded quotients of $O(1, n)$ [Od1], and Kudla–Millson studied more general cycles on quotients of $O(p, q)$ and $U(p, q)$ using the Weil representation and theta functions with values in closed differential forms [KM1, KM2, KM3], see also [Fu] for the case of non-compact quotients. The relationship of the Kudla–Millson lift and the regularized theta lift is clarified in [BrFu1].

Proof of Theorem 3.51. Using Borcherds products, Theorem 3.51 can be proved as follows (see [Bo6]). In view of Corollary 3.50 it suffices to show that

$$\tilde{c}(0)\, c_1(\mathcal{M}_{-1/2}) + \sum_{n<0} \tilde{c}(n)\pi^*(T_{-n}) = 0 \in \mathrm{CH}^1(\tilde{X})_{\mathbb{Q}}$$

for every $f = \sum_n c(n)q^n$ in $W_0^+(p, \chi_p)$ with integral Fourier coefficients. But this is an immediate consequence of Theorem 3.44: Up to torsion, the Borcherds lift of f is a rational section of $\mathcal{M}_{c(0)}$ with divisor $\sum_{n<0} \tilde{c}(n) \cdot \pi^*(T_{-n})$. $\qquad\square$

Notice that we have only used (i) and (ii) of Theorem 3.44. Using the product expansion (iii) in addition, one can prove an arithmetic version of Theorem 3.51, saying that certain arithmetic Hirzebruch–Zagier divisors are the coefficients of a modular form in $M_2^+(p, \chi_p)$ with values in an arithmetic Chow group, see [BBK], [Br3]. Finally, we mention that this argument generalizes to Heegner divisors on quotients of $O(2, n)$.

Remark 3.52. With some further work it can be proved that the dimension of the subspace of $\mathrm{CH}(\tilde{X})_{\mathbb{Q}}$ generated by the Hirzebruch–Zagier divisors is equal to $\dim M_2^+(p, \chi_p)$, see Corollary 3.62.

Examples

Recall that p is a prime congruent to 1 modulo 4. By a result due to Hecke [He] the dimension of $S_2^+(p, \chi_p)$ is equal to $[\frac{p-5}{24}]$. In particular there exist three such primes for which $S_2^+(p, \chi_p)$ is trivial, namely $p = 5, 13, 17$. In these cases $W_0^+(p, \chi_p)$ is a free module of rank $\frac{p+1}{2}$ over the ring $\mathbb{C}[j(p\tau)]$. Therefore it is not hard to compute explicit bases. For any $m \in \mathbb{Z}_{>0}$ with $\chi_p(m) \neq -1$ there is a unique $f_m = \sum_{n \geq -m} c_m(n)q^n \in W_0^+(p, \chi_p)$ whose Fourier expansion starts with

$$f_m = \begin{cases} q^{-m} + c_m(0) + O(q), & \text{if } p \nmid m, \\ \frac{1}{2}q^{-m} + c_m(0) + O(q), & \text{if } p \mid m. \end{cases}$$

The f_m ($m \in \mathbb{Z}_{>0}$) form a base of the space $W_0^+(p, \chi_p)$. The Borcherds lift Ψ_m of f_m is a Hilbert modular form for Γ_F of weight $c_m(0) = -B_2^+(m)/2$ with divisor T_m. Here $B_2^+(m)$ denotes the m-th coefficient of the Eisenstein series $E_2^+(\tau)$ as before.

The case $p = 5$. We consider the real quadratic field $F = \mathbb{Q}(\sqrt{5})$. The fundamental unit is given by $\varepsilon_0 = \frac{1}{2}(1 + \sqrt{5})$. Here the first few f_m were computed in [BB]. One obtains:

$$f_1 = q^{-1} + 5 + 11\,q - 54\,q^4 + 55\,q^5 + 44\,q^6 - 395\,q^9 + 340\,q^{10} + \ldots,$$

$$f_4 = q^{-4} + 15 - 216\,q + 4959\,q^4 + 22040\,q^5 - 90984\,q^6 + 409944\,q^9 + \ldots,$$

$$f_5 = \tfrac{1}{2}\,q^{-5} + 15 + 275\,q + 27550\,q^4 + 43893\,q^5 + 255300\,q^6 + \ldots,$$

$$f_6 = q^{-6} + 10 + 264\,q - 136476\,q^4 + 306360\,q^5 + 616220\,q^6 + \ldots,$$

$$f_9 = q^{-9} + 35 - 3555\,q + 922374\,q^4 + 7512885\,q^5 - 53113164\,q^6 + \ldots,$$

$$f_{10} = \tfrac{1}{2}\,q^{-10} + 10 + 3400\,q + 3471300\,q^4 + 9614200\,q^5 + 91620925\,q^6 + \ldots.$$

The Eisenstein series $E_2^+(\tau) \in M_2^+(5, \chi_5)$ has the Fourier expansion

$$E_2^+(\tau, 0) = 1 - 10q - 30q^4 - 30q^5 - 20q^6 - 70q^9 - 20q^{10} - 120q^{11} - 60q^{14} - \ldots.$$

One easily shows that the weight of any Borcherds product is divisible by 5. By a little estimate one concludes that there is just one holomorphic Borcherds product of weight 5, namely Ψ_1. There exist precisely 3 holomorphic Borcherds products in weight 10, namely Ψ_1^2, Ψ_6, and Ψ_{10}. In weight 15 there are the holomorphic Borcherds products Ψ_4, Ψ_5, Ψ_1^3, $\Psi_1\Psi_6$, and $\Psi_1\Psi_{10}$.

It follows from Lemma 3.37 that T_m does not go through the cusp ∞ when m is not the norm of some $\lambda \in \mathcal{O}_F$. In particular, T_6 and T_{10} do not meet ∞. This also implies that $S(6) = S(10) = \emptyset$. There is just one Weyl chamber of index 6 and 10 (namely $(\mathbb{R}_{>0})^2$) and the corresponding Weyl vector is 0. The divisor T_1 does meet ∞. As in Example 3.41, let W be the unique Weyl chamber of index 1 containing $(\varepsilon_0^{-1}, \varepsilon_0)$. The corresponding Weyl vector is $\rho_1 = \frac{\varepsilon_0}{\sqrt{D}}\frac{1}{\mathrm{tr}(\varepsilon_0)}$. We obtain the Borcherds product expansions

$$\Psi_1 = q_1^{\rho_1} q_2^{\rho_1'} \prod_{\substack{\nu \in \partial_F^{-1} \\ \varepsilon_0 \nu' - \varepsilon_0' \nu > 0}} \left(1 - q_1^{\nu} q_2^{\nu'}\right)^{\tilde{c}_1(5\nu\nu')},$$

$$\Psi_6 = \prod_{\substack{\nu \in \partial_F^{-1} \\ \nu \gg 0}} \left(1 - q_1^{\nu} q_2^{\nu'}\right)^{\tilde{c}_6(5\nu\nu')},$$

$$\Psi_{10} = \prod_{\substack{\nu \in \partial_F^{-1} \\ \nu \gg 0}} \left(1 - q_1^{\nu} q_2^{\nu'}\right)^{\tilde{c}_{10}(5\nu\nu')}.$$

Gundlach [Gu] constructed a Hilbert modular form s_5 for Γ_F with divisor T_1 as a product of 10 theta functions of weight $1/2$, see Section 1.5. We have $s_5 = \Psi_1$. Moreover, s_{15}, the symmetric cusp form of weight 15, is equal to Ψ_5. For further examples we refer to [Ma].

3.3 Automorphic Green Functions

By Theorem 3.49 of the previous section we know precisely which linear combinations of Hirzebruch–Zagier divisors occur as divisors of Borcherds products on $Y(\Gamma_F)$. It is natural to ask, whether every Hilbert modular form on $Y(\Gamma_F)$ whose divisor is a linear combination of Hirzebruch–Zagier divisors is a Borcherds product, i.e., in the image of the lift of Theorem 3.44. In this section we discuss this question in some detail. To answer it, we first simplify the problem. We extend the Borcherds lift to a larger space of "input modular forms" and answer the question for this extended lift. In that way we are led to automorphic Green functions associated with Hirzebruch–Zagier divisors.

Let k be an integer, let Γ be a subgroup of $SL_2(\mathbb{Q})$ which is commensurable with $SL_2(\mathbb{Z})$, and χ a character of Γ. A twice continuously differentiable function $f : \mathbb{H} \to \mathbb{C}$ is called a *weak Maass form* (of weight k and eigenvalue λ with respect to Γ and χ), if

(i) $f\left(\frac{a\tau+b}{c\tau+d}\right) = \chi(\gamma)(c\tau + d)^k f(\tau)$ for all $\left(\begin{smallmatrix} a & b \\ c & d \end{smallmatrix}\right) \in \Gamma$;

(ii) there is a $C > 0$ such that for any cusp $s \in \mathbb{Q} \cup \{\infty\}$ of Γ and $\delta \in SL_2(\mathbb{Z})$ with $\delta\infty = s$ the function $f_s(\tau) = j(\delta, \tau)^{-k} f(\delta\tau)$ satisfies $f_s(\tau) = O(e^{Cv})$ as $v \to \infty$;

(iii) $\Delta_k f = \lambda \Delta$ for some $\lambda \in \mathbb{C}$.

Here

$$\Delta_k = -v^2 \left(\frac{\partial^2}{\partial u^2} + \frac{\partial^2}{\partial v^2}\right) + ikv\left(\frac{\partial}{\partial u} + i\frac{\partial}{\partial v}\right) \tag{3.21}$$

denotes the usual hyperbolic Laplace operator in weight k and $\tau = u + iv$. In the special case where the eigenvalue λ is zero, f is called a *harmonic weak Maass form*. This is the case we are interested in here.

If we compare the definition of a harmonic weak Maass form with the definition of a weakly holomorphic modular form, we see that we simply replaced the condition that f be holomorphic on $\mathbb{H}$ by the weaker condition that f be annihilated by Δ_k, and the meromorphicity at the cusps by the corresponding growth condition. In particular, any weakly holomorphic modular form is a harmonic weak Maass form. The third condition implies that f is actually real analytic. Because of the transformation behavior, it has a Fourier expansion, which involves besides the exponential function a second type of Whittaker function. (See [BrFu1] Section 3 for more details.)

There are two fundamental differential operators on modular forms for Γ, the Maass raising and lowering operators

$$R_k = 2i\frac{\partial}{\partial \tau} + kv^{-1} \quad \text{and} \quad L_k = -2iv^2\frac{\partial}{\partial \bar{\tau}}.$$

If f is a differentiable function on $\mathbb{H}$ satisfying the transformation law (i) in weight k, then $L_k f$ transforms in weight $k - 2$, and $R_k f$ in weight $k + 2$. It can be shown that the assignment

$$f(\tau) \mapsto \xi_k(f)(\tau) := v^{k-2}\overline{L_k f(\tau)} = R_{-k} v^k \overline{f(\tau)}$$

defines an antilinear map ξ_k from harmonic weak Maass forms of weight k to weakly holomorphic modular forms of weight $2 - k$. Its kernel is precisely the space of weakly holomorphic modular forms in weight k.

We write $\mathcal{N}_k(p, \chi_p)$ for the space of harmonic weak Maass forms of weight k with respect to $\Gamma_0(p)$ and χ_p. Let us have a closer look at the map $\xi_k : \mathcal{N}_k(p, \chi_p) \to W_{2-k}(p, \chi_p)$. We denote by $N_k(p, \chi_p)$ the inverse image of $S_{2-k}(p, \chi_p)$ under ξ_k, and its plus subspace by $N_k^+(p, \chi_p)$. (Note that our notation differs from the notation of [BrFu1].)

Theorem 3.53. *We have the following exact sequence:*

$$0 \longrightarrow W_k^+(p, \chi_p) \longrightarrow N_k^+(p, \chi_p) \xrightarrow{\xi_k} S_{2-k}^+(p, \chi_p) \longrightarrow 0 .$$

Proof. This can be proved using Serre duality for the Dolbeault resolution of the structure sheaf on a modular curve (see [BrFu1] Theorem 3.7) or by means of Hejhal–Poincaré series (see [Br2] Chapter 1). □

Let $k \leq 0$. For every harmonic weak Maass form $f \in N_k^+(p, \chi_p)$ there is a unique Fourier polynomial $P(f) = \sum_{n<0} c(n) q^n \in \mathbb{C}[q^{-1}]$ (with $c(n) = 0$ if $\chi_p(n) = -1$) such that $f(\tau) - P(f)(\tau)$ is bounded as $v \to \infty$. It is called the *principal part* of f. This generalizes the notion of the principal part of a weakly holomorphic modular form.

Proposition 3.54. *Let $Q = \sum_{n<0} c(n) q^n \in \mathbb{C}[q^{-1}]$ be a Fourier polynomial satisfying $c(n) = 0$ if $\chi_p(n) = -1$. There exists a unique $f \in N_k^+(p, \chi_p)$ whose principal part is equal to Q.*

Proof. See [BrFu1] Proposition 3.11. □

This Proposition is a key fact, which suggests to study the regularized theta lift of harmonic weak Maass forms. If $f \in N_0^+(p, \chi_p)$, then we define its regularized theta lift $\Phi(z, f)$ by (3.18), in the same way as for weakly holomorphic modular forms.

Theorem 3.55. *Let $f \in N_0^+(p, \chi_p)$ be a harmonic weak Maass form with principal part $P(f) = \sum_{n<0} c(n) q^n$ and constant term $c(0)$.*

(i) *The regularized theta integral $\Phi(z, f)$ defines a Γ_F-invariant function on $\mathbb{H}^2$ with a logarithmic singularity along $-4Z(f)$, where*

$$Z(f) = \sum_{n<0} \tilde{c}(n) T_{-n} .$$

(ii) *It is a Green function for the divisor $2Z(f)$ on $Y(\Gamma_F)$ in the sense of [SABK], that is, it satisfies the identity of currents*

$$dd^c[\Phi(z, f)] + \delta_{2Z(f)} = [\omega(z, f)]$$

on $Y(\Gamma_F)$. Here δ_D denotes the Dirac current associated with a divisor D on $Y(\Gamma_F)$ and $\omega(z, f)$ is a smooth $(1, 1)$-form.

(iii) If $\Delta^{(j)} = -y_j^2 \left(\frac{\partial^2}{\partial x_j^2} + \frac{\partial^2}{\partial y_j^2} \right)$ denotes the $\mathrm{SL}_2(\mathbb{R})$-invariant Laplace operator on $\mathbb{H}^2$ in the variable z_j, then

$$\Delta^{(j)} \Phi(z, f) = -2c(0).$$

Proof. See [Br1] and [BBK]. □

In view of Proposition 3.54, for every positive integer m with $\chi_p(m) \neq -1$, there exists a unique harmonic weak Maass form $f_m \in N_0^+(p, \chi_p)$, whose principal part is given by

$$P(f_m) = \begin{cases} q^{-m}, & \text{if } p \nmid m, \\ \frac{1}{2} q^{-m}, & \text{if } p \mid m. \end{cases}$$

Its theta lift

$$\Phi_m(z) = \frac{1}{2} \Phi(z, f_m)$$

can be regarded as an *automorphic Green function* for T_m.

Let $\pi : \widetilde{X} \to X(\Gamma_F)$ be a desingularization. The Fourier expansion of $\Phi(z, f)$ can be computed explicitly. It can be used to determine the growth behavior at the boundary of $Y(\Gamma_F)$ in $\widetilde{X}$. It turns out that the boundary singularities are of log and log-log type. More precisely, one can view $\pi^* \Phi(z, f)$ as a pre-log-log Green function for the divisor $2\pi^*(Z(f))$ on $\widetilde{X}$ in the sense of [BKK] (see [BBK] Proposition 2.16). So the current equation in (ii) does not only hold for test forms with compact support on $Y(\Gamma_F)$, but also for test forms which are smooth on $\widetilde{X}$.

Moreover, one finds that $\Phi(z, f)$ can be split into a sum

$$\Phi(z, f) = -2 \log |\Psi(z, f)|^2 + \xi(z, f), \tag{3.22}$$

where $\xi(z, f)$ is real analytic on the whole domain $\mathbb{H}^2$ and $\Psi(z, f)$ is a meromorphic function on $\mathbb{H}^2$ whose divisor equals $Z(f)$. If f is weakly holomorphic, the function $\xi(z, f)$ is simply equal to $-2c(0) \left(\log(y_1 y_2) + \log(2\pi) + \Gamma'(1) \right)$, and we are back in the case of Borcherds' original lift. However, if f is an honest harmonic weak Maass form, then ξ is a complicated function and Ψ far from being modular.

The splitting (3.22) implies that the smooth form $\omega(z, f)$ in Theorem 3.55 is given by

$$\omega(z, f) = dd^c \xi(z, f).$$

By the usual Poincaré–Lelong argument, $\frac{1}{2} \omega(z, f)$ represents the Chern class of the divisor $Z(f)$ in the second cohomology $H^2(Y(\Gamma_F))$. One can further show that it is a square integrable harmonic representative. Moreover, $\frac{1}{2} \pi^* \omega(z, f)$ is a pre-log-log form on $\widetilde{X}$, representing the class of $\pi^* Z(f)$ in $H^2(\widetilde{X}, \mathbb{C})$.

We now discuss the relationship between the Borcherds lift (Theorem 3.44) and its generalization in the present section. For simplicity, we write N_k,

W_k, M_k, S_k for the spaces $N_k^+(p, \chi_p)$, $W_k^+(p, \chi_p)$, $M_k^+(p, \chi_p)$, $S_k^+(p, \chi_p)$, respectively. We denote by W_{k0} the subspace of elements of W_k with vanishing constant term. Moreover, we denote by $M_k^\vee$ the dual of the vector space M_k.

Theorem 3.56. *We have the following commutative diagram with exact rows:*

$$
\begin{array}{ccccccccc}
0 & \longrightarrow & W_{00} & \longrightarrow & N_0 & \longrightarrow & M_2^\vee & \longrightarrow & 0 \ . \\
& & \downarrow & & \downarrow & & \downarrow & & \\
0 & \longrightarrow & \mathrm{Rat}(\widetilde{X})_{\mathbb{C}} & \longrightarrow & \mathrm{Div}(\widetilde{X})_{\mathbb{C}} & \longrightarrow & \mathrm{CH}^1(\widetilde{X})_{\mathbb{C}} & \longrightarrow & 0
\end{array}
$$

Here the map $N_0 \to M_2^\vee$ is given by $f_m \mapsto a_m$, where a_m denotes the functional taking a modular form in M_2 to its m-th Fourier coefficient. The map $M_2^\vee \to \mathrm{CH}^1(\widetilde{X})_{\mathbb{C}}$ is defined by $a_m \mapsto \pi^* T_m$ for $m > 0$ and $a_0 \mapsto c_1(\mathcal{M}_{-1/2})$. The map $N_0 \to \mathrm{Div}(\widetilde{X})_{\mathbb{C}}$ is defined by $f \mapsto \pi^* Z(f)$.

Proof. The exactness of the first row is an immediate consequence of Theorem 3.53. Moreover, by Theorem 3.44, if $f \in W_{00}$, then $Z(f) \in \mathrm{Rat}(\widetilde{X})_{\mathbb{C}}$. $\qquad\square$

Remark 3.57. The map $N_0 \to \mathrm{Div}(\widetilde{X})_{\mathbb{C}}$ does not really depend on the analytic properties of the harmonic weak Maass forms. In particular the Green function $\Phi(z, f)$ associated to $f \in N_0$ does not play a role. However, there is an analogue of the above diagram in Arakelov geometry. If $\widetilde{\mathcal{X}}$ is a regular model of $\widetilde{X}$ over an arithmetic ring and $\mathcal{T}_m$ denotes the Zariski closure of $\pi^* T_m$ in $\widetilde{\mathcal{X}}$, then the pair

$$\widehat{\mathcal{T}}_m = (\mathcal{T}_m, \pi^* \Phi_m)$$

defines an arithmetic divisor in the sense of [BKK]. The map $N_0 \to \widehat{\mathrm{Div}}(\widetilde{\mathcal{X}})$, defined by $f_m \mapsto \widehat{\mathcal{T}}_m$, gives rise to a diagram as above for the first arithmetic Chow group of $\widetilde{\mathcal{X}}$. So the generalized Borcherds lift can be viewed as a map to the group of arithmetic divisors on $\widetilde{\mathcal{X}}$ (see [BBK], [Br3]).

Theorem 3.58. *Let h be a meromorphic Hilbert modular form of weight r for Γ_F, whose divisor $\mathrm{div}(h) = \sum_{n<0} \tilde{c}(n) T_{-n}$ is a linear combination of Hirzebruch–Zagier divisors. Then*

$$-2 \log |h(z)^2 (y_1 y_2)^r| = \Phi(z, f) + \text{constant},$$

where f is the unique harmonic weak Maass form in N_0 with principal part $\sum_{n<0} c(n) q^n$.

Proof. (See [Br2] Chapter 5.) Let f be the unique harmonic weak Maass form in N_0 with principal part $\sum_{n<0} c(n) q^n$. Then $\Phi(z, f)$ is real analytic on $\mathbb{H}^2 \setminus Z(f)$ and has a logarithmic singularity along $-4 Z(f)$. Hence

$$d(z) := \Phi(z, f) + 2 \log |h(z)^2 (y_1 y_2)^r|$$

is a smooth function on $Y(\Gamma_F)$. By Theorem 3.55 (iii), it is subharmonic.

One can show that $d(z)$ is in $L^{1+\varepsilon}(Y(\Gamma_F))$ for some $\varepsilon > 0$ (with respect to the invariant measure coming from the Haar measure). By results of Andreotti–Vesentini and Yau (see e.g. [Yau]) on sub-harmonic functions on complete Riemann manifolds that satisfy such integrability conditions it follows that $d(z)$ is constant.   □

The question regarding the surjectivity of the Borcherds lift raised at the beginning of this section is therefore reduced to the question whether the harmonic weak Maass form f in the Theorem is actually weakly holomorphic. It is answered affirmatively in Theorem 3.61 below.

Corollary 3.59. *The assignment* $\pi^* T_m \mapsto \frac{1}{2} dd^c \xi(z, f_m)$ *defines a linear map*

$$\mathrm{CH}^1_{HZ}(\widetilde{X})_{\mathbb{C}} \longrightarrow \mathcal{H}^{1,1}(Y(\Gamma_F))$$

from the subspace of $\mathrm{CH}^1(\widetilde{X})_{\mathbb{C}}$ *generated by the Hirzebruch–Zagier divisors to the space of square integrable harmonic* $(1,1)$*-forms on* $Y(\Gamma_F)$. □

Composing the map $M_2^{\vee} \to \mathrm{CH}^1(\widetilde{X})_{\mathbb{C}}$ with the map $\mathrm{CH}^1_{HZ}(\widetilde{X})_{\mathbb{C}} \to \mathcal{H}^{1,1}(Y(\Gamma_F))$ from Corollary 3.59, we obtain a linear map

$$M_2^{\vee} \longrightarrow \mathcal{H}^{1,1}(Y(\Gamma_F)).$$

On the other hand, we have the Doi–Naganuma lift $S_2 \to S_2(\Gamma_F)$, and there is a natural map from Hilbert cusp forms of weight 2 to harmonic $(1,1)$-forms on $Y(\Gamma_F)$ (see e.g. [Br1] Section 5). Summing up, we get the following diagram:

$$
\begin{array}{ccccc}
M_2^{\vee} & \longrightarrow & \mathrm{CH}^1_{HZ}(\widetilde{X})_{\mathbb{C}} & \longrightarrow & \mathcal{H}^{1,1}(Y(\Gamma_F)) \\
{\scriptstyle f \mapsto (\cdot, f)} \Big\uparrow & & & & \Big\uparrow \\
S_2 & & \longrightarrow & & S_2(\Gamma_F)
\end{array}
\qquad (3.23)
$$

Theorem 3.60. *The above diagram* (3.23) *commutes.*

Proof. See [Br1] Theorem 8.   □

So the above construction can be viewed as a different approach to the Doi–Naganuma lift, making its geometric properties quite transparent.

Using, for instance, the description of the Doi–Naganuma lifting in terms of Fourier expansions, it can be proved that $S_2 \to S_2(\Gamma_F)$ is injective. As a consequence, we obtain the following *converse theorem* for the Borcherds lift (see [Br1], [Br2] Chapter 5).

Theorem 3.61. *Let h be a meromorphic Hilbert modular form for Γ_F, whose divisor* $\mathrm{div}(F) = \sum_{n<0} \tilde{c}(n) T_{-n}$ *is given by Hirzebruch–Zagier divisors. Then there is a weakly holomorphic modular form $f \in W_0$ with principal part* $\sum_{n<0} c(n) q^n$, *and, up to a constant multiple, h is equal to the Borcherds lift of f in the sense of Theorem 3.44.*   □

Corollary 3.62. *The dimension of* $\mathrm{CH}^1_{HZ}(\widetilde{X})_{\mathbb{C}}$ *is equal to* $\dim(M_2)$.   □

Notice that the analogue of Theorem 3.58 holds for arbitrary congruence subgroups of Γ_F (more generally also for $O(2, n)$), whereas the analogue of Theorem 3.61 is related to the injectivity of a theta lift and therefore more complicated. So far it is only known for particular arithmetic subgroups of $O(2, n)$, see [Br2], [Br3]. For example, if we go to congruence subgroups of the Hilbert modular group Γ_F, it is not clear whether the analogue of Theorem 3.61 holds or not. See also [BrFu2] for this question.

A Second Approach

The regularized theta lift $\Phi_m(z) = \frac{1}{2}\Phi(z, f_m)$ of the weak Maass form $f_m \in N_0$ is real analytic on $\mathbb{H}^2 \setminus T_m$ and has a logarithmic singularity along $-2T_m$.

Here we present a different, more naive, construction of $\Phi_m(z)$. For details see [Br1]. The idea is to construct $\Phi_m(z)$ directly as a Poincaré series by summing over the logarithms of the defining equations of T_m. We consider the sum

$$\sum_{\substack{(a,b,\lambda)\in\mathbb{Z}^2\oplus\mathfrak{d}_F^{-1} \\ ab-\lambda\lambda'=m/D}} \log\left|\frac{az_1\bar{z}_2 + \lambda z_1 + \lambda'\bar{z}_2 + b}{az_1z_2 + \lambda z_1 + \lambda'z_2 + b}\right|. \tag{3.24}$$

The denominators of the summands ensure that this function has a logarithmic singularity along $-2T_m$ in the same way as $\Phi_m(z)$. The enumerators are smooth on the whole $\mathbb{H}^2$. They are included to make the sum formally Γ_F-invariant. Unfortunately, the sum diverges. However, it can be regularized in the following way. If we put $Q_0(z) = \frac{1}{2}\log\left(\frac{z+1}{z-1}\right)$, we may rewrite the summands as

$$\log\left|\frac{az_1\bar{z}_2 + \lambda z_1 + \lambda'\bar{z}_2 + b}{az_1z_2 + \lambda z_1 + \lambda'z_2 + b}\right| = Q_0\left(1 + \frac{|az_1z_2 + \lambda z_1 + \lambda'z_2 + b|^2}{2y_1y_2m/D}\right).$$

Now we replace Q_0 by the 1-parameter family Q_{s-1} of Legendre functions of the second kind (cf. [AbSt] §8), defined by

$$Q_{s-1}(z) = \int_0^\infty (z + \sqrt{z^2 - 1}\cosh u)^{-s}du. \tag{3.25}$$

Here $z > 1$ and $s \in \mathbb{C}$ with $\Re(s) > 0$. If we insert $s = 1$, we get back the above Q_0. Hence we consider

$$\phi_m(z, s) = \sum_{\substack{a,b\in\mathbb{Z} \\ \lambda\in\mathfrak{d}_F^{-1} \\ ab-\lambda\lambda'=m/D}} Q_{s-1}\left(1 + \frac{|az_1z_2 + \lambda z_1 + \lambda'z_2 + b|^2}{2y_1y_2m/D}\right). \tag{3.26}$$

It is easily seen that this series converges normally for $z \in \mathbb{H}^2 \setminus T_m$ and $\Re(s) > 1$ and therefore defines a Γ_F-invariant function, which has logarithmic growth along $-2T_m$. It is an eigenfunction of the hyperbolic Laplacians $\Delta^{(j)}$ with eigenvalue $s(s-1)$, because of the differential equation satisfied by Q_{s-1}. Notice that for $D = m = 1$ the function $\Phi_m(z,s)$ is simply the classical resolvent kernel for $\mathrm{SL}_2(\mathbb{Z})$ (cf. [Hej], [Ni]). One can compute the Fourier expansion of $\phi_m(z,s)$ explicitly and use it to obtain a meromorphic continuation to $s \in \mathbb{C}$. At $s = 1$ there is a simple pole, reflecting the divergence of the formal sum (3.24). We define the regularization $\phi_m(z)$ of (3.24) to be the constant term in the Laurent expansion of $\phi_m(z,s)$ at $s = 1$.

It turns out that ϕ_m is up to an additive constant equal to Φ_m. The Green functions ϕ_m can be used to give different proofs of the results of the previous section and of Theorem 3.44. Similar Green functions on $O(2,n)$ are investigated in the context of the theory of spherical functions on real Lie groups in [OT].

3.4 CM Values of Hilbert Modular Functions

In this section we consider the values of Borcherds products on Hilbert modular surfaces at certain CM cycles. We report on some results obtained in joint work with T. Yang, see [BY]. This generalizes work of Gross and Zagier on CM values of the j-function [GZ].

Singular Moduli

We review some of the results of Gross and Zagier on the j-function. We begin by recalling some background material (see also pp. 77–79).

Let k be a field and E/k an elliptic curve, that is, a non-singular projective curve over k of genus 1 together with a k-rational point. If char $k \neq 2, 3$, then by the Riemann–Roch theorem one finds that E has a Weierstrass equation of the form

$$y^2 = 4x^3 - g_2 x - g_3,$$

with $g_2, g_3 \in k$ and $g_2^3 - 27g_3^2 \neq 0$. The j-invariant of E is defined by

$$j(E) = 1728 \frac{g_2^3}{g_2^3 - 27g_3^2}.$$

A basic result of the theory of elliptic curves says that if k is algebraically closed then two elliptic curves over k are isomorphic if and only if they have the same j-invariant. Moreover, for every given $a \in k$ there is an elliptic curve with j-invariant a. So the assignment $E \mapsto j(E)$ defines a bijection

$$\{\text{elliptic curves over } k\}/\sim \longrightarrow k.$$

Over $\mathbb{C}$, the theory of the elliptic functions implies that any elliptic curve is complex analytically isomorphic to a complex torus $\mathbb{C}/L$, where $L \subset \mathbb{C}$ is a lattice. (Here $g_2 = 60G_4(L)$ and $g_3 = 140G_6(L)$ where G_4, G_6 are the usual Eisenstein series of weight 4 and 6.) Two elliptic curves E, E' over $\mathbb{C}$ are isomorphic if and only if the corresponding lattices L, L' satisfy

$$L = a\, L'$$

for some $a \in \mathbb{C}^*$. On the other hand it is easily seen that we have a bijection

$$\mathrm{SL}_2(\mathbb{Z})\backslash\mathbb{H} \longrightarrow \{\text{lattices in } \mathbb{C}\}/\mathbb{C}^*, \qquad [\tau] \mapsto [\mathbb{Z}\tau + \mathbb{Z}].$$

Summing up, we obtain a bijection

$$\mathrm{SL}_2(\mathbb{Z})\backslash\mathbb{H} \longrightarrow \{\text{elliptic curves over } \mathbb{C}\}/\sim, \qquad [\tau] \mapsto [\mathbb{C}/(\mathbb{Z}\tau + \mathbb{Z})]. \quad (3.27)$$

Hence, the j-invariant induces a function on $Y(1) := \mathrm{SL}_2(\mathbb{Z})\backslash\mathbb{H}$. A more detailed examination of the map in (3.27) shows that j is a holomorphic function on $Y(1)$ with the Fourier expansion $j(\tau) = q^{-1} + 744 + 196884q + \ldots$ at the cusp ∞.

So we may view the j-function as a function on the coarse moduli space of isomorphism classes of elliptic curves over $\mathbb{C}$. There are special points on $Y(1)$ which correspond to special elliptic curves, namely to elliptic curves with complex multiplication.

Let $K/\mathbb{Q}$ be an imaginary quadratic field with ring of integers $\mathcal{O}_K$. A point $\tau \in \mathbb{H}$ is called a CM point of type $\mathcal{O}_K$ if the corresponding elliptic curve $E_\tau = \mathbb{C}/(\mathbb{Z}\tau + \mathbb{Z})$ has complex multiplication $\mathcal{O}_K \hookrightarrow \mathrm{End}(E_\tau)$, or equivalently if $\mathbb{Z}\tau + \mathbb{Z} \subset K$ is a fractional ideal. We may consider the 0-cycle $\mathcal{CM}(K) \subset Y(1)$ given by the points τ for which E_τ has complex multiplication by $\mathcal{O}_K$.

The values of the j-function at CM points are classically known as *singular moduli*. If τ_0 is a CM point of type $\mathcal{O}_K$, then, by the theory of complex multiplication, $j(\tau_0)$ is an algebraic integer generating the Hilbert class field of K. Moreover, the Galois group $\mathrm{Gal}(H/K)$ acts transitively on $\mathcal{CM}(K) \subset Y(1)$. This implies that

$$j(\mathcal{CM}(K)) = \prod_{[\tau]\in\mathcal{CM}(K)} j(\tau)$$

is an integer. It is a natural question to ask for the shape of this number. At the beginning of the 20-th century, Berwick made extensive computations of these numbers and conjectured various congruences [Be]. We listed some values in Table 2.

In [GZ], Gross and Zagier found an explicit formula for the prime factorization of $j(\mathcal{CM}(K))$ and proved Berwick's conjectures. More precisely, they considered the function $j(z_1) - j(z_2)$ on $Y(1) \times Y(1)$.

Table 2. Some CM values of the j-function

$|\operatorname{disc}(K)|$	$h(K)$	$(j(\mathcal{CM}(K)))^{1/3}$
3	1	0
4	1	$2^2 \cdot 3$
7	1	$3 \cdot 5$
8	1	$2^2 \cdot 5$
11	1	2^5
19	1	$2^5 \cdot 3$
23	3	$5^3 \cdot 11 \cdot 17$
31	3	$3^3 \cdot 11 \cdot 17 \cdot 23$
43	1	$2^6 \cdot 3 \cdot 5$
47	5	$5^5 \cdot 11^2 \cdot 23 \cdot 29$
59	3	$2^{16} \cdot 11$
67	1	$2^5 \cdot 3 \cdot 5 \cdot 11$
71	7	$11^3 \cdot 17^2 \cdot 23 \cdot 41 \cdot 47 \cdot 53$

Let K_1 and K_2 be two imaginary quadratic fields of discriminants d_1 and d_2, respectively. Assume $(d_1, d_2) = 1$, and put $D = d_1 d_2$. We consider the CM cycle $\mathcal{CM}(K_1) \times \mathcal{CM}(K_2)$ on $Y(1) \times Y(1)$ and put

$$J(d_1, d_2) = \prod_{\substack{[\tau_1] \in \mathcal{CM}(K_1) \\ [\tau_2] \in \mathcal{CM}(K_2)}} (j(\tau_1) - j(\tau_2))^{\frac{4}{w_1 w_2}} ,$$

where w_i is the number of units in K_i.

Theorem 3.63 (Gross, Zagier). *We have*

$$J(d_1, d_2)^2 = \pm \prod_{\substack{x, n, n' \in \mathbb{Z}, \\ n, n' > 0 \\ x^2 + 4nn' = D}} n^{\epsilon(n')} . \tag{3.28}$$

Here ϵ is the genus character defined as follows: $\epsilon(n) = \prod \epsilon(l_i)^{a_i}$ if n has the prime factorization $n = \prod l_i^{a_i}$, and

$$\epsilon(l) = \begin{cases} \left(\frac{d_1}{l}\right) & \text{if } l \nmid d_1, \\ \left(\frac{d_2}{l}\right) & \text{if } l \nmid d_2, \end{cases}$$

for primes l with $\left(\frac{D}{l}\right) \neq -1$.

In particular, this result implies that the prime factors of $J(d_1, d_2)$ are bounded by $D/4$. Since $j(\mathcal{CM}(\mathbb{Q}(\sqrt{-3}))) = j(e^{2\pi i/3}) = 0$, we obtain an explicit formula for the CM values of j as a special case. It leads to the values in Table 2.

The surface $Y(1) \times Y(1)$ can be viewed as the Hilbert modular surface corresponding to the real quadratic "field" $\mathbb{Q} \oplus \mathbb{Q}$ of discriminant 1. Moreover, $j(z_1) - j(z_2)$ is a Borcherds product on this surface given by

$$j(z_1) - j(z_2) = q_1^{-1} \prod_{\substack{m>0 \\ n\in\mathbb{Z}}} (1 - q_1^m q_2^n)^{c(mn)} . \tag{3.29}$$

Here $q_j = e^{2\pi i z_j}$, and $c(n)$ is the n-th Fourier coefficient of $j(\tau) - 744$. In fact, this is the celebrated denominator identity of the monster Lie algebra, which is crucial in Borcherds' proof of the moonshine conjecture. From this viewpoint it is natural to ask if the formula of Gross and Zagier has a generalization to Hilbert modular surfaces. In the rest of this section we report on joint work with T. Yang on this problem [BY]. See also [Ya] for further motivation and background information.

CM Extensions

As before, let $F \subset \mathbb{R}$ be a real quadratic field. Let K be a CM extension of F, that is, $K = F(\sqrt{\Delta})$, where $\Delta \in F$ is totally negative. We view both K and $F(\sqrt{\Delta'})$ as subfields of $\mathbb{C}$ with $\sqrt{\Delta}, \sqrt{\Delta'} \in H$. The field $M = F(\sqrt{\Delta}, \sqrt{\Delta'})$ is Galois over $\mathbb{Q}$. There are three possibilities for the Galois group $\mathrm{Gal}(M/\mathbb{Q})$ of M over $\mathbb{Q}$:

$$\mathrm{Gal}(M/\mathbb{Q}) = \begin{cases} \mathbb{Z}/2\mathbb{Z} \times \mathbb{Z}/2\mathbb{Z}, & \text{if } K/\mathbb{Q} \text{ is biquadratic}, \\ \mathbb{Z}/4\mathbb{Z}, & \text{if } K/\mathbb{Q} \text{ is cyclic}, \\ D_4, & \text{if } K/\mathbb{Q} \text{ is non Galois}. \end{cases}$$

Lemma 3.64. *Let the notation be as above, and let $\tilde{F} = \mathbb{Q}(\sqrt{\Delta\Delta'})$.*

(i) $K/\mathbb{Q}$ is biquadratic if and only if $\tilde{F} = \mathbb{Q}$.
(ii) $K/\mathbb{Q}$ is cyclic if and only if $\tilde{F} = F$.
(iii) $K/\mathbb{Q}$ is non-Galois if and only if $\tilde{F} \neq F$ is a real quadratic field. □

Gross and Zagier considered a biquadratic case. Here we assume that K is non-biquadratic, i.e., $\tilde{F}$ is a real quadratic field. Then $M/\mathbb{Q}$ has an automorphism σ of order 4 such that

$$\sigma(\sqrt{\Delta}) = \sqrt{\Delta'}, \quad \sigma(\sqrt{\Delta'}) = -\sqrt{\Delta}. \tag{3.30}$$

Notice that K has four CM types, i.e., pairs of non complex conjugate complex embeddings: $\Phi = \{1, \sigma\}$, $\sigma\Phi = \{\sigma, \sigma^2\}$, $\sigma^2\Phi$, and $\sigma^3\Phi$. Since K is not biquadratic, these CM types are primitive. We write $(\tilde{K}, \tilde{\Phi})$ for the reflex of (K, Φ). Then $\tilde{K} = \mathbb{Q}(\sqrt{\Delta} + \sqrt{\Delta'})$ and $\tilde{F}$ is the real quadratic subfield of $\tilde{K}$. We refer to [Sh2] for details about CM types and reflex fields.

For the rest of this section we assume that the discriminant of F is a prime $p \equiv 1 \pmod 4$. Moreover, we suppose that the discriminant d_K of K is given

Table 3. CM extensions of $\mathbb{Q}(\sqrt{5})$

q	$K = F(\sqrt{\Delta})$	h_K	$\mathrm{Cl}(K)$
5	$\Delta = -\frac{5+\sqrt{5}}{2}$	1	$\mathcal{O}_K = \mathcal{O}_F + \sqrt{\Delta}\mathcal{O}_F$
41	$\Delta = -\frac{13+\sqrt{5}}{2}$	1	$\mathcal{O}_K = \mathcal{O}_F\frac{1}{2}\left(\sqrt{\Delta}+\frac{3+\sqrt{5}}{2}\right)\mathcal{O}_F$
61	$\Delta = -(9+2\sqrt{5})$	1	$\mathcal{O}_K = \mathcal{O}_F\frac{1}{2}\left(\sqrt{\Delta}+1\right)\mathcal{O}_F$
109	$\Delta = -\frac{21+\sqrt{5}}{2}$	1	$\mathcal{O}_K = \mathcal{O}_F\frac{1}{2}\left(\sqrt{\Delta}+\frac{3+\sqrt{5}}{2}\right)\mathcal{O}_F$
241	$\Delta = -\frac{33+5\sqrt{5}}{2}$	3	$\mathcal{O}_K = \mathcal{O}_F\frac{1}{2}\left(\sqrt{\Delta}+\frac{3+\sqrt{5}}{2}\right)\mathcal{O}_F,$
			$\mathfrak{A} = 2\mathcal{O}_F\frac{1}{2}\left(\sqrt{\Delta}+\frac{9+3\sqrt{5}}{2}\right)\mathcal{O}_F,$
			$\mathfrak{B} = 4\mathcal{O}_F\frac{1}{2}\left(\sqrt{\Delta}+\frac{9+3\sqrt{5}}{2}\right)\mathcal{O}_F$
281	$\Delta = -\frac{37+7\sqrt{5}}{2}$	3	$\mathcal{O}_K = \mathcal{O}_F\frac{1}{2}\left(\sqrt{\Delta}+\frac{1+\sqrt{5}}{2}\right)\mathcal{O}_F,$
			$\mathfrak{A} = 2\mathcal{O}_F\frac{1}{2}\left(\sqrt{\Delta}+\frac{1+\sqrt{5}}{2}\right)\mathcal{O}_F,$
			$\mathfrak{B} = 4\mathcal{O}_F\frac{1}{2}\left(\sqrt{\Delta}+\frac{9+\sqrt{5}}{2}\right)\mathcal{O}_F$
409	$\Delta = -\frac{41+3\sqrt{5}}{2}$	3	$\mathcal{O}_K = \mathcal{O}_F\frac{1}{2}\left(\sqrt{\Delta}+\frac{1+\sqrt{5}}{2}\right)\mathcal{O}_F,$
			$\mathfrak{A} = 2\mathcal{O}_F\frac{1}{2}\left(\sqrt{\Delta}+\frac{7+3\sqrt{5}}{2}\right)\mathcal{O}_F,$
			$\mathfrak{B} = 4\mathcal{O}_F\frac{1}{2}\left(\sqrt{\Delta}+\frac{-1+3\sqrt{5}}{2}\right)\mathcal{O}_F$

by $d_K = p^2 q$ for a prime $q \equiv 1 \pmod 4$. This assumption guarantees that the class number of K is odd, which is crucial in the argument of [BY]. It implies that $\tilde{F} = \mathbb{Q}(\sqrt{q})$ and $d_{\tilde{K}} = q^2 p$. In Table 3 we listed a few CM extensions of $F = \mathbb{Q}(\sqrt{5})$ satisfying the assumption, including the class number h_K, and a system of representatives for the ideal class group of K.

CM Cycles

We now define CM points on Hilbert modular surfaces analogously to the CM points on the modular curve $Y(1)$ above. Recall that the Hilbert modular surface $Y(\Gamma_F)$ corresponding to $\Gamma_F = \mathrm{SL}_2(\mathcal{O}_F)$ parameterizes isomorphism classes of triples $(A, \imath, m)$, where

(i) A is an abelian surface over $\mathbb{C}$,
(ii) $\imath : \mathcal{O}_F \to \mathrm{End}(A)$ is a real multiplication by $\mathcal{O}_F$,
(iii) and $m : (P_A, P_A^+) \to \left(\mathfrak{d}_F^{-1}, \mathfrak{d}_F^{-1,+}\right)$ is an $\mathcal{O}_F$-linear isomorphism between the polarization module $P_A = \mathrm{Hom}^{\mathrm{sym}}_{\mathcal{O}_F}(A, A^\vee)$ of A and $\mathfrak{d}_F^{-1}$, taking the subset of polarizations to totally positive elements of $\mathfrak{d}_F^{-1}$.

(See e.g. [Go], Theorem 2.17 and [BY] Section 3.) The moduli interpretation can be used to construct a model of the Hilbert modular surface $Y(\Gamma_F)$ over $\mathbb{Q}$, see [Rap], [DePa], [Ch].

Let $\Phi = (\sigma_1, \sigma_2)$ be a CM type of K. A point $z = (A, \imath, m) \in Y(\Gamma_F)$ is said to be a CM point of type (K, Φ) if one of the following equivalent conditions holds (see [BY] Section 3 for details):

(i) As a point $z \in \mathbb{H}^2$, there is $\tau \in K$ such that $\Phi(\tau) = (\sigma_1(\tau), \sigma_2(\tau)) = z$ and such that $\Lambda_\tau = \mathcal{O}_F \tau + \mathcal{O}_F$ is a fractional ideal of K.

(ii) $(A, \imath)$ is a CM abelian variety of type (K, Φ) with complex multiplication $\imath' : \mathcal{O}_K \hookrightarrow \mathrm{End}(A)$ such that $\imath = \imath'|_{\mathcal{O}_F}$.

We consider the CM type $\Phi = \{1, \sigma\}$ of K, where σ is defined by (3.30). Let $\mathcal{CM}(K, \Phi, \mathcal{O}_F)$ be the CM 0-cycle in $Y(\Gamma_F)$ of CM abelian surfaces of type (K, Φ). By the theory of complex multiplication [Sh2], the field of moduli for $\mathcal{CM}(K, \Phi, \mathcal{O}_F)$ is the reflex field $\tilde{K}$ of (K, Φ). In fact, one can show that the field of moduli for

$$\mathcal{CM}(K) = \mathcal{CM}(K, \Phi, \mathcal{O}_F) + \mathcal{CM}(K, \sigma^3 \Phi, \mathcal{O}_F)$$

is $\mathbb{Q}$ (see [BY], Remark 3.5). Therefore, if Ψ is a rational function on $Y(\Gamma_F)$, i.e., a Hilbert modular function for Γ_F over $\mathbb{Q}$, then $\Psi(\mathcal{CM}(K))$ is a rational number. The purpose of the following section is to find a formula for this number, when Ψ is given by a Borcherds product.

CM Values of Borcherds Products

We keep the above assumptions on F and K. We denote by W_K the number of roots of unity in K. For an ideal $\mathfrak{a}$ of $\tilde{F}$ we consider the representation number

$$\rho(\mathfrak{a}) = \#\{\mathfrak{A} \subset \mathcal{O}_{\tilde{K}}; \ N_{\tilde{K}/\tilde{F}}\mathfrak{A} = \mathfrak{a}\}$$

of $\mathfrak{a}$ by integral ideals of $\tilde{K}$. We briefly write $|\mathfrak{a}|$ for the norm of $\mathfrak{a}$. For a non-zero element $t \in d_{\tilde{K}/\tilde{F}}^{-1}$ and a prime ideal $\mathfrak{l}$ of $\tilde{F}$, we put

$$B_t(\mathfrak{l}) = \begin{cases} (\mathrm{ord}_{\mathfrak{l}}\, t + 1)\rho(td_{\tilde{K}/\tilde{F}}\mathfrak{l}^{-1}) \log |\mathfrak{l}| & \text{if } \mathfrak{l} \text{ is non-split in } \tilde{K}, \\ 0 & \text{if } \mathfrak{l} \text{ is split in } \tilde{K}, \end{cases}$$

and

$$B_t = \sum_{\mathfrak{l}} B_t(\mathfrak{l}).$$

We remark that $\rho(\mathfrak{a}) = 0$ for a non-integral ideal $\mathfrak{a}$, and that for every $t \neq 0$, there are at most finitely many prime ideals $\mathfrak{l}$ such that $B_t(\mathfrak{l}) \neq 0$. In fact, when $t > 0 > t'$, then $B_t = 0$ unless there is *exactly one* prime ideal $\mathfrak{l}$ such that $\chi_{\mathfrak{l}}(t) = -1$, in which case $B_t = B_t(\mathfrak{l})$ (see [BY], Remark 7.3). Here $\chi = \prod_{\mathfrak{l}} \chi_{\mathfrak{l}}$ is the quadratic Hecke character of $\tilde{F}$ associated to $\tilde{K}/\tilde{F}$. The following formula for the CM values of Borcherds products is proved in [BY].

Theorem 3.65. *Let* $f = \sum_{n \gg -\infty} c(n)q^n \in W_0^+(p, \chi_p)$, *and assume that* $\tilde{c}(n) \in \mathbb{Z}$ *for all* $n < 0$, *and* $c(0) = 0$. *Then the Borcherds lift* $\Psi = \Psi(z, f)$ *(see Theorem 3.44) is a rational function on* $Y(\Gamma_F)$, *whose value at the CM cycle* $\mathcal{CM}(K)$ *satisfies*

$$\log |\Psi(\mathcal{CM}(K))| = \frac{W_{\tilde{K}}}{4} \sum_{m>0} \tilde{c}(-m)b_m,$$

where

$$b_m = \sum_{\substack{t = \frac{n+m\sqrt{q}}{2p} \in d_{\tilde{K}/\tilde{F}}^{-1} \\ |n| < m\sqrt{q}}} B_t.$$

Observe that the number of roots of unity $W_{\tilde{K}}$ equals 2 unless $p = q = 5$, in which case $W_{\tilde{K}} = 10$. The theorem shows that the prime factorization of $\Psi(\mathcal{CM}(K))$ is determined by the arithmetic of the reflex field $\tilde{K}$.

Corollary 3.66. *Let the notation be as in Theorem 3.65. Then*

$$\Psi(\mathcal{CM}(K)) = \pm \prod_{l \text{ rational prime}} l^{e_l}, \tag{3.31}$$

where

$$e_l = \frac{W_{\tilde{K}}}{4} \sum_{m>0} \tilde{c}(-m)b_m(l),$$

and

$$b_m(l) \log l = \sum_{l|l} \sum_{\substack{t = \frac{n+m\sqrt{q}}{2p} \in d_{\tilde{K}/\tilde{F}}^{-1} \\ |n| < m\sqrt{q}}} B_t(l).$$

Moreover, when $K/\mathbb{Q}$ *is cyclic, the sign in (3.31) is positive.*

As in the case that Gross and Zagier considered, see Theorem 3.63, we find that the prime factors of the CM value are small.

Corollary 3.67. *Let the notation and assumption be as in Corollary 3.66. Then* $e_l = 0$ *unless* $4pl|m^2q - n^2$ *for some* $m \in M := \{m \in \mathbb{Z}_{>0}; \tilde{c}(-m) \neq 0\}$ *and some integer* $|n| < m\sqrt{q}$.

Corollary 3.68. *Let the notation and assumption be as in Corollary 3.66. Every prime factor of* $\Psi(\mathcal{CM}(K))$ *is less than or equal to* $\frac{N^2q}{4p}$, *where* $N = \max(M)$.

We now indicate the idea of the proof of Theorem 3.65. It roughly follows the analytic proof of Theorem 3.63 given in [GZ], although each step requires some new ideas. By the construction of the Borcherds lift and by the results of Section 3.3, we have

$$-4\log|\Psi(z,f)| = \Phi(z,f) = \sum_{m>0} \tilde{c}(-m)\phi_m(z),$$

where $\phi_m(z)$ denotes the automorphic Green function for T_m. Consequently, it suffices to compute $\phi_m(\mathcal{CM}(K))$. Using a CM point, the lattice $\mathbb{Z}^2 \oplus \mathfrak{d}_F^{-1}$ defining the automorphic Green function can be related to some ideal of the reflex field $\tilde{K}$ of (K,Φ). In that way, one derives an expression for $\phi_m(\mathcal{CM}(K))$ as an infinite sum involving arithmetic data of $\tilde{K}/\tilde{F}$.

To come up with a finite sum for the CM value $\phi_m(\mathcal{CM}(K))$, we consider an auxiliary function. It is constructed using an incoherent Eisenstein series (see e.g. [Kul]) of weight 1 on $\tilde{F}$ associated to $\tilde{K}/\tilde{F}$. We consider the central derivative of this Eisenstein series, take its restriction to $\mathbb{Q}$, and compute its holomorphic projection.

In that way we obtain a holomorphic cusp form $h \in S_2^+(p,\chi_p)$ of weight 2. Its m-th Fourier coefficient is the sum of two parts. One part is the infinite sum for $\phi_m(\mathcal{CM}(K))$, the other part is a linear combination of the quantity b_m (what we want) and the logarithmic derivative of the Hecke L-series of $\tilde{K}/\tilde{F}$. Finally, the duality between $W_0^+(p,\chi_p)$ and $S_2^+(p,\chi_p)$ of Theorem 3.49, applied to f and h, implies a relation for the Fourier coefficients of h, which leads to the claimed formula.

Notice that the assumption in Theorem 3.65 that the constant term of f vanishes can be dropped. Then the Borcherds lift of f is a meromorphic modular form of non-zero weight, and one can prove a formula for $\log\|\Psi(\mathcal{CM}(K),f)\|_{\mathrm{Pet}}$, where $\|\cdot\|_{\mathrm{Pet}}$ denotes the Petersson metric on the line bundle of modular forms (see [BY] Theorem 1.4).

In a recent preprint [Scho], Schofer obtained a formula for the evaluation of Borcherds products on $O(2,n)$ at CM 0-cycles associated with biquadratic CM fields by means of a different method. It would be interesting to use his results to derive explicit formulas as in Theorem 3.65 for the values of Hilbert modular functions at CM cycles associated to biquadratic CM fields. Finally, notice that Goren and Lauter have recently proved results on the CM values of Igusa genus two invariants using arithmetic methods [GL].

Examples

We first consider the real quadratic field $F = \mathbb{Q}(\sqrt{5})$ and the cyclic CM extension $K = \mathbb{Q}(\zeta_5)$, where $\zeta_5 = e^{2\pi i/5}$. So $p = q = 5$. If σ denotes the complex embedding of K taking ζ_5 to ζ_5^2 then $\Phi = \{1,\sigma\}$ is a CM type of K. We have $\mathcal{O}_K = \mathcal{O}_F + \mathcal{O}_F\zeta_5$, and the corresponding CM cycle $\mathcal{CM}(K,\Phi)$ is represented by the point $(\zeta_5,\zeta_5^2) \in \mathbb{H}^2$.

In Section 3.2 we constructed some Borcherds products for Γ_F. Using the basis (f_m) of $W_0^+(p, \chi_p)$ we see that the Borcherds products

$$R_1(z) = \Psi(z, f_6 - 2f_1) = \frac{\Psi_6}{\Psi_1^2},$$

$$R_2(z) = \Psi(z, f_{10} - 2f_1) = \frac{\Psi_{10}}{\Psi_1^2}$$

are rational functions on $Y(\Gamma_F)$ with divisors $T_6 - 2T_1$ and $T_{10} - 2T_1$, respectively. Let us see what the above results say about $R_1(\mathcal{CM}(K))$. We have $M = \{1, 6\}$ and $N = 6$. According to Corollary 3.68, the prime divisors of $R_1(\mathcal{CM}(K))$ are bounded by 9. Consequently, only the primes $2, 3, 5, 7$ can occur in the factorization. The divisibility criterion given in Corollary 3.67 actually shows that only $2, 3, 5$ can occur. The exact value is given by Corollary 3.66. It is equal to $R_1(\mathcal{CM}(K)) = 2^{20} \cdot 3^{10}$.

In Table 4 we listed some further CM values of R_1 and R_2.

Table 4. The case $F = \mathbb{Q}(\sqrt{5})$

q	$R_1(\mathcal{CM}(K))$	$R_2(\mathcal{CM}(K))$
5 (cyclic)	$2^{20} \cdot 3^{10}$	$2^{20} \cdot 5^{10}$
41	$2^{14} \cdot 3^{10} \cdot 61 \cdot 73$	$2^{14} \cdot 5^9 \cdot 37 \cdot 41$
61	$2^{20} \cdot 3^6 \cdot 13 \cdot 97 \cdot 109$	$2^{20} \cdot 5^9 \cdot 61$
109	$2^{20} \cdot 3^8 \cdot 61 \cdot 157 \cdot 193$	$2^{20} \cdot 5^{12} \cdot 73$
149	$2^{20} \cdot 3^{10} \cdot 31^2 \cdot 37 \cdot 229$	$2^{20} \cdot 5^{12} \cdot 17 \cdot 113$
269	$2^{20} \cdot 3^{10} \cdot 13^{-2} \cdot 37^2 \cdot 61 \cdot 97 \cdot 349 \cdot 433$	$2^{20} \cdot 5^{14} \cdot 13^{-1} \cdot 53 \cdot 73 \cdot 233$

References

[AbSt]   M. Abramowitz and I. Stegun, Pocketbook of Mathematical Functions, Verlag Harri Deutsch, Thun (1984).

[Be]   W. H. Berwick, Modular Invariants, Proc. Lond. Math. Soc. **28** (1927), 53–69.

[Bo1]   R. E. Borcherds, Automorphic forms on $O_{s+2,2}(\mathbb{R})$ and infinite products, Invent. Math. **120** (1995), 161–213.

[Bo2]   R. E. Borcherds, Automorphic forms on $O_{s+2,2}(\mathbb{R})^+$ and generalized Kac-Moody algebras, Proceedings of the ICM 1994, Birkhäuser, Basel (1995), 744-752.

[Bo3]   R. E. Borcherds, Automorphic forms and Lie algebras, Current Developments in mathematics 1996, International Press (1998).

[Bo4]   R. E. Borcherds, Automorphic forms with singularities on Grassmannians, Invent. Math. **132** (1998), 491–562.

[Bo5]   R. E. Borcherds, What is moonshine?, Proceedings of the International congress of Mathematicians, Doc. Math., Extra Vol. I (1998), 607-615.

[Bo6]   R. E. Borcherds, The Gross-Kohnen-Zagier theorem in higher dimensions, Duke Math. J. **97** (1999), 219–233.

[Br1]   J. H. Bruinier, Borcherds products and Chern classes of Hirzebruch–Zagier divisors, Invent. math. **138** (1999), 51–83.

[Br2]   J. H. Bruinier, Borcherds products on $O(2, l)$ and Chern classes of Heegner divisors, Lect. Notes Math. **1780**, Springer-Verlag, Berlin (2002).

[Br3]   J. H. Bruinier, Infinite propducts in number theory and geometry, Jahresber. Dtsch. Math. Ver. **106** (2004), Heft 4, 151–184.

[BB]   J. H. Bruinier and M. Bundschuh, On Borcherds products associated with lattices of prime discriminant, Ramanujan J. **7** (2003), 49–61.

[BrFr]   J. H. Bruinier and E. Freitag, Local Borcherds products, Annales de l'Institut Fourier **51.1** (2001), 1–26.

[BrFu1]   J. H. Bruinier and J. Funke, On two geometric theta lifts, Duke Math. J. **125** (2004), 45–90.

[BrFu2]   J. H. Bruinier and J. Funke, On the injectivity of the Kudla–Millson lift and surjectivity of the Borcherds lift, preprint (2006). http://arxiv.org/math.NT/0606178

[BY]   J. H. Bruinier and T. Yang, CM values of Hilbert modular functions, Invent. Math. **163** (2006), 229–288.

[BBK]   J. H. Bruinier, J. Burgos, and U. Kühn, Borcherds products and arithmetic intersection theory on Hilbert modular surfaces, Duke Math. J. **139** (2007), 1–88.

[BKK]   J. Burgos, J. Kramer, and U. Kühn, Cohomological Arithmetic Chow groups, J. Inst. Math. Jussieu. **6**, 1–178 (2007).

[Ch]   C.-L. Chai, Arithmetic minimal compactification of the Hilbert-Blumenthal moduli spaces, Ann. Math. **131** (1990), 541–554.

[DePa]   P. Deligne and G. Pappas, Singularités des espaces de modules de Hilbert, en les caractéristiques divisant le discriminant, Compos. Math. **90** (1994), 59–79.

[DI]   F. Diamond and J. Im, Modular forms and modular curves, Canadian Mathematical Society Conference Proceedings **17** (1995), 39–133.

[DN]   K. Doi and H. Naganuma, On the functional equation of certain Dirichlet series, Invent. Math. **9** (1969), 1–14.

[EZ]   M. Eichler and D. Zagier, The Theory of Jacobi Forms, Progress in Math. **55** (1985), Birkhäuser.

[Fra]   H.-G. Franke, Kurven in Hilbertschen Modulflächen und Humbertsche Flächen im Siegelraum, Bonner Math. Schriften **104** (1978).

[Fr]   E. Freitag, Hilbert Modular Forms, Springer-Verlag, Berlin (1990).

[Fu]   J. Funke, Heegner divisors and nonholomorphic modular forms, Compos. Math. **133** (2002), 289-321.

[Ga]   P. Garrett, Holomorphic Hilbert modular forms. Wadsworth and Brooks Advanced Books, Pacific Grove, (1990).

[Ge1]   G. van der Geer, Hilbert Modular Surfaces, Springer-Verlag, Berlin (1988).

[Ge2]   G. van der Geer, On the geometry of a Siegel modular threefold, Math. Ann. **260** (1982), 317–350.

[Go]   E. Z. Goren, Lectures on Hilbert Modular Varieties and Modular Forms, CRM Monograph Series **14**, American Mathematical Society, Providence (2002).

178   J. H. Bruinier

[GL]   *E. Z. Goren and K. E. Lauter*, Class invariants for quartic CM fields, Ann. Inst. Fourier **57** (2007), 457–480.

[GZ]   *B. Gross and D. Zagier*, On singular moduli, J. Reine Angew. Math. **355** (1985), 191–220.

[Gu]   *K.-B. Gundlach*, Die Bestimmung der Funktionen zur Hilbertschen Modulgruppe des Zahlkörpers $\mathbb{Q}(\sqrt{5})$, Math. Annalen **152** (1963), 226–256.

[HM]   *J. Harvey and G. Moore*, Algebras, BPS states, and strings, Nuclear Phys. B **463** (1996), no. 2-3, 315–368.

[Ha]   *W. Hausmann*, Kurven auf Hilbertschen Modulflächen, Bonner Math. Schriften **123** (1980).

[He]   *E. Hecke*, Analytische Arithmetik der positiv definiten quadratischen Formen, Kgl. Danske Vid. Selskab. Math. fys. Med. XIII **12** (1940). Werke, 789–918.

[Hej]   *D. A. Hejhal*, The Selberg Trace Formula for $PSL(2, \mathbb{R})$, Lect. Notes Math. **1001**, Springer-Verlag, Berlin (1983).

[Hi]   *H. Hironaka*, Resolution of singularities of an algebraic variety over a field of characteristic zero I, II, Ann. Math. **79** (1964), 109–203 , 205–326.

[Hz]   *F. Hirzebruch*, The Hilbert modular group, resolution of the singularities at the cusps and related problems. Sem. Bourbaki 1970/71, Lect. Notes Math. **244** (1971), 275–288.

[HZ]   *F. Hirzebruch and D. Zagier*, Intersection Numbers of Curves on Hilbert Modular Surfaces and Modular Forms of Nebentypus, Invent. Math. **36** (1976), 57–113.

[Ho]   *R. Howe*, θ-series and invariant theory, Proceedings of Symposia in Pure Mathematics **33**, part 1, American Mathematical Society (1979), 275–285.

[Ki]   *Y. Kitaoka*, Arithmetic of quadratic forms, Cambridge Tracts in Mathematics **106**, Cambridge University Press, Cambridge (1993).

[Kon]   *M. Kontsevich*, Product formulas for modular forms on $O(2, n)$, Séminaire Bourbaki **821** (1996).

[Ku1]   *S. Kudla*, Central derivatives of Eisenstein series and height pairings, Ann. Math. **146** (1997), 545–646.

[Ku2]   *S. Kudla*, Algebraic cycles on Shimura varieties of orthogonal type, Duke Math. J. **86** (1997), 39–78.

[Ku3]   *S. Kudla*, Integrals of Borcherds forms, Compos. Math. **137** (2003), 293–349.

[KM1]   *S. Kudla and J. Millson*, The theta correspondence and harmonic forms I, Math. Ann. **274**, (1986), 353–378.

[KM2]   *S. Kudla and J. Millson*, The theta correspondence and harmonic forms II, Math. Ann. **277**, (1987), 267–314.

[KM3]   *S. Kudla and J. Millson*, Intersection numbers of cycles on locally symmetric spaces and Fourier coefficients of holomorphic modular forms in several complex variables, IHES Publi. Math. **71** (1990), 121–172.

[KR]   *S. Kudla and M. Rapoport*, Arithmetic Hirzebruch–Zagier divisors, J. Reine Angew. Math. **515** (1999), 155–244.

[Lo]   *E. Looijenga*, Compactifications defined by arrangements. II. Locally symmetric varieties of type IV. Duke Math. J. **119** (2003), 527–588.

[Ma]   *S. Mayer*, Hilbert Modular Forms for the Fields $\mathbb{Q}(\sqrt{5})$, $\mathbb{Q}(\sqrt{13})$, and $\mathbb{Q}(\sqrt{17})$, Thesis, University of Aachen (2007).

[Mü]   *R. Müller*, Hilbertsche Modulfunktionen zu $\mathbb{Q}(\sqrt{5})$, Arch. Math. **45**, 239–251 (1985).

[Na]   H. *Naganuma*, On the coincidence of two Dirichlet series associated with cusp forms of Hecke's "Neben"-type and Hilbert modular forms over a real quadratic field, J. Math. Soc. Japan **25** (1973), 547–555.

[Ne]   J. *Neukirch*, Algebraic number theory, Grundlehren der Mathematischen Wissenschaften **322**, Springer-Verlag, Berlin (1999).

[Ni]   D. *Niebur*, A class of nonanalytic automorphic functions, Nagoya Math. J. **52** (1973), 133–145.

[Od1]   T. *Oda*, On Modular Forms Associated with Indefinite Quadratic Forms of Signature $(2, n-2)$, Math. Ann. **231** (1977), 97–144.

[Od2]   T. *Oda*, A note on a geometric version of the Siegel formula for quadratic forms of signature $(2, 2k)$, Sci. Rep. Niigata Univ. Ser. A No. **20** (1984), 13–24.

[OT]   T. *Oda and M. Tsuzuki*, Automorphic Green functions associated with the secondary spherical functions. Publ. Res. Inst. Math. Sci. **39** (2003), 451–533.

[Ral]   S. *Rallis*, Injectivity properties of liftings associated to Weil representations, Compositio Math. **52** (1984), 139–169.

[RS]   S. *Rallis and G. Schiffmann*, On a relation between SL_2 cusp forms and cusp forms on tube domains associated to orthogonal groups, Trans. AMS **263** (1981), 1–58.

[Rap]   M. *Rapoport*, Compactifications de l'espace de modules de Hilbert-Blumenthal, Compos. Math. **36** (1978), 255–335.

[Scha]   W. *Scharlau*, Quadratic and Hermitian forms, Grundlehren der Mathematischen Wissenschaften **270**, Springer-Verlag, Berlin (1985).

[Scho]   J. *Schofer*, Borcherds Forms and Generalizations of Singular Moduli, preprint (2006).

[Sh1]   G. *Shimura*, Introduction to the Arithmetic Theory of Automorphic Functions, Princeton University Press, Princeton (1971).

[Sh2]   G. *Shimura*, Abelian varieties with complex multiplication and modular functions, Princeton Univ. Press (1998).

[SABK]   C. *Soulé, D. Abramovich, J.-F. Burnol, and J. Kramer*, Lectures on Arakelov Geometry, Cambridge Studies in Advanced Mathematics **33**, Cambridge University Press, Cambridge (1992).

[Ya]   T. *Yang*, Hilbert modular functions and their CM values, preprint (2005).

[Yau]   S.-T. *Yau*, Some function-theoretic properties of complete Riemannian Manifolds and their applications to geometry, Indiana Univ. Math. J. **25** (1976), 659–670.

[Za1]   D. *Zagier*, Modular Forms Associated to Real Quadratic Fields, Invent. Math. **30** (1975), 1–46.

[Za2]   D. *Zagier*, Modular forms whose Fourier coefficients involve zeta-functions of quadratic fields. In: Modular Functions of One Variable VI, Lecture Notes in Math. **627**, Springer-Verlag (1977), 105–169.

Siegel Modular Forms and Their Applications

Gerard van der Geer

Korteweg-de Vries Instituut, Universiteit van Amsterdam,
Plantage Muidergracht 24, 1018 TV Amsterdam, The Netherlands
E-mail: geer@science.uva.nl

Summary. These are the lecture notes of the lectures on Siegel modular forms at the Nordfjordeid Summer School on Modular Forms and their Applications. We give a survey of Siegel modular forms and explain the joint work with Carel Faber on vector-valued Siegel modular forms of genus 2 and present evidence for a conjecture of Harder on congruences between Siegel modular forms of genus 1 and 2.

1 Introduction

Siegel modular forms generalize the usual modular forms on $\mathrm{SL}(2, \mathbb{Z})$ in that the group $\mathrm{SL}(2, \mathbb{Z})$ is replaced by the automorphism group $\mathrm{Sp}(2g, \mathbb{Z})$ of a unimodular symplectic form on $\mathbb{Z}^{2g}$ and the upper half plane is replaced by the Siegel upper half plane $\mathcal{H}_g$. The integer $g \geq 1$ is called the degree or genus.

Siegel pioneered the generalization of the theory of elliptic modular forms to the modular forms in more variables now named after him. He was motivated by his work on the Minkowski–Hasse principle for quadratic forms over the rationals, cf., [96]. He investigated the geometry of the Siegel upper half plane, determined a fundamental domain and its volume and proved a central result equating an Eisenstein series with a weighted sum of theta functions.

No doubt, Siegel modular forms are of fundamental importance in number theory and algebraic geometry, but unfortunately, their reputation does not match their importance. And although vector-valued rather than scalar-valued Siegel modular forms are the natural generalization of elliptic modular forms, their reputation amounts to even less. A tradition of ill-chosen notations may have contributed to this, but the lack of attractive examples that can be handled decently seems to be the main responsible. Part of the beauty of elliptic modular forms is derived from the ubiquity of easily accessible examples. The accessible examples that we have of Siegel modular forms are scalar-valued Siegel modular forms given by Fourier series and for $g > 1$ it is difficult to extract the arithmetic information (e.g., eigenvalues of Hecke operators) from the Fourier coefficients.

The general theory of automorphic representations provides a generalization of the theory of elliptic modular forms. But despite the obvious merits of this approach some of the attractive explicit features of the $g = 1$ theory are lost in the generalization.

The elementary theory of elliptic modular forms ($g = 1$) requires little more than basic function theory, while a good grasp of the elementary theory of Siegel modular forms requires a better understanding of the geometry involved, in particular of the compactifications of the quotient space $\mathrm{Sp}(2g, \mathbb{Z}) \backslash \mathcal{H}_g$. A singular compactification was provided by Satake and Baily-Borel and a smooth compactification by Igusa in special cases and by Mumford c.s. by an intricate machinery in the general case.

The fact that $\mathrm{Sp}(2g, \mathbb{Z}) \backslash \mathcal{H}_g$ is the moduli space of principally polarized abelian varieties plays an important role in the arithmetic theory of modular forms. Even for $g = 1$ one needs the understanding of the geometry of moduli space as a scheme (stack) over the integers and its cohomology as Deligne's proof of the estimate $|a(p)| \leq 2p^{(k-1)/2}$ (the Ramanujan conjecture) for the Fourier coefficients of a Hecke eigenform of weight k showed. For quite some time the lack of a well-developed theory of moduli spaces of principally polarized abelian varieties over the integers formed a serious hurdle for the development of the arithmetic theory. Fortunately, Faltings' work on the moduli spaces of abelian varieties has provided us with the first necessary ingredients of the arithmetic theory, both the smooth compactification over $\mathbb{Z}$ as well as the Satake compactification over $\mathbb{Z}$. It also gives the analogue of the Eichler–Shimura theorem which expresses Siegel modular forms in terms of the cohomology of local systems on $\mathrm{Sp}(2g, \mathbb{Z}) \backslash \mathcal{H}_g$. The fact that the vector-valued Siegel modular forms are the natural generalization of the classical elliptic modular forms becomes apparent if one studies the cohomology of the universal abelian variety.

Examples of modular forms for $\mathrm{SL}(2, \mathbb{Z})$ are easily constructed using Eisenstein series or theta series. These methods are much less effective when dealing with the case $g \geq 2$, especially if one is interested in vector-valued Siegel modular forms. Some examples can be constructed using theta series, but it is not always easy to calculate the Fourier coefficients and more difficult to extract the eigenvalues of the Hecke operators.

We show that there is an alternative approach that uses the analogue of the classical Eichler–Shimura theorem. Since cohomology of a variety over a finite field can be calculated by determining the number of rational points over extension fields one can count curves over finite fields to calculate traces of Hecke operators on spaces of vector-valued cusp forms for $g = 2$. This is joint work with Carel Faber. It has the pleasant additional feature that our forms all live in level 1, i.e. on the full Siegel modular group.

We illustrate this by providing convincing evidence for a conjecture of Harder on congruences between the eigenvalues of Siegel modular forms of genus 2 and elliptic modular forms.

In these lectures we concentrate on modular forms for the full Siegel modular group $\mathrm{Sp}(2g, \mathbb{Z})$ and leave modular forms on congruence subgroups aside. We start with the elementary theory and try to give an overview of the various interesting aspects of Siegel modular forms. An obvious omission are the Galois representations associated to Siegel modular forms.

A good introduction to the Siegel modular group and Siegel modular forms is Freitag's book [30]. The reader may also consult the introductory book by Klingen [61]. Two other references to the literature are the two books [94, 95] by Shimura. Vector-valued Siegel modular forms are also discussed in a paper by Harris, [46].

Acknowledgements. I would like to thank Carel Faber, Alex Ghitza, Christian Grundh, Robin de Jong, Winfried Kohnen, Sam Grushevsky, Martin Weissman and Don Zagier for reading the manuscript and/or providing helpful comments. Finally I would like to thank Kristian Ranestad for inviting me to lecture in Nordfjordeid in 2004.

2 The Siegel Modular Group

The ingredients of the definition of 'elliptic modular form' are the group $\mathrm{SL}(2, \mathbb{Z})$, the upper half plane $\mathcal{H}$, the action of $\mathrm{SL}(2, \mathbb{Z})$ on $\mathcal{H}$, the concept of a holomorphic function and the factor of automorphy $(cz + d)^k$. So if we want to generalize the concept 'modular form' we need to generalize these notions. But the upper half plane can be expressed in terms of the group as $\mathrm{SL}(2, \mathbb{Z})/\mathrm{SO}(2)$, where $\mathrm{SO}(2) = \mathrm{U}(1)$, a maximal compact subgroup, is the stabilizer of the point $i = \sqrt{-1}$. Therefore, the group is the central object and we start by generalizing the group. The group $\mathrm{SL}(2, \mathbb{Z})$ is the automorphism group of the lattice $\mathbb{Z}^2$ with the standard alternating form $\langle \, , \, \rangle$ with

$$\langle (a, b), (c, d) \rangle = ad - bc.$$

This admits an obvious generalization by taking for $g \in \mathbb{Z}_{\geq 1}$ the lattice $\mathbb{Z}^{2g}$ of rank $2g$ with basis $e_1, \ldots, e_g, f_1, \ldots, f_g$ provided with the symplectic form $\langle \, , \, \rangle$ with

$$\langle e_i, e_j \rangle = 0, \ \langle f_i, f_j \rangle = 0 \quad \text{and} \quad \langle e_i, f_j \rangle = \delta_{ij},$$

where δ_{ij} is Kronecker's delta. The *symplectic group* $\mathrm{Sp}(2g, \mathbb{Z})$ is by definition the automorphism group of this symplectic lattice

$$\mathrm{Sp}(2g, \mathbb{Z}) := \mathrm{Aut}(\mathbb{Z}^{2g}, \langle \, , \, \rangle).$$

By using the basis of the e's and the f's we can write the elements of this group as matrices

$$\begin{pmatrix} A & B \\ C & D \end{pmatrix},$$

where A, B, C and D are $g \times g$ integral matrices satisfying $AB^t = BA^t$, $CD^t = DC^t$ and $AD^t - BC^t = 1_g$. Here we write 1_g for the $g \times g$ identity matrix. For $g = 1$ we get back the group $SL(2, \mathbb{Z})$. The group $Sp(2g, \mathbb{Z})$ is called the *Siegel modular group* (of degree g) and often denoted Γ_g.

Exercise 1. Show that the conditions on A, B, C and D are equivalent to $C^t \cdot A - A^t \cdot C = 0$, $D^t \cdot B - B^t \cdot D = 0$ and $D^t \cdot A - B^t \cdot C = 1_g$.

The upper half plane $\mathcal{H}$ can be given as a coset space $SL(2, \mathbb{R})/K$ with $K = U(1)$ a maximal compact subgroup, and this admits a generalization, but the desired generalization also admits a description as a half plane and with this we start: the *Siegel upper half plane* $\mathcal{H}_g$ is defined as

$$\mathcal{H}_g = \{\tau \in \mathrm{Mat}(g \times g, \mathbb{C}) \colon \tau^t = \tau, \, \mathrm{Im}(\tau) > 0\},$$

consisting of $g \times g$ complex symmetric matrices which have positive definite imaginary part (obtained by taking the imaginary part of every matrix entry). Clearly, we have $\mathcal{H}_1 = \mathcal{H}$.

An element $\gamma = \begin{pmatrix} A & B \\ C & D \end{pmatrix}$ of the group $Sp(2g, \mathbb{Z})$, sometimes denoted by $(A, B; C, D)$, acts on the Siegel upper half plane by

$$\tau \mapsto \gamma(\tau) = (A\tau + B)(C\tau + D)^{-1}. \tag{1}$$

Of course, we must check that this is well-defined, in particular that $C\tau + D$ is invertible. For this we use the identity

$$(C\bar{\tau} + D)^t(A\tau + B) - (A\bar{\tau} + B)^t(C\tau + D) = \tau - \bar{\tau} = 2iy, \tag{2}$$

where we write $\tau = x + iy$ with x and y symmetric real $g \times g$ matrices. We claim that $\det(C\tau + D) \neq 0$. Indeed, if the equation $(C\tau + D)\xi = 0$ has a solution $\xi \in \mathbb{C}^g$ then equation (2) implies $\bar{\xi}^t y \xi = 0$ and by the assumed positive definiteness of y that $\xi = 0$.

One can also check directly the identity

$$(C\tau + D)^t(\gamma(\tau) - \gamma(\tau)^t)(C\tau + D)$$
$$= (C\tau + D)^t(A\tau + B) - (A\tau + B)^t(C\tau + D) = \tau - \tau^t = 0$$

that shows that $\gamma(\tau)$ is symmetric. Moreover, again by (2) and this last identity we find the relation between $y' = \mathrm{Im}(\gamma(\tau))$ and y

$$(C\bar{\tau} + D)^t y'(C\tau + D) = \frac{1}{2i}(C\bar{\tau} + D)^t(\gamma(\tau) - (\overline{\gamma(\tau)})^t)(C\tau + D) = y$$

and this shows that $y' = \mathrm{Im}(\gamma(\tau))$ is positive definite. Using these details one easily checks that (1) defines indeed an action of $Sp(2g, \mathbb{Z})$, and even of $Sp(2g, \mathbb{R})$ on $\mathcal{H}_g$.

The group $\mathrm{Sp}(2g, \mathbb{R})/\{\pm 1\}$ acts effectively on $\mathcal{H}_g$ and it is the biholomorphic automorphism group of $\mathcal{H}_g$. The action is transitive and the stabilizer of $i\,1_g$ is

$$U(g) := \left\{ \begin{pmatrix} A & B \\ -B & A \end{pmatrix} \in \mathrm{Sp}(2g, \mathbb{R}) \colon A \cdot A^t + B \cdot B^t = 1_g \right\},$$

the unitary group. We may thus view $\mathcal{H}_g$ as the coset space $\mathrm{Sp}(2g, \mathbb{R})/U(g)$ of a simple Lie group by a maximal compact subgroup (which is unique up to conjugation).

The disguise of $\mathcal{H}_1$ as the unit disc $\{z \in \mathbb{C} \colon |z| < 1\}$ also has an analogue for $\mathcal{H}_g$. The space $\mathcal{H}_g$ is analytically equivalent to a bounded symmetric domain

$$D_g := \{Z \in \mathrm{Mat}(g \times g, \mathbb{C}) \colon Z^t = Z,\, Z^t \cdot Z < 1_g\}$$

and the generalized Cayley transform

$$\tau \mapsto z = (\tau - i1_g)(\tau + i1_g)^{-1}, \qquad z \mapsto \tau = i \cdot (1_g + z)(1_g - z)^{-1}$$

makes the correspondence explicit. The 'symmetric' in the name refers to the existence of an involution on $\mathcal{H}_g$ (or D_g)

$$\tau \mapsto -\tau^{-1} \qquad (z \mapsto -z)$$

having exactly one isolated fixed point. Note that we can write $\mathcal{H}_g$ also as $S_g + iS_g^+$ with S_g (resp. S_g^+) the $\mathbb{R}$-vector space (resp. cone) of real symmetric (resp. real positive definite symmetric) matrices of size $g \times g$.

The group $\mathrm{Sp}(2g, \mathbb{Z})$ is a discrete subgroup of $\mathrm{Sp}(2g, \mathbb{R})$ and acts properly discontinuously on $\mathcal{H}_g$, i.e., for every $\tau \in \mathcal{H}_g$ there is an open neighborhood U of τ such that $\{\gamma \in \mathrm{Sp}(2g, \mathbb{Z}) \colon \gamma(U) \cap U \neq \emptyset\}$ is finite. In fact, this follows immediately from the properness of the map $\mathrm{Sp}(2g, \mathbb{R}) \to \mathrm{Sp}(2g, \mathbb{R})/U(g)$.

For $g = 1$ usually one proceeds after these introductory remarks on the action to the construction of a fundamental domain for the action of $\mathrm{SL}(2, \mathbb{Z})$ and all the texts display the following archetypical figure.

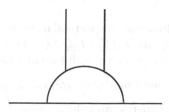

Siegel (see [97]) constructed also a fundamental domain for $g \geq 2$, namely the set of $\tau = x + iy \in \mathcal{H}_g$ satisfying the following three conditions:

1. We have $|\det(C\tau + D)| \geq 1$ for all $(A, B; C, D) \in \Gamma_g$;
2. the matrix y is reduced in the sense of Minkowski;
3. the entries x_{ij} of x satisfy $|x_{ij}| \leq 1/2$.

Here Minkowski reduced means that y satisfies the two properties 1) $h^t y h \geq y_{kk}$ $(k = 1, \ldots, g)$ for all primitive vectors h in $\mathbb{Z}^g$ and 2) $y_{k,k+1} \geq 0$ for $0 \leq k \leq g - 1$. Already for $g = 2$ the boundary of this fundamental domain is complicated; Gottschling found that it posesses 28 boundary pieces, cf., [39], and the whole thing does not help much to understand the nature of the quotient space $\mathrm{Sp}(2g, \mathbb{Z}) \backslash \mathcal{H}_g$.

The group $\mathrm{Sp}(2g, \mathbb{Z})$ does not act freely on $\mathcal{H}_g$, but the subgroup

$$\Gamma_g(n) := \{\gamma \in \mathrm{Sp}(2g, \mathbb{Z}) : \gamma \equiv 1_{2g} \,(\mathrm{mod}\, n)\}$$

acts freely if $n \geq 3$ as is easy to check, cf. [89]. The quotient space (orbit space)

$$Y_g(n) := \Gamma_g(n) \backslash \mathcal{H}_g$$

is then for $n \geq 3$ a complex manifold of dimension $g(g + 1)/2$. Note that the finite group $\mathrm{Sp}(2g, \mathbb{Z}/n\mathbb{Z})$ acts on $Y_g(n)$ as a group of biholomorphic automorphisms and we can thus view

$$\mathrm{Sp}(2g, \mathbb{Z}) \backslash \mathcal{H}_g$$

as an orbifold (quotient of a manifold by a finite group).

The Poincaré metric on the upper half plane also generalizes to the Siegel upper half plane. The corresponding volume form is given by

$$(\det y)^{-(g+1)} \prod_{i \leq j} dx_{ij}\, dy_{ij}$$

which is $\partial\bar{\partial} \log \det \mathrm{Im}(\tau)^g$. The volume of the fundamental domain was calculated by Siegel, [98]. If we normalize the volume such that it gives the orbifold Euler characteristic the result is (cf. Harder [44])

$$\mathrm{vol}(\mathrm{Sp}(2g, \mathbb{Z}) \backslash \mathcal{H}_g) = \zeta(-1)\zeta(-3) \cdots \zeta(1 - 2g)$$

with $\zeta(s)$ the Riemann zeta function. In particular, for $n \geq 3$ the Euler number of the manifold $\Gamma_g(n) \backslash \mathcal{H}_g$ equals $[\Gamma_g(1) : \Gamma_g(n)]\zeta(-1) \cdots \zeta(1 - 2g)$.

We first present two exercises for the solution of which we refer to [30].

Exercise 2. Show that the Siegel modular group Γ_g is generated by the elements $\begin{pmatrix} 1_g & s \\ 0 & 1_g \end{pmatrix}$ with $s = s^t$ symmetric and the element $\begin{pmatrix} 0 & 1_g \\ -1_g & 0 \end{pmatrix}$.

Exercise 3. Show that $\mathrm{Sp}(2g, \mathbb{Z})$ is contained in $\mathrm{SL}(2g, \mathbb{Z})$.

We close with another model of the domain $\mathcal{H}_g$ that can be obtained as follows. Extend scalars of our symplectic lattice $(\mathbb{Z}^{2g}, \langle \, , \, \rangle)$ to $\mathbb{C}$ and let Y_g be the Lagrangian Grassmann variety parametrizing totally isotropic subspaces of dimension g:

$$Y_g := \{L \subset \mathbb{C}^{2g} : \dim(L) = g, \langle x, y \rangle = 0 \quad \text{for all } x, y \in L\}.$$

Since the group $\mathrm{Sp}(2g, \mathbb{C})$ acts transitively on the set of totally isotropic subspaces we may identify Y_g with the compact manifold $\mathrm{Sp}(2g, \mathbb{C})/Q$, where Q is the parabolic subgroup that fixes the first summand $\mathbb{C}^g$. Consider now in Y_g the open set Y_g^+ of Lagrangian subspaces L such that $-i\langle x, \bar{x} \rangle > 0$ for all non-zero x in L. Then Y_g^+ is stable under the action of $\mathrm{Sp}(2g, \mathbb{R})$ and the stabilizer of a point is isomorphic to the unitary group $U(g)$. A basis of such an L is given by the columns of a unique $2g \times g$ matrix $\begin{pmatrix} -1_g \\ \tau \end{pmatrix}$ with $\tau \in \mathcal{H}_g$ and this embeds $\mathcal{H}_g$ in Y_g as the open subset Y_g^+; for $g = 1$ we get the upper half plane in $\mathbb{P}^1$. The manifold Y_g is called the *compact dual* of $\mathcal{H}_g$.

Remark 1. Just as for $g = 1$ we could consider congruence subgroups of $\mathrm{Sp}(2g, \mathbb{Z})$, like for example $\Gamma_g(n)$, the kernel of the natural homomorphism $\mathrm{Sp}(2g, \mathbb{Z}) \to \mathrm{Sp}(2g, \mathbb{Z}/n\mathbb{Z})$ for natural numbers n. We shall stick to the full symplectic group $\mathrm{Sp}(2g, \mathbb{Z})$ here.

3 Modular Forms

To generalize the notion of modular form as we know it for $g = 1$ we still have to generalize the 'automorphy factor' $(cz + d)^k$. To do this we consider a representation

$$\rho \colon \mathrm{GL}(g, \mathbb{C}) \to \mathrm{GL}(V)$$

with V a finite-dimensional $\mathbb{C}$-vector space.

For reasons that become clear later, it is useful to provide V with a hermitian metric $(\, , \,)$ such that $(\rho(g)v_1, v_2) = (v_1, \rho(g^t)v_2)$ and we shall put $\|v\| = (v, v)^{1/2}$. Such a hermitian metric can always be found and is unique up to a scalar for irreducible representations.

Definition 1. *A holomorphic map* $f \colon \mathcal{H}_g \to V$ *is called a Siegel modular form of weight* ρ *if*

$$f(\gamma(\tau)) = \rho(C\tau + D)f(\tau)$$

for all $\gamma = \begin{pmatrix} A & B \\ C & D \end{pmatrix} \in \mathrm{Sp}(2g, \mathbb{Z})$ *and all* $\tau \in \mathcal{H}_g$, *plus for* $g = 1$ *the requirement that* f *is holomorphic at* ∞.

Before we proceed, a word about notations. The subject has been plagued with unfortunate choices of notations, and the tradition of using capital letters for the matrix blocks of elements of the symplectic group is one of them.

I propose to use lower case letters, so I will write $f(\gamma(\tau)) = \rho(c\tau + d)f(\tau)$ for all $\gamma = (a, b; c, d) \in \Gamma_g$ for our condition.

The modular forms we consider here are vector-valued modular forms. As it turns out, the holomorphicity condition is not necessary for $g > 1$, see the Koecher principle hereafter.

Modular forms of weight ρ form a $\mathbb{C}$-vector space $M_\rho = M_\rho(\Gamma_g)$ and we shall see later (in Section 13) that all the M_ρ are finite-dimensional. If ρ is a direct sum of two representations $\rho = \rho_1 \oplus \rho_2$ then M_ρ is isomorphic to the direct sum $M_{\rho_1} \oplus M_{\rho_2}$ and this allows us to restrict ourselves to studying M_ρ for the irreducible representations of $\mathrm{GL}(g, \mathbb{C})$.

As is well-known (see [34], but see also the later Section 12), the irreducible finite-dimensional representations of $\mathrm{GL}(g, \mathbb{C})$ correspond bijectively to the g-tuples $(\lambda_1, \ldots, \lambda_g)$ of integers with $\lambda_1 \geq \lambda_2 \geq \cdots \geq \lambda_g$, the highest weight of the representation ρ. That is, for each irreducible V there exists a unique 1-dimensional subspace $\langle v_\rho \rangle$ of V such that $\rho(\mathrm{diag}(a_1, \ldots, a_g))$ acts on v_ρ by multiplication by $\prod_{i=1}^g a_i^{\lambda_i}$. For example, the g-tuple $(1, 0, \ldots, 0)$ corresponds to the tautological representation $\rho(x) = x$ for $x \in \mathrm{GL}(g, \mathbb{C})$, while the determinant representation corresponds to $\lambda_1 = \ldots = \lambda_g = 1$. Tensoring a given irreducible representation with the k-th power of the determinant changes the λ_i to $\lambda_i + k$. We thus can arrange that $\lambda_g = 0$ or that $\lambda_g \geq 0$ (i.e. that the representation is 'polynomial'). Let R be the set of isomorphism classes of representations of $\mathrm{GL}(g, \mathbb{C})$. This set forms a ring with $\oplus$ as addition and $\otimes$ as multiplication. It is called the representation ring of $\mathrm{GL}(g, \mathbb{C})$.

For $g = 1$ one usually forms a graded ring of modular forms by taking $M_*(\Gamma_1) = \oplus M_k(\Gamma_1)$. We can try do something similar for $g > 1$ and try to make the direct sum $\oplus_{\rho \in R} M_\rho(\Gamma_g)$ into a graded ring. But of course, this is a huge ring, even for $g = 1$ much larger than $M_*(\Gamma_1)$ since it involves also the reducible representations and it is not really what we want.

The classes of the irreducible representations of $\mathrm{GL}(g, \mathbb{C})$ form a subset of all classes of representations. For $g = 1$ and $g = 2$ the fact is that the tensor product of two irreducible representations is a direct sum of irreducible representations with multiplicity 1. In fact, for $g = 1$ the tensor product of the irreducible representations ρ_{k_1} and ρ_{k_2} of degree $k_1 + 1$ and $k_2 + 1$ is the irreducible representation $\rho_{k_1+k_2}$. For $g = 2$, a case that will play a prominent role in these lecture notes, we let $\rho_{j,k}$ denote the irreducible representation of $\mathrm{GL}(2, \mathbb{C})$ that is $\mathrm{Sym}^j(W) \otimes \det(W)^k$ with W the standard 2-dimensional representation; it corresponds to highest weight $(\lambda_1, \lambda_2) = (j + k, k)$. Then there is the formula

$$\rho_{j_1,k_1} \otimes \rho_{j_2,k_2} \cong \sum_{r=0}^{\min(j_1,j_2)} \rho_{j_1+j_2-2r,k_1+k_2+r} \cdot$$

So we can decompose $M_{\rho_{j_1,k_1}}$ as a direct sum $\sum_{r=0}^{\min(j_1,j_2)} M_{\rho_{j_1+j_2-2r,k_1+k_2+r}}$, but this is not canonical as it depends upon a choice of isomorphism in the above formula. Nevertheless, this decomposition is useful to construct modular forms in new weights by multiplying modular forms.

To make $\oplus_{\rho \in \mathrm{Irr}} M_\rho(\Gamma_2)$ into a ring requires a consistent choice for all these identifications. We can avoid this by viewing the symmetric power $\mathrm{Sym}^j(W)$ as a space of polynomials of degree j in two variables and then by remarking that multiplication of polynomials defines a canonical map $\mathrm{Sym}^{j_1}(W) \otimes \mathrm{Sym}^{j_2}(W) \to \mathrm{Sym}^{j_1+j_2}(W)$. Using this and the obvious map $\det(W)^{k_1} \otimes \det(W)^{k_2} \to \det(W)^{k_1+k_2}$ the direct sum $\oplus_{\rho \in \mathrm{Irr}} M_\rho(\Gamma_2)$ becomes a ring; we just 'forgot' the terms in the above sum with $r > 0$. For $g \geq 3$ the tensor products come in general with multiplicities, given by Littlewood–Richardson numbers. Nevertheless, one can define a ring structure on $\oplus_{\rho \in \mathrm{Irr}} M_\rho(\Gamma_g)$ that extends the multiplication of modular forms for $g = 1$ and the one given here for $g = 2$ as Weissman shows. We refer to his interesting paper, [105].

For every g one obtains a subring of the representation ring by taking the powers of the determinant $\det : \mathrm{GL}(g, \mathbb{C}) \to \mathbb{C}^*$. This leads to a ring of 'classical' modular forms.

Definition 2. *A classical Siegel modular form of weight k (and degree g) is a holomorphic function $f \colon \mathcal{H}_g \to \mathbb{C}$ such that*

$$f(\gamma(\tau)) = \det(c\tau + d)^k f(\tau)$$

for all $\gamma = (a, b; c, d) \in Sp(2g, \mathbb{Z})$ (with for $g = 1$ the usual holomorphicity requirement at ∞).

Classical Siegel modular forms are also known as scalar-valued Siegel modular forms.

Let $M_k = M_k(\Gamma_g)$ be the vector space of classical Siegel modular forms of weight k. Together these spaces form a graded ring $M^{\mathrm{cl}} := \oplus M_k$ of M of classical Siegel modular forms. Of course, for $g = 1$ the notion of classical modular form reduces to the usual notion of modular form on $\mathrm{SL}(2, \mathbb{Z})$.

4 The Fourier Expansion of a Modular Form

The classical Fourier expansion of a modular form on $\mathrm{SL}(2, \mathbb{Z})$ has an analogue. To define it we need the following definition.

Definition 3. *A symmetric $g \times g$-matrix $n \in \mathrm{GL}(g, \mathbb{Q})$ is called half-integral if $2n$ is an integral matrix the diagonal entries of which are even.*

Every half-integral $g \times g$-matrix n defines a linear form with integral coefficients in the coordinates τ_{ij} with $1 \leq i \leq j \leq g$ of $\mathcal{H}_g$, namely

$$\mathrm{Tr}(n\tau) = \sum_{i=1}^{g} n_{ii}\tau_{ii} + 2 \sum_{1 \leq i < j \leq g} n_{ij}\tau_{ij}$$

and every linear integral combination of the coordinates is of this form.

Let us write $\tau = x + iy$ with x and y symmetric $g \times g$ matrices. A function $f : \mathcal{H}_g \to \mathbb{C}$ that is periodic in the sense that $f(\tau + s) = f(\tau)$ for all integral symmetric $g \times g$-matrices s admits a Fourier expansion

$$f(\tau) = \sum_{n \text{ half-integral}} a(n) e^{2\pi i \operatorname{Tr}(n\tau)}$$

with $a(n) \in \mathbb{C}$ given by

$$a(n) = \int_{x \bmod 1} f(\tau) e^{-2\pi i \operatorname{Tr}(n\tau)} dx$$

with dx the Euclidean volume of the space of x-coordinates and the integral runs over $-1/2 \le x_{ij} \le 1/2$. This is a series which is uniformly convergent on compact subsets. If f is a vector-valued modular form in M_ρ we have a similar *Fourier series*

$$f(\tau) = \sum_{n \text{ half-integral}} a(n) e^{2\pi i \operatorname{Tr}(n\tau)}$$

with $a(n) \in V$. One could also use the suggestive notation

$$f(\tau) = \sum_{n \text{ half-integral}} a(n) q^n,$$

where we write q^n for $e^{2\pi i \operatorname{Tr}(n\tau)}$. Moreover, we have the property

$$a(u^t n u) = \rho(u^t) a(n) \qquad \text{for all } u \in \operatorname{GL}(g, \mathbb{Z}). \tag{3}$$

Indeed, we have

$$\begin{aligned}
a(u^t n u) &= \int_{x \bmod 1} f(\tau) e^{-2\pi i \operatorname{Tr}(u^t n u \tau)} dx \\
&= \rho(u^t) \int_{x \bmod 1} f(u\tau u^t) e^{-2\pi i \operatorname{Tr}(n\, u\tau u^t)} dx \\
&= \rho(u^t) a(n).
\end{aligned}$$

A direct corollary of formula (3) (proof left to the reader) restricts the weight of non-zero forms.

Corollary 1. *A classical Siegel modular form of weight k with $kg \equiv 1 (\bmod 2)$ vanishes.*

A basic result is the following theorem.

Theorem 1. *Let $f \in M_\rho(\Gamma_g)$. Then f is bounded on any subset of $\mathcal{H}_g$ of the form $\{\tau \in \mathcal{H}_g : \operatorname{Im}(\tau) > c \cdot 1_g\}$ with $c > 0$.*

Proof. For $g = 1$ the boundedness comes from the requirement in the definition that the Fourier expansion $f = \sum_n a(n) q^n$ has no negative terms. So suppose that $g \ge 2$ and let $f = \sum_n a(n) e^{2\pi i \operatorname{Tr} n\tau} \in M_\rho(\Gamma_g)$. Since f

converges absolutely on $\mathcal{H}_g$ we see by substitution of $\tau = i \cdot 1_g$ that there exists a constant $c > 0$ such that for all half-integral matrices we have $|a(n)| \leq ce^{2\pi \mathrm{Tr} n\tau}$. We first will show that $a(n)$ vanishes for n that are not positive semi-definite.

Suppose that n is not positive semi-definite. Then there exists a primitive integral (column) vector ξ such that $\xi^t n \xi < 0$. We can complete ξ to a unimodular matrix u. Using the relation $a(u^t n u) = \rho(u^t) a(n)$ and replacing n by $u^t n u$ we may assume that entry n_{11} of n is negative. Consider now for $m \in \mathbb{Z}$ the matrix

$$v = \begin{pmatrix} 1 & m & \\ 0 & 1 & \\ & & 1_{g-2} \end{pmatrix} \in \mathrm{GL}(g, \mathbb{Z}),$$

where the omitted entries are zero. We have

$$|a(n)| = |\rho(v^t)^{-1}| \, |a(v^t n v)| \leq ce^{2\pi \mathrm{Tr} v^t n v}.$$

But $\mathrm{Tr}(v^t n v) = \mathrm{Tr}(v) + n_{11} m^2 + 2n_{12} m$ and if $m \to \infty$ then this expression goes to $-\infty$, so $|a(n)| = 0$.

We conclude that $f = \sum_{n \geq 0} a(n) e^{2\pi i \mathrm{Tr} n \tau}$. We can now majorize by the value at $ci \cdot 1_g$ of f, viz. $\sum_{n \geq 0} |a(n)| e^{-2\pi \mathrm{Tr} n c}$, uniformly in τ on $\{\tau \in \mathcal{H}_g : \mathrm{Im}(\tau) > c \cdot 1_g\}$.

The proof of this theorem shows the validity of the so-called Koecher principle announced above.

Theorem 2. *(Koecher Principle) Let* $f = \sum_n a(n) q^n \in M_\rho(\Gamma_g)$ *with* $q^n = e^{2\pi i \mathrm{Tr}(n\tau)}$ *be a modular form of weight* ρ. *Then* $a(n) = 0$ *if the half-integral matrix* n *is not positive semi-definite.*

The Koecher principle was first observed in 1928 by Götzky for Hilbert modular forms and in general by Koecher in 1954, see [63] and Bruinier's lectures.

Corollary 2. *A classical Siegel modular form of negative weight vanishes.*

Proof. Let $f \in M_k(\Gamma_g)$ with $k < 0$. Then the function $h = \det(y)^{k/2} |f(\tau)|$ is invariant under Γ_g since $\mathrm{Im}(\gamma(\tau)) = (c\tau + d)^{-t}(\mathrm{Im}(\tau))\overline{(c\tau + d)}^{-1}$. It is not difficult to see that a fundamental domain is contained in $\{\tau \in \mathcal{H}_g : \mathrm{Tr}(x^2) < 1/c, y > c \cdot 1_g\}$ for some suitable c. This implies that for negative k the expression $\det(y)^{k/2}$ is bounded on a fundamental domain, and by the Koecher principle f is bounded on $\{\tau \in \mathcal{H}_g : \det y \geq c\}$. It follows that h is bounded on $\mathcal{H}_g$, say $h \leq c'$ and with

$$a(n) e^{-2\pi \mathrm{Tr} n y} = \int_{x \bmod 1} f(\tau) e^{-2\pi \mathrm{Tr} n x} dx$$

we get

$$|a(n)| e^{-2\pi \mathrm{Tr} n y} \leq \sup_{x \bmod 1} |f(x+iy)| \leq c' \det y^{-k/2}.$$

If we let $y \to 0$ then for $k < 0$ we see $|a(n)| = 0$ for all $n \geq 0$.

This corollary admits a generalization for vector-valued Siegel modular forms, cf., [32]:

Proposition 1. *Let ρ be a non-trivial irreducible representation of $\mathrm{GL}(g, \mathbb{C})$ with highest weight $\lambda_1 \geq \ldots \geq \lambda_g$. If $M_\rho \neq \{0\}$ then we have $\lambda_g \geq 1$.*

One proves this by taking a totally real field K of degree g over $\mathbb{Q}$ and by identifying the symplectic space $O_K \oplus O_K^\vee$ (with $O_K^\vee$ the dual of O_K with respect to the trace) with our standard symplectic space $(\mathbb{Z}^{2g}, \langle, \rangle)$. This induces an embedding $\mathrm{SL}(2, O_K) \to \mathrm{Sp}(2g, \mathbb{Z})$ and a map $\mathrm{SL}(2, O_K) \backslash \mathcal{H}_1^g \to \mathrm{Sp}(2g, \mathbb{Z}) \backslash \mathcal{H}_g$. Pulling back Siegel modular forms yields Hilbert modular forms on $\mathrm{SL}(2, O_K)$. Now use that a Hilbert modular form of weight $(k_1, \ldots, k_g)$ vanishes if one of the $k_i \leq 0$, cf., [35]. By varying K one sees that if $\lambda_g \leq 0$ then a non-constant f vanishes on a dense subset of $\mathcal{H}_g$.

5 The Siegel Operator and Eisenstein Series

Since modular forms $f \in M_\rho(\Gamma_g)$ are bounded in the sets of the form $\{\tau \in \mathcal{H}_g : \mathrm{Im}(\tau) > c \cdot 1_g\}$ we can take the limit.

Definition 4. *We define an operator Φ on $M_\rho(\Gamma_g)$ by*

$$\Phi f = \lim_{t \to \infty} f \begin{pmatrix} \tau' & 0 \\ 0 & it \end{pmatrix} \qquad \text{with } \tau' \in \mathcal{H}_{g-1}, t \in \mathbb{R}.$$

In view of the convergence we can also apply this limit to all terms in the Fourier series and get

$$(\Phi f)(\tau') = \sum_{n' \geq 0} a \begin{pmatrix} n' & 0 \\ 0 & 0 \end{pmatrix} e^{2\pi i \mathrm{Tr}(n'\tau')}.$$

The values of Φf generate a subspace $V' \subseteq V$ that is invariant under the action of the subgroup of matrices $\{(a, 0; 0, 1) : a \in \mathrm{GL}(g-1, \mathbb{C})\}$ and that defines a representation ρ' of $\mathrm{GL}(g-1, \mathbb{C})$. The operator Φ defined on Siegel modular forms of degree g is called the *Siegel operator* and defines a linear map $M_\rho(\Gamma_g) \to M_{\rho'}(\Gamma_{g-1})$. If ρ is the irreducible representation with highest weight $(\lambda_1, \ldots, \lambda_g)$ then Φ maps $M_\rho(\Gamma_g)$ to $M_{\rho'}(\Gamma_{g-1})$ with ρ' the irreducible representation of $\mathrm{GL}(g-1, \mathbb{C})$ with highest weight $(\lambda_1, \ldots, \lambda_{g-1})$.

Definition 5. *A modular form $f \in M_\rho$ is called a cusp form if $\Phi f = 0$. The subspace of M_ρ of cusp forms is denoted by $S_\rho = S_\rho(\Gamma_g)$.*

Exercise 4. Show that a modular $f = \sum a(n) e^{2\pi i \mathrm{Tr}(n\tau)} \in M_\rho$ is a cusp form if and only if $a(n) = 0$ for all semi-definite n that are not definite.

We can apply the Siegel operator repeatedly (say $r \leq g$ times) to a Siegel modular form on Γ_g and one thus obtains a Siegel modular form on Γ_{g-r}. If ρ is irreducible with highest weight $(\lambda_1, \ldots, \lambda_g)$ and $\Phi F = f \neq 0$ for some $F \in M_\rho(\Gamma_g)$ then necessarily $\lambda_g \equiv 0 \pmod 2$ because with γ also $-\gamma$ lies in Γ_g.

Let now f_1 and f_2 be modular forms of weight ρ, one of them a cusp form. Then we define the *Petersson product* of f_1 and f_2 by

$$\langle f_1, f_2 \rangle = \int_F (\rho(\mathrm{Im}(\tau)) f_1(\tau), f_2(\tau)) d\tau ,$$

where $d\tau = \det(y)^{-(g+1)} \prod_{i \leq j} dx_{ij} dy_{ij}$ is an invariant measure on $\mathcal{H}_g$, F is a fundamental domain for the action of Γ_g on $\mathcal{H}_g$ and the brackets $(\,,\,)$ refer to the Hermitian product defined in Section 3. One checks that it converges exactly because at least one of the two forms is a cusp form. Furthermore, we define

$$N_\rho = S_\rho^\perp ,$$

for the orthogonal complement of S_ρ and then have an orthogonal decomposition $M_\rho = S_\rho \oplus N_\rho$.

Just as in the case $g = 1$ one can construct modular forms explicitly using Eisenstein series. We first deal with the case of classical Siegel modular forms. Let $g \geq 1$ be the degree and let r be a natural number with $0 \leq r \leq g$. Suppose that $f \in S_k(\Gamma_r)$ is a (classical Siegel modular) cusp form of even weight k. For a matrix $\begin{pmatrix} \tau_1 & z \\ z & \tau_2 \end{pmatrix}$ with $\tau_1 \in \mathcal{H}_r$ and $\tau_2 \in \mathcal{H}_{g-r}$ we write $\tau^* = \tau_1 \in \mathcal{H}_r$. (For $r = 0$ we let τ^* be the unique point of $\mathcal{H}_0$.) If k is positive and even we define the *Klingen Eisenstein series*, a formal series,

$$E_{g,r,k}(f) := \sum_{A=(a,b;c,d) \in P_r \backslash \Gamma_g} f((a\tau + b)(c\tau + d)^{-1})^*) \det(c\tau + d)^{-k} ,$$

where P_r is the subgroup

$$P_r := \left\{ \begin{pmatrix} a' & 0 & b' & * \\ * & u & * & * \\ c' & 0 & d' & * \\ 0 & 0 & 0 & u^{-t} \end{pmatrix} \in \Gamma_g : \begin{pmatrix} a' & b' \\ c' & d' \end{pmatrix} \in \Gamma_r, u \in \mathrm{GL}(g - r, \mathbb{Z}) \right\} .$$

For an interpretation of this subgroup we refer to Section 11. In case $r = 0$, f constant, say $f = 1$, we get the old Eisenstein series

$$E_{g,0,k} = \sum_{(a,b;c,d)} \det(c\tau + d)^{-k} ,$$

where the summation is over a full set of representatives for the cosets $\mathrm{GL}(g, \mathbb{Z}) \backslash \Gamma_g$.

Theorem 3. *Let $g \geq 1$ and $0 \leq r \leq g$ and $k > g + r + 1$ be integers with k even. Then for every cusp form $f \in S_k(\Gamma_r)$ the series $E_{g,r,k}(f)$ converges to a classical Siegel modular form of weight k in $M_k(\Gamma_g)$ and $\Phi^{g-r}E_{g,r,k}(f) = f$.*

This theorem was proved by Hel Braun[1] in 1938 for $r = 0$ and $k > g + 1$.

The Fourier coefficients of these Eisenstein series were determined by Maass, see [70]. Often we shall restrict the summation over co-prime (c, d) in order to avoid an unnecessary factor.

Corollary 3. *The Siegel operator $\Phi : M_k(\Gamma_g) \to M_k(\Gamma_{g-1})$ is surjective for even $k > 2g$.*

Weissauer improved the above result and proved that Φ^r is surjective if $k > (g + r + 3)/2$, see [107]. He also treated the case of vector-valued modular forms and showed that the image $\Phi(M_\rho(\Gamma_g))$ contains the space of cusp forms $S_{\rho'}(\Gamma_{g-1})$ if $k = \lambda_g \geq g + 2$, see loc. cit. p. 87.

If k is odd we have no good Eisenstein series; for example look at the Siegel operator $M_k(\Gamma_g) \to M_k(\Gamma_{g-1})$ for $k \equiv g \equiv 1 \,(\mathrm{mod}\, 2)$. Then $M_k(\Gamma_g) = (0)$ while the target space $M_k(\Gamma_{g-1})$ is non-zero for sufficiently large k (e.g. $M_{35}(\Gamma_2) \neq (0)$ as we shall see later).

Just as for $g = 1$ one can construct Poincaré series and use these to generate the spaces of cusp forms if the weight is sufficiently high. These Poincaré series behave well with respect to the Petersson product. We refer to [61], Ch. 6, or [8] for the general setting.

6 Singular Forms

A particularity of $g > 1$ are the so-called singular modular forms.

Definition 6. *A modular form $f = \sum_n a(n)e^{2\pi i \,\mathrm{Tr} n\tau} \in M_k(\Gamma_g)$ is called singular if $a(n) \neq 0$ implies that n is a singular matrix $(\det(n) = 0)$.*

Modular forms of small weight are singular as the following theorem shows, see [106].

Theorem 4. *(Freitag, Saldaña, Weissauer) Let ρ be irreducible with highest weight $(\lambda_1, \ldots, \lambda_g)$. A non-zero modular form $f \in M_\rho(\Gamma_g)$ is singular if and only if $2\lambda_g < g$.*

In particular, there are no cusp forms of weight $2\lambda_g < g$. One defines the *co-rank* of an irreducible representation as $\#\{1 \leq i \leq g \colon \lambda_i = \lambda_g\}$. For a modular form $f = \sum_n a(n)\exp(2\pi i \,\mathrm{Tr} n\tau) \in M_\rho(\Gamma_g)$, Weissauer introduced the *rank* and *co-rank* of f by

---

[1] Hel Braun was a student of Carl Ludwig Siegel (1896–1981), the mathematician after whom our modular forms are named. She sketches an interesting portrait of Siegel in [16]

$$\mathrm{rank}(f) = \max\{\mathrm{rank}(n)\colon a(n) \neq 0\}$$

and

$$\text{co-rank}(f) = g - \min\{\mathrm{rank}(n)\colon a(n) \neq 0\}.$$

In particular, modular forms of rank $<g$ are singular while cusp forms have co-rank 0 and Siegel–Eisenstein forms $E_{g,0,k}$ have co-rank g; Φ applied $k+1$ times to forms of co-rank k should be zero. Weissauer proved (see [106]) for irreducible ρ that co-rank$(f) \leq$ co-rank(ρ) and also that $M_\rho(\Gamma_g) = (0)$ if $\lambda_g \leq g/2 -$ co-rank(ρ). More precisely, he proved

Theorem 5. *Let $\rho = (\lambda_1, \ldots, \lambda_g)$ be an irreducible representation of co-rank $< g - \lambda_g$. If $\#\{i\colon 1 \leq i \leq g, \lambda_i = \lambda_g + 1\} < 2(g - \lambda_g -$ co-rank$(\rho))$ then $M_\rho = (0)$.*

Finally, Duke and Imamoğlu prove in [24] that there are no cusp forms of small weights; for example, $S_6(\Gamma_g) = (0)$ for all g.

7 Theta Series

Besides Eisenstein series one can construct Siegel modular forms using theta series. We begin with the so-called *theta-constants*. Let $\epsilon = \begin{pmatrix} \epsilon' \\ \epsilon'' \end{pmatrix}$ with $\epsilon', \epsilon'' \in \{0,1\}^g$ and consider the rapidly converging series

$$\theta[\epsilon] = \sum_{m \in \mathbb{Z}^g} \exp 2\pi i \left\{ \left(m + \frac{1}{2}\epsilon'\right)^t \tau \left(m + \frac{1}{2}\epsilon'\right) + \frac{1}{2}\left(m + \frac{1}{2}\epsilon'\right)^t (\epsilon'') \right\}.$$

This vanishes identically if ϵ is odd, that is, if $\epsilon'(\epsilon'')^t$ is odd. The other $2^{g-1}(2^g+1)$ cases (the 'even' ones) yield the so-called even theta characteristics. These are modular forms on a level 2 congruence subgroup of $\mathrm{Sp}(2g,\mathbb{Z})$ of weight $1/2$, cf. [56]. These can be used to construct classical Siegel modular forms on $\mathrm{Sp}(2g,\mathbb{Z})$. For example, for $g = 1$ one has

$$\left(\theta\genfrac[]{0pt}{}{0}{0}\theta\genfrac[]{0pt}{}{0}{1}\theta\genfrac[]{0pt}{}{1}{0}\right)^8 = 2^8 \Delta \in S_{12}(\Gamma_1).$$

For $g = 2$ the product $-2^{-14} \prod \theta[\epsilon]^2$ of the squares of the ten even theta characteristics gives a cusp form χ_{10} of weight 10 on $\mathrm{Sp}(4,\mathbb{Z})$, cf. [52–54]. Similarly, an expression

$$\left(\prod \theta[\epsilon]\right) \sum \pm (\theta[\epsilon_1]\theta[\epsilon_2]\theta[\epsilon_3])^{20},$$

where the product is over the even theta characteristics and the sum is over so-called azygous triples of theta characteristics (i.e., triples such that $\epsilon_1 + \epsilon_2 + \epsilon_3$ is odd) defines (up to a normalization $-2^{-39}5^{-3}i$) a cusp form χ_{35} of weight 35 on $\mathrm{Sp}(4,\mathbb{Z})$. Similarly, for $g = 3$ the product of the 36 even theta characteristics defines a cusp form of weight 18 on $\mathrm{Sp}(6, \mathbb{Z})$. The reason why one needs such a complicated expression is that the theta characteristics are modular forms on a subgroup $\Gamma_g(4, 8)$ of $\mathrm{Sp}(2g, \mathbb{Z})$ and the quotient group $\mathrm{Sp}(2g, \mathbb{Z})/\Gamma_g(4, 8)$ permutes them and creates signs in addition so that we need a sort of symmetrization to get something invariant.

Another source of Siegel modular forms are theta series associated to even unimodular lattices. Let B be a positive definite symmetric even unimodular matrix of size $r \equiv 0 \pmod 8$. We denote by $H_k(r, g)$ the space of harmonic polynomials $P : \mathbb{C}^{r \times g} \to \mathbb{C}$ satisfying for $M \in \mathrm{GL}(g, \mathbb{C})$ the identity $P(zM) = \det(M)^k P(z)$. Recall that harmonic means that $\sum_{i,j} \partial^2/\partial z_{ij}^2 \, P(z) = 0$ if z_{ij} are the coordinates on $\mathbb{C}^{r \times g}$. For a pair (B, P) with $P \in H_k(r, g)$ we set

$$\theta_{B,P}(\tau) = \sum_{A \in \mathbb{Z}^{r \times g}} P(\sqrt{B}A)e^{\pi i \mathrm{Tr}(A^t B A \tau)},$$

where $\sqrt{B}$ is a positive matrix with square B. Then $\theta_{B,P}$ is a classical Siegel modular form in $M_{k+r/2}(\Gamma_g)$, see [30]. Such theta series for $P \in H_{k-r/2,g}$ and B as above span a subspace of $M_k(\Gamma_g)$ that is invariant under the Hecke-operators that will be introduced later, cf. Section 16. There are analogues of these that give vector-valued Siegel modular forms if we require that P is a vector-valued polynomial satisfying the relation $P(zM) = \rho(M)P(z)$. See also Section 25 and [48, 49] for an example.

Finally, we would like to make a reference to Siegel's Hauptsatz [96] (or [30], p. 285) on representations of quadratic forms by quadratic forms which can be viewed as an identity between an Eisenstein series and a weighted sum of theta series, and to its far-reaching generalizations, cf. [68].

8 The Fourier–Jacobi Development of a Siegel Modular Form

As we saw above, just as for $g = 1$ we have a Fourier expansion of a Siegel modular form $f = \sum_{n \geq 0} a(n)e^{2\pi i \mathrm{Tr}(n\tau)}$. But for $g > 1$ there are other developments that provide more information, like the so-called Fourier–Jacobi development, a concept due to Piatetski–Shapiro.

We consider classical Siegel modular forms of weight k on Γ_g. We write $\tau \in \mathcal{H}_g$ as

$$\tau = \begin{pmatrix} \tau' & z \\ z^t & \tau'' \end{pmatrix} \text{ with } \tau' \in \mathcal{H}_1 \,, \; z \in \mathbb{C}^{g-1} \text{and } \tau'' \in \mathcal{H}_{g-1} \,. \tag{4}$$

From the definition of modular form it is clear that f is invariant under $\tau' \mapsto \tau' + b$ for $b \in \mathbb{Z}$ (given by an element of $\mathrm{Sp}(2g, \mathbb{Z})$), hence we have a Fourier series

$$f = \sum_{m=0}^{\infty} \phi_m(\tau'', z)e^{2\pi i m \tau'} \,.$$

Here the function ϕ_m is a holomorphic function on $\mathcal{H}_{g-1} \times \mathbb{C}^{g-1}$ satisfying certain transformation rules. More generally, if we split τ as in (4) but with $\tau' \in \mathcal{H}_r$, $z \in \mathbb{C}^{r(g-r)}$ and $\tau'' \in \mathcal{H}_{g-r}$ we find a development

$$\sum_m \phi_m(\tau'', z) e^{2\pi i \mathrm{Tr}(m\tau')},$$

where the sum is over positive semi-definite half-integral matrices $r \times r$ matrices m and the functions ϕ_m are holomorphic on $\mathcal{H}_r \times \mathbb{C}^{r(g-r)}$. For $r = g$ we get back the Fourier expansion and for $r = 1$ we get what is called the Fourier–Jacobi development.

For ease of explanation and to simplify matters we start with $g = 2$. Then the function $\phi_m(\tau', z)$ turns out to be a Jacobi form of weight k and index m, i.e., $\phi_m \in J_{k,m}$ which amounts to saying that it satisfies

1. $\phi_m((a\tau' + b)/(\tau' + d), z/(c\tau' + d)) = (c\tau' + d)^k e^{2\pi i m c z^2/(c\tau'+d)} \phi_m(\tau', z)$,
2. $\phi_m(\tau', z + \lambda\tau' + \mu) = e^{-2\pi i m(\lambda^2 \tau' + 2\lambda z)} \phi_m(\tau', z)$,
3. ϕ_m has a Fourier expansion of the form

$$\phi_m = \sum_{n=0}^{\infty} \sum_{r \in \mathbb{Z},\, r^2 \le 4mn} c(n, r) e^{2\pi(n\tau' + rz)}.$$

This gives a relation between Siegel modular forms for genus 2 and Jacobi forms (see [26]) that we shall exploit later. In the general case, if we split τ as

$$\tau = \begin{pmatrix} \tau' & z \\ z^t & \tau'' \end{pmatrix} \quad \text{with } \tau' \in \mathcal{H}_r, z \in \mathbb{C}^{r(g-r)} \text{ and } \tau'' \in \mathcal{H}_{g-r}$$

and a symmetric matrix n as $\begin{pmatrix} n' & \nu \\ \nu^t & n'' \end{pmatrix}$ and if we use the fact that $\mathrm{Tr}(n\tau) = \mathrm{Tr}(n'\tau') + 2\mathrm{Tr}(\nu z) + \mathrm{Tr}(n''\tau'')$ then we can decompose the Fourier series of $f \in M_\rho(\Gamma_g)$ as

$$\sum_{n'' \ge 0} \phi_{n''}(\tau', z) e^{2\pi i \mathrm{Tr}(n''\tau'')}$$

with V-valued holomorphic functions $\phi_{n''}(\tau', z)$ that satisfy the rules

1. For $\lambda, \mu \in \mathbb{Z}^g$ we have

$$\phi_{n''}(\tau', z + \tau'\lambda + \mu) = \rho\left(\begin{pmatrix} 1_r & -\lambda \\ 0 & 1_{g-r} \end{pmatrix}\right) e^{-2\pi i \mathrm{Tr}(2\lambda^t z + \lambda^t \tau'\lambda)} \phi_{n''}(\tau', z).$$

2. For $\gamma' = (a', b; c', d') \in \Gamma_{g-1}$ we have

$$\phi_{n''}(\gamma'(\tau'), (c'\tau' + d')^{-t}z) =$$
$$e^{2\pi i \mathrm{Tr}(n'' z^t (c'\tau'+d')^{-1} c'z)} \rho\left(\begin{pmatrix} c'\tau' + d' & c'z \\ 0 & 1_{g-r} \end{pmatrix}\right) \phi_{n''}(\tau', z).$$

3. $\phi_{n''}(\tau', z)$ is regular at infinity.

The last condition means that $\phi_{n''}(\tau', z)$ has a Fourier expansion $\phi_{n''}(\tau', z) = \sum c(m, r) \exp(2\pi i \mathrm{Tr}(m\tau' + 2r^t z))$ for which $c(m, r) \neq 0$ implies that $\begin{pmatrix} m & r \\ r^t & n'' \end{pmatrix}$ is positive semi-definite. A holomorphic V-valued function $\phi(\tau', z)$ satisfying 1), 2) and 3) is called a Jacobi form of weight (ρ', n''). The sceptical reader may frown upon this unattractive set of transformation formulas, but there is a natural geometric explanation for this transformation behavior that we shall see in Section 11.

9 The Ring of Classical Siegel Modular Forms for Genus Two

So far we have not met any striking examples of Siegel modular forms. To convince the reader that the subject is worthy of his attention we turn to the first non-trivial case: classical Siegel modular forms of genus 2.

For $g = 1$ we know the structure of the graded ring $M_*(\Gamma_1) = \oplus_k M_k(\Gamma_1)$. It is a polynomial ring generated by the Eisenstein series $e_4 = E_4^{(1)}$ and $e_6 = E_6^{(1)}$ and the ideal of cusp forms is generated by the famous cusp form $\Delta = (e_4^3 - e_6^2)/1728$ of weight 12.

In comparison to this our knowledge of the graded ring $\oplus_{\rho \in \mathrm{Irr}} M_\rho$ of Siegel modular forms for $g = 2$ is rather restricted and most of what we know concerns classical Siegel modular forms. A first basic result was the determination by Igusa [52] of the ring of classical Siegel modular forms for $g = 2$. We now know also the structure of the ring of classical Siegel modular forms for $g = 3$, a result of Tsuyumine, [102].

Recall that we have the Eisenstein Series $E_k^{(g)} \in M_k(\Gamma_g)$ for $k > g + 1$. In particular, for $g = 2$ we have $E_4 = E_4^{(2)} \in M_4(\Gamma_2)$ and $E_6 = E_6^{(2)} \in M_6(\Gamma_2)$. Let us normalize them here so that

$$E_k = \sum_{(c,d)} \det(c\tau + d)^{-k},$$

where the sum is over non-associated pairs of *co-prime* symmetric integral matrices (non-associated w.r.t. to the multiplication on the left by $\mathrm{GL}(g, \mathbb{Z})$). The Fourier expansion of these modular forms is known. If we write $\tau = \begin{pmatrix} \tau_1 & z \\ z & \tau_2 \end{pmatrix}$ then

$$E_k = \sum_N a(N) e^{2\pi i \mathrm{Tr}(N\tau)},$$

with constant term 1 and for non-zero $N = \begin{pmatrix} n & r/2 \\ r/2 & m \end{pmatrix}$ the coefficient $a(N)$ given as

$$a(N) = \sum_{d|(n,r,m)} d^{k-1} H\left(k - 1, \frac{4mn - r^2}{d^2}\right)$$

with $H(k-1, D)$ Cohen's function, i.e., $H(k-1, D) = L_{-D}(2-k)$, where $L_D(s) = L(s, (\frac{D}{\cdot}))$ is the Dirichlet L-series associated to D if D is 1 or a discriminant of a real quadratic field, cf., [26], p. 21. (This $H(k-1, D)$ is essentially a class number.) Explicitly we have with $q_j = e^{2\pi i \tau_j}$ and $\zeta = e^{2\pi i z}$ the developments (cf., [26])

$$E_4 = 1 + 240(q_1 + q_2) + 2160\left(q_1^2 + q_2^2\right)$$
$$(240\,\zeta^{-2} + 13440\,\zeta^{-1} + 30240 + 13440\,\zeta + 240\,\zeta^2)q_1 q_2 + \ldots$$

and

$$E_6 = 1 - 504(q_1 + q_2) - 16632\left(q_1^2 + q_2^2\right) +$$
$$+(-504\,\zeta^{-2} + 44352\,\zeta^{-1} + 166320 + 44352\,\zeta - 504\,\zeta^2)q_1 q_2 + \ldots .$$

Under Siegel's operator $\Phi \colon M_k(\Gamma_2) \to M_k(\Gamma_1)$ the Eisenstein series E_k on Γ_2 maps to the Eisenstein series e_k on Γ_1 for $k \geq 4$. In particular, the modular form $E_{10} - E_4 E_6$ maps to $e_{10} - e_4 e_6$, and this is zero since $\dim M_{10}(\Gamma_1) = 1$ and the e_k are normalized so that their Fourier expansions have constant term 1. We thus find a cusp form. Similarly, $E_{12} - E_6^2$ defines a cusp form of weight 12 on Γ_2. To see that these are not zero we restrict to the 'diagonal' locus as follows.

Consider the map $\delta \colon \mathcal{H}_1 \times \mathcal{H}_1 \to \mathcal{H}_2$ given by $(\tau_1, \tau_2) \mapsto \begin{pmatrix} \tau_1 & 0 \\ 0 & \tau_2 \end{pmatrix}$. There is a corresponding map $\mathrm{SL}(2, \mathbb{Z}) \times \mathrm{SL}(2, \mathbb{Z}) \to \mathrm{Sp}(4, \mathbb{Z})$ by sending $\begin{pmatrix} a & b \\ c & d \end{pmatrix}, \begin{pmatrix} a' & b' \\ c' & d' \end{pmatrix}$ to $(A, B; C, D)$ (difficult to avoid capital letters here) with $A = \begin{pmatrix} a & 0 \\ 0 & a' \end{pmatrix}$, etc. that induces δ (on $(\mathrm{SL}(2, \mathbb{R})/U(1))^2 \to \mathrm{Sp}(4, \mathbb{R})/U(2))$. If we use the coordinates

$$\tau = \begin{pmatrix} \tau_1 & z \\ z & \tau_2 \end{pmatrix} \in \mathcal{H}_2$$

then the image of the map δ is given by $z = 0$ and it is the fixed point locus of the involution on $\mathcal{H}_2$ given by $(\tau_1, z, \tau_2) \mapsto (\tau_1, -z, \tau_2)$ induced by the element $(A, B; C, D)$ with $A = (1, 0; 0, -1) = D$ and $B = C = 0$.

An element $F \in M_k(\Gamma_2)$ can be developed around this locus $z = 0$

$$F = f(\tau_1, \tau_2)z^n + O(z^{n+1}) \qquad \text{for some } n \in \mathbb{Z}_{\geq 0}. \tag{5}$$

It is now easy to check that

1. $f(\tau_1, \tau_2) \in M_{k+n}(\Gamma_1) \otimes M_{k+n}(\Gamma_1)$;
2. $f(\tau_2, \tau_1) = (-1)^k f(\tau_1, \tau_2)$;
3. $f(\tau_1, -z, \tau_2) = (-1)^k f(\tau_1, z, \tau_2)$.

the first by looking at the action of $\mathrm{SL}(2, \mathbb{Z}) \times \mathrm{SL}(2, \mathbb{Z})$ and the second by applying the involution $(A, B; C, D)$ with $A = D = (0, 1; 1, 0)$ and $B = C = 0$

which interchanges τ_1 and τ_2 and the last by using the involution $z \mapsto -z$. The idea of developing along the diagonal locus was first used by Witt, [108].

Developing $E_{10} - E_4 E_6$ along $z = 0$ and writing $q_j = e^{2\pi i \tau_j}$ one finds $c q_1 q_2 z^2 + O(z^3)$, with $c \neq 0$, so we normalize to get a cusp form $\chi_{10} = E_{10}^{(1)}(\tau_1) \otimes E_{10}^{(1)}(\tau_2) z^2 + O(z^3)$. Similarly, the form $E_{12}^{(2)} - (E_6^{(2)})^2$ gives after normalization a non-zero cusp form $\chi_{12} = \Delta(\tau_1) \otimes \Delta(\tau_2) z^2 + O(z^3)$.

As we saw above in Section 7 we also know the existence of a cusp form χ_{35} of odd weight 35.

We now describe the structure of the ring of classical Siegel modular forms for $g = 2$. The theorem is due to Igusa and various proofs have been recorded in the literature, cf. [5, 33, 43, 52–54]. Here is another variant.

Theorem 6. *(Igusa) The graded ring $M = \oplus_k M_k(\Gamma_2)$ of classical Siegel modular forms of genus 2 is generated by E_4, E_6, χ_{10}, χ_{12} and χ_{35} and*

$$M \cong \mathbb{C}[E_4, E_6, \chi_{10}, \chi_{12}, \chi_{35}]/(\chi_{35}^2 = R),$$

where R is an explicit (isobaric) polynomial in E_4, E_6, χ_{10} and χ_{12} (given on [53], p. 849).

Proof. (Isobaric means that every monomial has the same weight (here 70) if E_4, E_6, χ_{10} and χ_{12} are given weights 4, 6, 10 and 12.) We start by introducing the vector spaces of modular forms:

$$M_k^{\geq n}(\Gamma_1) = \{ f \in M_k(\Gamma_1) : f = O(q^n) \text{ at } \infty \} = \Delta^n M_{k-12n}(\Gamma_1)$$

and

$$M_k^{\geq n}(\Gamma_2) = \{ F \in M_k(\Gamma_2) : F = O(z^n) \text{ near } \delta(\mathcal{H}_1 \times \mathcal{H}_1) \}$$

We distinguish two cases depending on the parity of k.

k **even.** As we saw above (use properties (1), (2), (5)) any element $F \in M_k^{\geq 2n}(\Gamma_2)$ can be written as $F(\tau_1, z, \tau_2) = f(\tau_1, \tau_2) z^{2n} + O(z^{2n+2})$ with $f \in M_{k+2n}(\Gamma_1) \otimes M_{k+2n}(\Gamma_1)$ symmetric (i.e. $f(\tau_1, \tau_2) = f(\tau_2, \tau_1)$) and $f = O(q_1^n, q_2^n)$. This last fact follows from the observation that each Fourier–Jacobi coefficient $\phi_m(\tau_1, z)$ of F is also $O(z^{2n})$, so is zero if $2n > 2m$. We find an exact sequence

$$0 \to M_k^{\geq 2n+2}(\Gamma_2) \to M_k^{\geq 2n}(\Gamma_2) \xrightarrow{r} \mathrm{Sym}^2\left(M_{k+2n}^{\geq n}(\Gamma_1)\right) \to 0,$$

where the surjectivity of r is a consequence of the fact that

$$\mathrm{Sym}^2\left(M_{k+2n}^{\geq n}(\Gamma_1)\right) = \mathbb{C}[e_4 \otimes e_4, e_6 \otimes e_6, \Delta \otimes \Delta]$$

and $\chi_{10} = \Delta(\tau_1)\Delta(\tau_2) z^2 + O(z^4)$ so that a modular form $\chi_{10}^n P(E_4, E_6, \chi_{12})$ with P an isobaric polynomial maps to $P(e_4 \otimes e_4, e_6 \otimes e_6, \Delta \otimes \Delta)$. It follows that

$$\dim M_k(\Gamma_2) = \sum_{n=0}^{\infty} \dim \operatorname{Sym}^2(M_{k+2n}^{\geq n}(\Gamma_1)) = \sum_{0 \leq n \leq k/10} \dim \operatorname{Sym}^2(M_{k-10n}(\Gamma_1)),$$

i.e., we get

$$\sum_{k \text{ even}} \dim M_k(\Gamma_2) t^k = \frac{1}{1 - t^{10}} \sum_{k \geq 0} \dim \operatorname{Sym}^2(M_k(\Gamma_1)) t^k$$

$$= \frac{1}{1 - t^{10}} \text{ Hilbert series of } \mathbb{C}[e_4 \otimes e_4, e_6 \otimes e_6, \Delta \otimes \Delta]$$

$$= \frac{1}{(1 - t^4)(1 - t^6)(1 - t^{10})(1 - t^{12})}.$$

k **odd.** For $F \in M_k^{\geq 2n+1}(\Gamma_2)$ we find $f = O(q_1^{n+2}, q_2^{n+2})$. Since our Fourier–Jacobi coefficients $\phi_m(\tau_1, z)$ have a zero of order $2n + 1$ at $z = 0$ and another three at the 2-torsion points we see $2m \geq (2n + 1) + 3$ for non-zero ϕ_m. Also we know that f is anti-symmetric now, so $\dim M_k(\Gamma_2) \leq \sum_{n \geq 0} \dim \wedge^2(M_{k+2n+1}^{\geq n+2}(\Gamma_1))$ and this shows that for odd $k < 35$ $\dim M_k(\Gamma_2) = 0$. Since we have a non-trivial form of weight 35 we see that

$$\sum_{k \text{ odd}} \dim M_k(\Gamma_2) t^k = \frac{t^{35}}{(1 - t^4)(1 - t^6)(1 - t^{10})(1 - t^{12})}.$$

The square χ_{35}^2 is a modular form of even weight, hence can be expressed as an polynomial R in E_4, E_6, χ_{10} and χ_{12}. This was done by Igusa in [53]. This completes the proof.

10 Moduli of Principally Polarized Complex Abelian Varieties

For $g = 1$ the quotient space $\Gamma_1 \backslash \mathcal{H}_1$ has an interpretation as the moduli space of elliptic curves over the complex numbers (complex tori of dimension 1). To a point $\tau \in \mathcal{H}_1$ we associate the complex torus $\mathbb{C}/\mathbb{Z} + \mathbb{Z}\tau$. Then to a point $(a\tau + b)/(c\tau + d)$ in the Γ_1-orbit of τ we associate the torus $\mathbb{C}/\mathbb{Z} + \mathbb{Z}(a\tau + b)/(c\tau + d)$, and the homothety $z \mapsto (c\tau + d)z$ defines an isomorphism of this torus with $\mathbb{C}/\mathbb{Z}(c\tau + d) + \mathbb{Z}(a\tau + b) = \mathbb{C}/\mathbb{Z} + \mathbb{Z}\tau$ since $(c\tau + d, a\tau + b)$ is a basis of $\mathbb{Z} + \mathbb{Z}\tau$ as well. Conversely, every 1-dimensional complex torus can be represented as $\mathbb{C}/\mathbb{Z} + \mathbb{Z}\tau$. This can be generalized to $g > 1$ as follows. A point $\tau \in \mathcal{H}_g$ determines a complex torus $\mathbb{C}^g/\mathbb{Z}^g + \mathbb{Z}^g\tau$, but we do not get all complex g-dimensional tori. The following lemma, usually ascribed to Lefschetz, tells us what conditions this imposes.

Lemma 1. *The following conditions on a complex torus $X = V/\Lambda$ are equivalent:*

1. *X admits an embedding into a complex projective space;*
2. *X is the complex manifold associated to an algebraic variety;*
3. *There is a positive definite Hermitian form H on V such that $\mathrm{Im}(H)$ takes integral values on $\Lambda \times \Lambda$.*

A complex torus satisfying these requirements is called a complex *abelian variety*. For $g = 1$ we could take $H(z, w) = z\bar{w}/\mathrm{Im}(\tau)$ on $\Lambda = \mathbb{Z} + \mathbb{Z}\tau$ and indeed, the map $\mathbb{C}/\Lambda \to \mathbb{P}^2$ given by $z \mapsto (\wp(z) : \wp'(z) : 1)$ for $z \notin \Lambda$ with $\wp$ the Weierstrass $\wp$-function defines the embedding. For $g > 1$ we can take $H(z, w) = z^t(\mathrm{Im}(\tau))^{-1}\bar{w}$. An H as in the lemma is called a *polarization*. It is called a *principal polarization* if the map $\mathrm{Im}(H) : \Lambda \times \Lambda \to \mathbb{Z}$ is unimodular. We shall write $E = \mathrm{Im}(H)$ for the alternating form that is the imaginary part of H. Given a complex torus $X = V/\Lambda$ and a principal polarization on Λ we can normalize things as follows. We choose an isomorphism $V \cong \mathbb{C}^g$ and choose a symplectic basis $e_1, \ldots, e_{2g}$ of the lattice Λ such that E takes the standard form

$$J = \begin{pmatrix} 0_g & 1_g \\ -1_g & 0_g \end{pmatrix}$$

with respect to this basis. These two bases yield us a period matrix $\Omega \in \mathrm{Mat}(g \times 2g, \mathbb{C})$ expressing the e_i in terms of the chosen $\mathbb{C}$-basis of V. A natural question is which period matrices occur. For this we note that E is the imaginary part of a Hermitian form $H(x, y) = E(ix, y) + \sqrt{-1}E(x, y)$ if and only if E satisfies the condition $E(iz, iw) = E(z, w)$ for all $z, w \in V$ and this translates into (Exercise!)

$$\Omega\, J^{-1} \Omega^t = 0$$

while the positive definiteness of H translates into the condition

$$2i(\bar{\Omega}J^{-1}\Omega^t)^{-1}\text{is positive definite}\,.$$

These conditions were found by Riemann in his brilliant 1857 paper [81]. If we now associate to $\Omega = (\Omega_1\, \Omega_2)$ with Ω_i complex $g \times g$ matrices we see that the two conditions just found say that if we put $\tau = \Omega_2^{-1}\Omega_1$ we have $\tau = \tau^t$, $\mathrm{Im}(\tau) > 0$ i.e., τ lies in $\mathcal{H}_g$. A change of basis of Λ changes $(\tau\, 1_g)$ into $(\tau a + c, \tau b + d)$ with $(a, b; c, d) \in \mathrm{Sp}(2g, \mathbb{Z})$, but the corresponding torus is isomorphic to $\mathbb{C}^g/\mathbb{Z}^g(\tau b + d)^{-1}(\tau a + c) + \mathbb{Z}^g$. In this way we see that the isomorphism classes of complex tori with a principal polarization are in 1–1 correspondence with the points of the orbit space $\mathcal{H}_g/\mathrm{Sp}(2g, \mathbb{Z})$. If we transpose we can identify this orbit space with the orbit space $\mathrm{Sp}(2g, \mathbb{Z})\backslash\mathcal{H}_g$ for the usual action $\tau \mapsto (a\tau + b)(c\tau + d)^{-1}$.

Proposition 2. *There is a canonical bijection between the set of isomorphism classes of principally polarized abelian varieties of dimension g and the orbit space $\Gamma_g\backslash\mathcal{H}_g$.*

If we try to construct the whole family of abelian varieties we encounter a difficulty. The action of the semi-direct product $\Gamma_g \ltimes \mathbb{Z}^{2g}$ on $\mathcal{H}_g \times \mathbb{C}^g$ given by the usual action of Γ_g on $\mathcal{H}_g$ and the action of $(\lambda, \mu) \in \mathbb{Z}^{2g}$ on a fibre $\{\tau\} \times \mathbb{C}^{2g}$ by $z \mapsto z + \tau\lambda + \mu$ forces $-1_{2g} \in \Gamma_g$ to act by -1 on a fibre, so instead of finding the complex torus $\mathbb{C}^g/\mathbb{Z}^g + \tau\mathbb{Z}^g$ we get its quotient by the action $z \mapsto -z$. However, if we replace Γ_g by the congruence subgroup $\Gamma_g(n)$ with $n \geq 3$ (see [89]) then we get an honest family $\mathcal{X}_g(n) = \Gamma_g(n) \ltimes \mathbb{Z}^{2g} \backslash \mathcal{H}_g \times \mathbb{C}^g$ of abelian varieties over $\Gamma_g(n) \backslash \mathcal{H}_g$. If we insist on using Γ_g then we have to work with orbifolds or stacks to have a universal family available; the orbifold in question is the quotient of $\mathcal{X}_g(n)$ under the action of the finite group $\mathrm{Sp}(2g, \mathbb{Z}/n\mathbb{Z})$.

The cotangent bundle of the family of abelian varieties over $\mathcal{A}_g(n) = \Gamma_g(n) \backslash \mathcal{H}_g$ along the zero section defines a vector bundle of rank g on $\mathcal{A}_g(n)$. It can be constructed explicitly as a quotient $\Gamma_g(n) \backslash \mathcal{H}_g \times \mathbb{C}^g$ under the action of $\gamma \in \Gamma_g(n)$ by $(\tau, z) \mapsto (\gamma(\tau), (c\tau + d)^{-t}z)$. The bundle is called the *Hodge bundle* and denoted by $\mathbb{E} = \mathbb{E}_g$. The finite group $\mathrm{Sp}(2g, \mathbb{Z}/n\mathbb{Z})$ acts on the bundle $\mathbb{E}$ on $\mathcal{A}_g(n)$. A section of $\det(\mathbb{E})^{\otimes k}$ that is $\mathrm{Sp}(2g, \mathbb{Z}/n\mathbb{Z})$-invariant comes from a holomorphic function on $\mathcal{H}_g$ that is a classical Siegel modular form of weight k. Classical modular forms thus get a geometric interpretation. In particular, the determinant of the cotangent bundle of $\mathcal{A}_g(n)$, i.e., the canonical bundle, is isomorphic to $\det(\mathbb{E})^{\otimes g+1}$; so to a modular form f of weight $g + 1$ we can associate a top differential form on $\mathcal{H}_g$ that is Γ_g-invariant via $f \mapsto f(\tau) \prod_{i \leq j} d\tau_{ij}$. In a similar way one can construct for each ρ a vectorbundle over $\mathcal{A}_g(n)$ whose $\mathrm{Sp}(2g, \mathbb{Z}/n\mathbb{Z})$-invariant sections are the Siegel modular forms of weight ρ by taking the quotient $\mathcal{H}_g \times V$ by Γ_g under $(\tau, z) \mapsto (\gamma(\tau), \rho(c\tau + d)z)$, see Section 13.

The Hermitian form H on the lattice $\Lambda \subset \mathbb{C}^g$ can be viewed as the first Chern class (in $H^2(X, \mathbb{Z}) \cong \wedge^2(H_1(X, \mathbb{Z})^\vee) \cong (\wedge^2\Lambda)^\vee$) of a line bundle L on $X = \mathbb{C}^g/\Lambda$ with $\dim_{\mathbb{C}} H^0(X, L) = 1$. A non-zero section determines an effective divisor Θ on X. The line bundle L and the corresponding divisor Θ are determined by H up to translation over X. If we require that Θ be invariant under $z \mapsto -z$ then Θ is unique up to translation over a point of order 2 on X and then 2Θ is unique.

If we pull a non-zero section s of L back to the universal cover $\mathbb{C}^g$ then we obtain a holomorphic function with a certain transformation behavior under translations by elements of Λ. An example of such a function is provided by Riemann's theta function

$$\theta(\tau, z) = \sum_{n \in \mathbb{Z}^g} e^{\pi i (n^t \tau n + 2n^t z)}, \qquad (\tau \in \mathcal{H}_g, z \in \mathbb{C}^g)$$

a series that converges very rapidly and defines a holomorphic function that satisfies for all $\lambda, \mu \in \mathbb{Z}^g$

$$\theta(\tau, z + \tau\lambda + \mu) = e^{-\pi i (\lambda^t \tau \lambda + 2\lambda^t z)} \theta(\tau, z).$$

Conversely, a holomorphic function f on $\mathbb{C}^g$ that satisfies for all $\lambda, \mu \in \mathbb{Z}^g$

$$f(z + \tau\lambda + \mu) = e^{-\pi i({}^t\lambda \tau \lambda + 2{}^t\lambda^t z)} f(z)$$

is up to a multiplicative constant precisely $\theta(\tau, z)$ as one sees by developing f in a Fourier series $f = \sum_n c(n) \exp(2\pi i n^t z)$ and observing that addition of a column τ_k of τ to z produces

$$f(z + \tau_k) = \sum_n c(n) \exp(2\pi i n^t(z + \tau_k)) = \sum_n c(n) \exp(2\pi i n^t \tau_k) \exp(2\pi i n^t z)$$

from which one obtains $c(n + e_k) = c(n) \exp(2\pi i n^t \tau_k + \pi i \tau_{kk})$ and gets that f is completely determined by $c(0)$.

If S is a compact Riemann surface of genus g it determines a Jacobian variety $\mathrm{Jac}(S)$ which is a principally polarized complex abelian variety of dimension g. Sending S to $\mathrm{Jac}(S)$ provides us with a map $\mathcal{M}_g(\mathbb{C}) \to \Gamma_g \backslash \mathcal{H}_g$ from the moduli space of compact Riemann surfaces of genus g to the moduli of complex principally polarized abelian varieties of dimension g which is injective by a theorem of Torelli. The geometric interpretation given for Siegel modular forms thus pulls back to the moduli of compact Riemann surfaces.

11 Compactifications

It is well known that $\Gamma_1 \backslash \mathcal{H}_1$ is not compact, but can be compactified by adding the cusp, that is, the orbit of Γ_1 acting on $\mathbb{Q} \subset \bar{\mathcal{H}}_1$. Or if we use the equivalence of $\mathcal{H}_1$ with the unit disc D_1 given by $\tau \mapsto (\tau - i)/(\tau + i)$ then we add to D_1 the rational points of the boundary of the unit disc and take the orbit space of this enlarged space. We can do something similar for $g > 1$ by considering the bounded symmetric domain

$$D_g = \{z \in \mathrm{Mat}(g \times g, \mathbb{C}) : z^t = z, z^t \cdot \bar{z} < 1_g\}$$

which is analytically equivalent to $\mathcal{H}_g$. We now enlarge this space by adding not the whole boundary but only part of it as follows. Let

$$D_r = \left\{ \begin{pmatrix} z' & 0 \\ 0 & 1_{g-r} \end{pmatrix} : z' \in D_r \right\} \subset \bar{D}_g$$

and define now D_g^* to be the union of all Γ_g-orbits of these D_r for $0 \le r \le g$. Note that Γ_g acts on D_g and on its closure $\bar{D}_g$. Then Γ_g acts in a natural way on D_g^* and the orbit space decomposes naturally as a disjoint union

$$\Gamma_g \backslash D_g^* = \sqcup_{i=0}^g \Gamma_i \backslash D_i .$$

Going back to the upper half plane model this means that we consider

$$\Gamma_g \backslash \mathcal{H}_g^* = \sqcup_{i=0}^g \Gamma_i \backslash \mathcal{H}_i$$

Satake has shown how to make this space into a normal analytic space, the Satake compactification. One first defines a topology on $\mathcal{H}_g^*$ and then a sheaf of holomorphic functions. The quotient $\Gamma_g \backslash \mathcal{H}_g^*$ then becomes a normal analytic space. By using explicitly constructed modular forms one then shows that classical modular forms of a suitably high weight separate points and tangent vectors and thus define an embedding of $\Gamma_g \backslash \mathcal{H}_g^*$ into projective space. By Chow's lemma it is then a projective variety. The following theorem is a special case of a general theorem due to Baily and Borel, [8].

Theorem 7. *Scalar Siegel modular forms of an appropriately high weight define an embedding of $\Gamma_g \backslash \mathcal{H}_g$ into projective space and the image of $\Gamma_g \backslash \mathcal{H}_g$ (resp. $\Gamma_g \backslash \mathcal{H}_g^*$) is a quasi-projective (resp. a projective) variety.*

The resulting Satake or Baily–Borel compactification is for $g > 1$ very singular. As a first attempt at constructing a smooth compactification we reconsider the case $g = 1$. In $\mathcal{H}_{1,c} = \{\tau \in \mathcal{H}_1 : \mathrm{Im}(\tau) \geq c\}$ with $c > 1$ the action of Γ_1 reduces to the action of $\mathbb{Z}$ by translations $\tau \mapsto \tau + b$. So consider the map $\mathcal{H}_{1,c} \to \mathbb{C}^*$, $\tau \mapsto q = \exp 2\pi i \tau$. It is clear how to compactify $\mathcal{H}_{1,c}/\mathbb{Z}$: just add the origin $q = 0$ to the image in $\mathbb{C}^* \subset \mathbb{C}$. In other words, glue $\Gamma_1 \backslash \mathcal{H}_1$ with $\mathbb{Z} \backslash \mathcal{H}_{1,c}^*$ over $\mathbb{Z} \backslash \mathcal{H}_{1,c}$. To do something similar for $g > 1$ we consider the subset (for a suitable real symmetric $g \times g$-matrix $c \gg 0$ which is sufficiently positive definite)

$$\mathcal{H}_{g,c} = \left\{ \tau = \begin{pmatrix} \tau_1 & z \\ z^t & \tau_2 \end{pmatrix} \in \mathcal{H}_g : \mathrm{Im}(\tau_2) - \mathrm{Im}(z^t)\mathrm{Im}(\tau_1)^{-1}\mathrm{Im}(z) \geq c \right\}$$

The action of Γ_g in $\mathcal{H}_{g,c}$ reduces to the action of the subgroup P

$$\left\{ \begin{pmatrix} a & 0 & b & * \\ * & \pm 1 & * & * \\ c & 0 & d & * \\ 0 & 0 & 0 & \pm 1 \end{pmatrix} \in \Gamma_g, \begin{pmatrix} a & b \\ c & d \end{pmatrix} \in \Gamma_{g-1} \right\},$$

the normalizer of the 'boundary component' $\mathcal{H}_{g-1}$. We now make a map

$$\mathcal{H}_{g,c} \to \mathcal{H}_{g-1} \times \mathbb{C}^{g-1} \times \mathbb{C}^*, \quad \tau \mapsto (\tau_1, z, q_2 = \exp(2\pi i \tau_2)).$$

The associated parabolic subgroup P acts on $\mathcal{H}_{g-1} \times \mathbb{C}^{g-1} \times \mathbb{C}^*$ and this action can be extended to an action on $\mathcal{H}_{g-1} \times \mathbb{C}^{g-1} \times \mathbb{C}$, where

$$\begin{pmatrix} 1_{g-1} & 0 & 0 & 0 \\ 0 & 1 & 0 & b \\ 0 & 0 & 1_{g-1} & 0 \\ 0 & 0 & 0 & 1 \end{pmatrix}$$

acts now by $(\tau_1, z, q_2) \mapsto (\tau_1, z, e^{2\pi i b} q_2)$ while the matrix

$$\begin{pmatrix} 1_{g-1} & 0 & 0 & l \\ m & 1 & l & 0 \\ 0 & 0 & 1_{g-1} & -m \\ 0 & 0 & 0 & 1 \end{pmatrix}$$

acts by $(\tau_1, z, q_2) \mapsto (\tau_1, z + \tau_1 m + l, e^{2\pi i(m\tau_1 m + 2mz + lm)}q_2)$, and the diagonal matrix with entries $(1, \ldots, -1, 1, \ldots, 1, -1)$ acts by $(\tau_1, z, \zeta) \mapsto (\tau_1, -z, \zeta)$ and finally $(a, b; c, d) \in \Gamma_{g-1}$ acts on $\mathcal{H}_{g-1} \times \mathbb{C}^{g-1} \times \mathbb{C}$ by

$$(\tau_1, z, q_2) \mapsto (\gamma(\tau_1), (a - (\gamma(\tau_1)c) z, \tau_2 - z^t(c\tau_1 + d)^{-1}cz)$$

and this action can be extended similarly.

We now have an embedding $\Gamma_g \backslash \mathcal{H}_{g,c} \longrightarrow P \backslash \mathcal{H}_{g-1} \times \mathbb{C}^{g-1} \times \mathbb{C}$ and by taking the closure of the image we obtain a 'partial compactification'. The quotient of $\mathcal{H}_{g-1} \times \mathbb{C}^{g-1} \times \{0\}$ by this action is the 'dual universal abelian variety $\hat{\mathcal{X}}_{g-1} = \Gamma_{g-1} \ltimes \mathbb{Z}^{2g-2} \backslash \mathcal{H}_{g-1} \times \mathbb{C}^{g-1}$ over $\Gamma_{g-1} \backslash \mathcal{H}_{g-1}$ (in the orbifold sense). Note that a principally polarized abelian variety is isomorphic to its dual, so we can enlarge our orbifold $\Gamma_g \backslash \mathcal{H}_g$ by adding this orbifold quotient $\mathcal{X}_{g-1} = \Gamma_{g-1} \times \mathbb{Z}^{2g-2} \backslash \mathcal{H}_{g-1} \times \mathbb{C}^{2g-2}$. The result is a partial compactification

$$\mathcal{A}_g^{(1)} = \mathcal{A}_g \sqcup \mathcal{X}_{g-1}',$$

where the prime refers to the fact that we are dealing with orbifolds and have to divide by (at least) an extra involution since a semi-abelian variety generically has $\mathbb{Z}/2 \times \mathbb{Z}/2$ as its automorphism group, while a generic abelian variety has only $\mathbb{Z}/2$.

This space parametrizes principally polarized complex abelian varieties of dimension g or degenerations of such (so-called semi-abelian varieties of torus rank 1) that are extensions

$$1 \to \mathbb{G}_m \to \tilde{X} \to X \to 0$$

of a $g-1$-dimensional principally polarized complex abelian variety by a rank 1 torus $\mathbb{G}_m = \mathbb{C}^*$. Such extension classes are classified by the dual abelian variety $\hat{X} \cong X$ (associate to a line bundle on X the $\mathbb{G}_m$-bundle obtained by deleting the zero section) which explains why we find the universal abelian variety of dimension $g - 1$ in the 'boundary' of $\mathcal{A}_g$. (There is the subtlety whether one allows isomorphisms to be -1 on $\mathbb{G}_m$ or not.) This partial compactification is canonical. If we wish to construct a full smooth compactification one can use Mumford's theory of toroidal compactifications, but unfortunately there is (for $g \geq 4$) no unique such compactification. We refer e.g. to [77].

This partial compactification enables us to reinterpret the Fourier–Jacobi series of a Siegel modular form. In particular, the formulas in Section 8 tell us that the pull back of f to a fibre of $\mathcal{X}_{g-1} \to \mathcal{A}_{g-1}$ is an abelian function and that f restricted to the zero-section of $\mathcal{X}_{g-1} \to \mathcal{A}_{g-1}$ is a Siegel modular form of weight $k - 1$ on Γ_{g-1}.

We can be more precise. We work with a group $\Gamma_g(n)$ with $n \geq 3$ or interpret everything in the orbifold sense. The normal bundle of $\mathcal{X}_{g-1}$ is the line bundle $O(-2\Theta)$, as one can deduce from the action given above.

We can also extend the Hodge bundle $\mathbb{E} = \mathbb{E}_g$ to a vector bundle on $\mathcal{A}_g^{(1)}$. On the boundary divisor $\mathcal{X}'_{g-1}$ it is the extension of the pull back $\pi^* \mathbb{E}_{g-1}$ from $\mathcal{A}_{g-1}$ to $\mathcal{X}_{g-1}$ by a line bundle.

So if we are given a classical Siegel modular form of weight k we can interpret it as a section of $\det(\mathbb{E})^{\otimes k}$ and develop (the pull back of) f along the boundary $\mathcal{X}_{g-1}$ where the m-th term in the development is a section of

$$(\det(\mathbb{E})_{|\mathcal{X}_{g-1}})^{\otimes k} \otimes O(-2m\Theta)$$

on $\mathcal{X}_{g-1}$. This gives us a geometric interpretation of the Fourier–Jacobi development.

Of course, it is useful to have not only a partial compactification, but a smooth compactification. The theory of toroidal compactifications developed by Mumford and his co-workers Ash, Rapoport and Tai provides such compactifications $\tilde{\mathcal{A}}_g$. They depend on the choice of a certain cone decomposition of the cone of positive definite bilinear forms in g variables, cf. [7]. The 'boundary' $\tilde{\mathcal{A}}_g - \mathcal{A}_g$ is a divisor with normal crossings and one has a universal semi-abelian variety over $\tilde{\mathcal{A}}_g$ in the orbifold sense.

12 Intermezzo: Roots and Representations

Here we record a few concepts and notations that we shall need in the later sections. The reader may want to skip this on a first reading.

Recall that we started out in Section 2 with a symplectic lattice $(\mathbb{Z}^{2g}, \langle , \rangle)$ with a basis $e_1, \ldots, e_g, f_1, \ldots, f_g$ with $\langle e_i, f_j \rangle = \delta_{ij}$ and $\langle e_1, \ldots, e_g \rangle$ and $\langle f_1, \ldots, f_g \rangle$ isotropic subspaces. We let $G := \mathrm{GSp}(2g, \mathbb{Q})$ be the group of rational symplectic similitudes (transformations that preserve the form up to a scalar), viz.,

$$G := \mathrm{GSp}(2g, \mathbb{Q}) = \{\gamma \in \mathrm{GL}(\mathbb{Q}^{2g}) : \gamma^t J \gamma = \eta(\gamma) J\}$$

and $G^+ = \{\gamma \in G : \eta(\gamma) > 0\}$. Note that $\det(\gamma) = \eta(\gamma)^g$ for $\gamma \in G$ and that $G^0 = \mathrm{Sp}(2g, \mathbb{Z})$ is the kernel of the map that sends γ to $\eta(\gamma)$ on $G^+(\mathbb{Z})$. For $\gamma \in G$ the element $\eta(\gamma)$ is called the *multiplier*. Note that we view elements of $\mathbb{Z}^{2g}$ as column vectors and G acts from the left.

There are several important subgroups that play a role in the sequel. Given our choice of basis there is a natural Borel subgroup B respecting the symplectic flag $\langle e_1 \rangle \subset \langle e_1, e_2 \rangle \subset \ldots \subset \langle e_1, e_2 \rangle^\perp \subset \langle e_1 \rangle^\perp$. It consists of the matrices $(a, b; 0, d)$ with a upper triangular and d lower triangular.

Other natural subgroups are: the subgroup M of elements respecting the decomposition $\mathbb{Z}^g \oplus \mathbb{Z}^g$ of our symplectic space. It is isomorphic to $\mathrm{GL}(g) \times \mathbb{G}_m$

and consists of the matrices $\gamma = (a, 0; 0, d)$ with $ad^t = \eta(\gamma)1_g$. Furthermore, we have the Siegel (maximal) parabolic subgroup Q of elements that stabilize the first summand $\mathbb{Z}^g = \langle e_1, \ldots, e_g \rangle$; it consists of the matrices $(a, b; 0, d)$. It contains the subgroup U (unipotent radical) of matrices of the form $(1_g, b; 0, 1_g)$ with b symmetric that act as the identity of the first summand $\mathbb{Z}^g$.

Another important subgroup of G is the diagonal torus $\mathbb{T}$ isomorphic to $\mathbb{G}_m^{g+1}$ of matrices $\gamma = \mathrm{diag}(a_1, \ldots, a_g, d_1, \ldots, d_g)$ with $a_i d_i = \eta(\gamma)$. Let X be the character group of $\mathbb{T}$; it is generated by the characters $\epsilon_i : \gamma \mapsto a_i$ for $i = 1, \ldots, g$ and $\epsilon_0(\gamma) = \eta(\gamma)$. Let Y be the co-character group of $\mathbb{T}_m$, i.e., $Y = \mathrm{Hom}(\mathbb{G}_m, \mathbb{T})$. This group is isomorphic to the group $\mathbb{Z}^{g+1}$ of $g + 1$-tuples with $(\alpha_1, \ldots, \alpha_g, c)$ corresponding to the co-character $t \mapsto \mathrm{diag}(t^{\alpha_1}, \ldots, t^{\alpha_g}, t^{c-\alpha_1}, \ldots, t^{c-\alpha_g})$. We fix a basis of Y by letting χ_i for $i = 1, \ldots, g$ correspond to $\alpha_j = \delta_{ij}$ and $c = 0$ and χ_0 to $\alpha_j = 0$ and $c = 1$. Then the characters and co-characters pair via $\langle \epsilon_i, \chi_j \rangle = \delta_{ij}$.

The adjoint action of $\mathbb{T}$ on the Lie algebras of M and G defines root systems Φ_M and Φ_G in X. Concretely, we may take as simple roots $\alpha_i = \epsilon_i - \epsilon_{i+1}$ for $i = 1, \ldots, g-1$ and $\alpha_g = 2\epsilon_g - \epsilon_0$ and coroots $\alpha_i^\vee = \chi_i - \chi_{i+1}$ for $i = 1, \ldots, g-1$ and $\alpha_g^\vee = \chi_g$.

The set Φ_G^+ of positive roots (those occuring in the Lie algebra of the nilpotent radical of B) consists of the so-called compact roots $\Phi_M^+ = \{\epsilon_i - \epsilon_j : 1 \le i < j \le g\}$ and the non-compact roots $\Phi_{nc}^+ = \{\epsilon_i + \epsilon_j - \epsilon_0 : 1 \le i, j \le g\}$. We let $2\varrho = 2\varrho_G$ (resp. $2\varrho_M$) be the sum of the positive roots in Φ_G^+ (res. Φ_M^+). When viewed as characters $2\varrho_M$ corresponds to $\gamma \mapsto \prod_{i=1}^g a_i^{g+1-2i}$ and $2\varrho_G$ to $\gamma \mapsto \eta(\gamma)^{-g(g+1)/2} \prod_{i=1}^g a_i^{2g+2-2i}$.

There is a symmetry group acting on our situation, the Weyl group $W_G = N(\mathbb{T})/\mathbb{T}$, with $N(\mathbb{T})$ the normalizer of $\mathbb{T}$ in G. This group W_G is isomorphic to $S_g \ltimes (\mathbb{Z}/2\mathbb{Z})^g$, where the generator of the i-th factor $\mathbb{Z}/2\mathbb{Z}$ acts on a matrix of the form $\mathrm{diag}(a_1, \ldots, a_g, d_1, \ldots, d_g)$ by interchanging a_i and d_i and the symmetric group S_g acts by permuting the a's and d's. The Weyl group of M (normalizer this time in M) is isomorphic to the symmetric group S_g. We have positive Weyl chambers $P_G^+ = \{\chi \in Y : \langle \chi, \alpha \rangle \ge 0 \text{ for all } \alpha \in \Phi_G^+\}$ and similarly for M: $P_M^+ = \{\chi \in Y : \langle \chi, \alpha \rangle \ge 0 \text{ for all } \alpha \in \Phi_M^+\}$ giving the dominant weights.

Lemma 2. *The irreducible complex representations of G (resp. M) correspond to integral weights in the chamber P_G^+ (res. P_M^+) that come from characters of $\mathbb{T}$.*

Sometimes we just work with G^0 and $M^0 = M \cap G^0$. This means that we forget about the action of the multiplier η.

We can give a set W_0 of 2^g canonical coset representatives of $W_M \backslash W_G$, the Kostant representatives, which are characterized by the conditions

$$W_0 = \{w \in W_G : \Phi_M^+ \subset w(\Phi_G^+)\} = \{w \in W_G : w(\varrho) - \varrho \in P_M^+\}.$$

With our normalizations we have $\varrho = (g, g-1, \ldots, 1, 0)$ and $2\varrho_M = (g+1, \ldots, g+1, -g(g+1)/2)$. If we restrict to G^0 and M^0 then dominant weights for $M^0 \cong \mathrm{GL}(g)$ are given by g-tuples $(\lambda_1, \ldots, \lambda_g)$ with $\lambda_i \geq \lambda_{i+1}$ for $i = 1, \ldots, g - 1$. A coset in $W_M \backslash W_G$ is given by a vector s (in $\{\pm 1\}^g$) of g signs. The Kostant representative of s is the element σs such that $(s_{\sigma(1)} \lambda_{\sigma(1)}, \ldots, s_{\sigma(g)} \lambda_{\sigma(g)})$ is in P_M^+, i.e., $s_{\sigma(i)} \lambda_{\sigma(i)} \geq s_{\sigma(i+1)} \lambda_{\sigma(i+1)}$ for $i = 1, \ldots, g - 1$ for all $(\lambda_1 \geq \ldots \geq \lambda_g)$.

13 Vector Bundles Defined by Representations

Let $\pi \colon \mathcal{X}_g \to \mathcal{A}_g$ be the universal family of abelian varieties over $\mathcal{A}_g$. The Hodge bundle $\mathbb{E} = \pi_* \Omega_{\mathcal{X}_g / \mathcal{A}_g}$, a holomorphic bundle of rank g, and the de Rham bundle $R^1 \pi_* \mathbb{C}$ on $\mathcal{A}_g$, a locally constant sheaf of rank $2g$, are examples of vector bundles associated to representations of $\mathrm{GL}(g)$ and $\mathrm{GSp}(2g)$. Their fibres at a point $[X] \in \mathcal{A}_g$ are $H^0(X, \Omega_X^1)$ and $H^1(X, \mathbb{C})$. The first is a holomorphic vector bundle, the second a local system. Both are important for understanding Siegel modular forms.

To define these bundles recall the description of $\mathcal{H}_g$ as an open part Y_g^+ of the symplectic Grassmann variety Y_g given in Section 2. We can identify Y_g with $G(\mathbb{C})/Q(\mathbb{C})$ with Q the subgroup fixing the totally isotropic first summand $\mathbb{C}^g$ of our complexified symplectic lattice $(\mathbb{Z}^g, \langle \, , \, \rangle) \otimes \mathbb{C}$. If $\rho \colon Q^0 \to \mathrm{End}(V)$ is a complex representation (with $Q^0 = Q \cap G^0$) then we can define a $G^0(\mathbb{C})$-equivariant vector bundle $\mathcal{V}_\rho$ on Y_g by $\mathcal{V}_\rho = G^0(\mathbb{C}) \times^{Q^0(\mathbb{C})} V$ as the quotient of $G^0(\mathbb{C}) \times V$ under the equivalence relation $(g, v) \sim (g\, q, \rho(q)^{-1} v)$ for all $g \in G^0(\mathbb{C})$ and $q \in Q^0(\mathbb{C})$. Then Γ_g (or any finite index subgroup Γ') acts on $\mathcal{V}_\rho$ and the quotient is a vector bundle V_ρ on $\mathcal{A}_g$ in the orbifold sense (or a true one if Γ' acts freely on $\mathcal{H}_g$).

Recall that M is the subgroup of $\mathrm{GSp}(2g, \mathbb{Q})$ respecting the decomposition $\mathbb{Q}^g \oplus \mathbb{Q}^g$ of our symplectic space and $M^0 = M \cap \mathrm{Sp}(2g, \mathbb{Q})$. If we are given a complex representation of $M^0(\mathbb{C}) \cong \mathrm{GL}(g)$ (or of $M \cong \mathrm{GL}(g) \times \mathbb{G}_m$) we can obtain a vector bundle by extending the representation to a representation on $Q^0(\mathbb{C})$ by letting it be trivial on the unipotent radical U of Q. (Note that $Q = M \cdot U$.) If we do this with the tautological representation of M^0 we get the Hodge bundle $\mathbb{E}$. But there is a subtle point here. If we work with M instead of M^0 then the Hodge bundle is given by the representation of M that acts by $\eta(\gamma)^{-1} a$ on $\mathbb{C}^g$ for $\gamma = (a, 0; 0, d)$.

In any case we thus get a holomorphic vector bundle $\mathcal{W}(\lambda)$ associated to each dominant weight $(\lambda_1 \geq \ldots \geq \lambda_g)$ of $\mathrm{GL}(g)$. Another way of getting these vector bundles thus associated to the irreducible representations of M^0 (or M) is by starting from the Hodge bundle and applying Schur operators (idempotents) to the symmetric powers of $\mathbb{E}$ analogously to the way one gets the corresponding representations from the standard one. Since the Hodge bundle $\mathbb{E}$ extends over a toroidal compactification $\tilde{\mathcal{A}}_g$ this makes it clear that these vector bundles $\mathcal{W}(\lambda)$ can be extended over any toroidal compactification as constructed by Mumford (or Faltings–Chai). The space of sections can be identified with a space of modular forms M_ρ and it thus follows from general theorems in algebraic geometry that these spaces of Siegel modular forms M_ρ are finite dimensional.

Another important vector bundle is the bundle associated to the first co-homology of the universal abelian variety $\mathcal{X}_g$ with fibre $H^1(X, \mathbb{C})$; more pre-cisely, it is given by $\mathbb{V} := R^1\pi_*\mathbb{C}$ with $\pi: \mathcal{X}_g \to \mathcal{A}_g$ the universal abelian variety. It can be gotten from the construction just given by taking the dual or contragredient of the standard or tautological representation of $\mathrm{Sp}(2g, \mathbb{C})$ and restricting it to $Q^0(\mathbb{C})$. (Again, if one takes the multiplier into account – as one should – then $R^1\pi_*\mathbb{C}$ corresponds to η^{-1} times the standard representa-tion.) In this case we find a flat bundle: all the bundles $\mathcal{V}_\rho$ on Y_g come with a trivialization given by $[(g, v)] \mapsto \rho(g)v$. So the quotient bundle carries a nat-ural integrable connection. The resulting $\mathbb{V}$ is a local system (locally constant sheaf). We thus find for each dominant weight $\lambda = (\lambda_1 \geq \ldots \geq \lambda_g, c)$ of G a local system $\mathbb{V}_\lambda(c)$ on $\mathcal{A}_g$. The multiplier representation defines a local sys-tem of rank 1 denoted by $\mathbb{C}(1)$ and we can twist $\mathbb{V}_\lambda(c)$ by the nth power of $\mathbb{C}(1)$ to change c, cf. Section 12.

14 Holomorphic Differential Forms

Let $\Gamma' \subset \Gamma_g$ be a subgroup of finite index which acts freely on $\mathcal{H}_g$, e.g., $\Gamma' = \ker\{\mathrm{Sp}(2g, \mathbb{Z}) \to \mathrm{Sp}(2g, \mathbb{Z}/n\mathbb{Z}\}$ for $n \geq 3$. Let Ω^i be the sheaf of holomorphic i-forms on $\mathcal{H}_g$. A section of Ω^1 can be written as

$$\omega = \mathrm{Tr}(f(\tau)d\tau),$$

where $d\tau = (d\tau_{ij})$ and f is a symmetric matrix of holomorphic functions on $\mathcal{H}_g$. Then ω is invariant under the action of Γ' if and only if $f(\gamma(\tau)) = (c\tau + d)f(\tau)(c\tau + d)^t$ for all $\gamma = (a, b; c, d) \in \Gamma'$. Note that if r is the standard representation of $\mathrm{GL}(g, \mathbb{C})$ on $V = \mathbb{C}^g$ then the action on symmetric bilinear forms $\mathrm{Sym}^2(V)$ is given by $b \mapsto r(g)\, b\, r(g)^t$. So the space of holomorphic 1-forms on $\Gamma'\backslash\mathcal{H}_g$ can be identified with $M_\rho(\Gamma')$, with ρ the second symmetric power of the standard representation and the space of holomorphic i-forms with $M_{\rho'}(\Gamma')$ with ρ' equal to the ith exterior power of $\mathrm{Sym}^2 V$. So we find an isomorphism $\Omega^1_{\Gamma'\backslash\mathcal{H}_g} \cong \mathrm{Sym}^2\mathbb{E}$ and this can be extended over a toroidal compactification $\tilde{\mathcal{A}}$ to an isomorphism

$$\Omega^1_{\tilde{\mathcal{A}}}(\log D) \cong \mathrm{Sym}^2(\mathbb{E})$$

with D the divisor at infinity. (But again, one should be aware of the action of the multiplier: if one looks at the action of $\mathrm{GSp}(2g, \mathbb{R})^+$ one has $d((a\tau + b)(c\tau + d)^{-1}) = \eta(\gamma)(c\tau + d)^{-1}\, d\tau\, (c\tau + d)^{-1}.$)

The question arises which representations occur in $\wedge^i\mathrm{Sym}^2(V)$.

The answer is given in terms of roots. A theorem of Kostant [66] tells that the irreducible representations ρ of $\mathrm{GL}(g, \mathbb{C})$ that occur in the exterior algebra $\wedge^*\mathrm{Sym}^2(V)$ with V the standard representation of $\mathrm{GL}(g,\mathbb{C})$ are those ρ for which the dual $\hat{\rho}$ is of the form $w\delta - \delta$ with $\delta = (g, g-1, \ldots, 1)$ the half-sum of the positive roots and w in the set W_0 of Kostant representatives. Now if $\hat{\rho} = (\lambda_1 \geq \lambda_2 \ldots \geq \lambda_g)$ occurs in this exterior algebra then $w\delta$ is of the form $(g - \lambda_g, g - 1 - \lambda_{g-1}, \ldots, 1 - \lambda_1)$. If α is the largest integer that occurs among the entries of $w\delta$ then either $\alpha = -1$ or $1 \leq \alpha \leq g$. In the latter case $w\delta$ is of the form $(\alpha, *, \ldots, *, -\alpha - 1, -\alpha - 2, \ldots, -g)$ and it follows that $\lambda_{g-\alpha} = g + 1$. This implies that the number of λ_j with $\lambda_j = \lambda_g$ (the co-rank of $\hat{\rho}$, cf., Section 5) plus the number of those with $\lambda_j = \lambda_g + 1$ is at most α. The vanishing theorem of Weissauer (Thm. 5) now implies that non-zero differentials can only come from representations that are of the form

$$\rho = (g + 1, g + 1, \ldots, g + 1)$$

which corresponds to top differentials $(\wedge^{g(g+1)/2}\Omega^1)$ and classical Siegel modular forms of weight $g + 1$, or of the form

$$\rho = (g + 1, g + 1, \ldots, g + 1, g - \alpha, \ldots, g - \alpha),$$

with $1 \leq \alpha \leq g$ and these occur in $\wedge^p\Omega^1$ with $p = g(g+1)/2 - \alpha(\alpha+1)/2$. For the following theorem of Weissauer we refer to [106].

Theorem 8. *Let $\tilde{\mathcal{A}}_g$ be a smooth compactification of $\mathcal{A}_g$. If p is an integer $0 \leq p < g(g+1)/2$ then the space of holomorphic p-forms on $\tilde{\mathcal{A}}_g$ is zero unless p is of the form $g(g+1)/2 - \alpha(\alpha+1)/2$ with $1 \leq \alpha \leq g$ and then $H^0(\tilde{\mathcal{A}}_g, \Omega^p_{\tilde{\mathcal{A}}_g}) \cong M_\rho(\Gamma_g)$ with $\rho = (g+1, \ldots, g+1, g-\alpha, \ldots, g-\alpha)$ with $g - \alpha$ occuring α times.*

If f is a classical Siegel modular form of weight $k = g + 1$ on the group Γ_g then $f(\tau) \prod_{i \leq j} d\tau_{ij}$ is a top differential on the smooth part of quotient space $\Gamma_g \backslash \mathcal{H}_g = \mathcal{A}_g$. It can be extended over the smooth part of the rank-1 compactification $\mathcal{A}_g^{(1)}$ if and only if f is a cusp form. It is not difficult to see that this form can be extended as a holomorphic form to the whole smooth compactification $\tilde{\mathcal{A}}_g$.

Proposition 3. *The map that associates to a classical cusp form $f \in S_{g+1}(\Gamma_g)$ of weight $g+1$ the top differential $\omega = f(\tau) \prod_{i \leq j} d\tau_{ij}$ gives an isomorphism between $S_{g+1}(\Gamma_g)$ and the space of holomorphic top differentials $H^0(\tilde{\mathcal{A}}_g, \Omega^{\frac{g(g+1)}{2}})$ on any smooth compactification $\tilde{\mathcal{A}}_g$.*

For this and an analysis of when the other forms extend over the singularities in these cases we refer to [30, 106].

Finally we refer to two papers of Salvati–Manni where he proves the existence of differential forms of some weights, [83, 84] and a paper of Igusa, [57], where Igusa discusses the question whether certain Nullwerte of jacobians of odd thetafunctions can be expressed as polynomials or rational functions in theta Nullwerte.

15 Cusp Forms and Geometry

The very first cusp forms that one encounters often have a beautiful geometric interpretation. We give some examples.

For $g = 1$ the first cusp form is $\Delta = \sum \tau(n) q^n \in S_{12}(\Gamma_1)$. It is up to a normalization the discriminant $g_2(\tau)^3 - 27 g_3^2$ of the equation $y^2 = 4x^3 - g_2 x - g_3$ for the Riemann surface $\mathbb{C}/\mathbb{Z}\tau + \mathbb{Z}$ and does not vanish on $\mathcal{H}_1$. Here $g_2 = (4\pi^4/3) E_4(\tau)$ and $g_3 = (8\pi^6/27) E_6(\tau)$ are the suitably normalized Eisenstein series.

For $g = 2$ there is a similar cusp form χ_{10} of weight 10 with development

$$\chi_{10} \begin{pmatrix} \tau_1 & z \\ z & \tau_2 \end{pmatrix} = (\exp(2\pi i \tau_1) \exp(2\pi i \tau_2) + \ldots)(\pi z)^2 + \ldots$$

which vanishes (with multiplicity 2) along the 'diagonal' $z = 0$. So its zero divisor in $\mathcal{A}_2$ is the divisor of abelian surfaces that are products of elliptic curves with multiplicity 2. There is the Torelli map $\mathcal{M}_2 \to \mathcal{A}_2$ that associates to a hyperelliptic complex curve of genus 2 given by $y^2 = f(x)$ its Jacobian. Then the pull back of χ_{10} to $\mathcal{M}_2$ is related to the discriminant of f, cf. Igusa's paper [53] or [60], Prop. 2.2.

For $g = 3$ the ring of classical modular forms is generated by 34 elements, cf. [102]. As we saw above, there is a cusp form of weight 18, namely the product of the 36 even theta constants $\theta[\epsilon]$ and its zero divisor is the closure of the hyperelliptic locus. This expresses the fact that a genus 3 Riemann surface with a vanishing theta characteristic is hyperelliptic.

For $g = 4$ there is the following beautiful example. There is up to isometry only one isomorphism class of even unimodular positive definite quadratic forms in 8 variables, namely E_8. In 16 variables there are exactly two such classes, $E_8 \oplus E_8$ and E_{16}. To each of these quadratic forms in 16 variables we can associate a Siegel modular form on Γ_4 by means of a theta series: $\theta_{E_8 \oplus E_8}$ and $\theta_{E_{16}}$. The difference $\theta_{E_8 \oplus E_8} - \theta_{E_{16}}$ is a cusp form of weight 8. Its zero divisor is the closure of the locus of Jacobians of Riemann surfaces of genus 4 in $\mathcal{A}_4$ as shown by Igusa, cf., [55]. Here also refer to [80] for a proof. We shall encounter this form again in Section 21.

Similarly, the theta series associated to the 24 different Niemeier lattices (even, positive definite) of rank 24 produce in genus 12 a linear subspace of $M_{12}(\Gamma_{12})$ of dimension 12. It intersects the space of cusp forms in a 1-dimensional subspace, as was proved in [14]. We thus find a cusp form of weight 12. As we shall see later, it is an Ikeda lift of the cusp form Δ for $g = 1$ (proven in [15]).

Question 1. What is the geometric meaning of this cusp form?

The paper [78] contains explicit results on Siegel modular forms of weight 12 obtained from lattices in dimension 24. For example, it gives a non-zero cusp form of weight 12 on Γ_{11}, hence one has a top differential on $\tilde{\mathcal{A}}_{11}$, cf., Prop. 3, implying that this modular variety is not rational or unirational. It is a well-known result of Mumford ([75]) that $\tilde{\mathcal{A}}_g$ is of general type for $g > 6$.

16 The Classical Hecke Algebra

In the arithmetic theory of elliptic modular forms Hecke operators play a pivotal role. They enable one to extract arithmetic information from the Fourier coefficients of a modular form: if $f = \sum_n a(n)q^n$ is a common eigenform of the Hecke operators which is normalized $(a(1) = 1)$ then the eigenvalue $\lambda(p)$ of f under the Hecke operator $T(p)$ equals the Fourier coefficient $a(p)$.

The classical theory of Hecke operators as for example exposed in Shimura's book ([92]) can be generalized to the setting of $g > 1$ as Shimura showed in [93], though the larger size of the matrices involved is a discouraging aspect of it. It is worked out in the books [1, 4, 30], of which the last, by Freitag, is certainly the most accessible. In this section we sketch this approach, in the next section we give another approach. We refer to loc. cit. for details.

Recall the group $G := \mathrm{GSp}(2g, \mathbb{Q}) = \{\gamma \in \mathrm{GL}(\mathbb{Q}^{2g}) : \gamma^t J \gamma = \eta(\gamma)J, \eta(\gamma) \in \mathbb{Q}^*\}$ of symplectic similitudes of the symplectic vector space $(\mathbb{Q}^{2g}, \langle, \rangle)$ and $G^+ = \{\gamma \in G : \eta(\gamma) > 0\}$.

We start by defining the abstract *Hecke algebra* $H(\Gamma, G)$ for the pair (Γ, G) with $\Gamma = \Gamma_g$ and $G = \mathrm{GSp}(2g, \mathbb{Q})$. Its elements are finite formal sums (with $\mathbb{Q}$-coefficients) of double cosets $\Gamma\gamma\Gamma$ with $\gamma \in G^+$. Each such double coset $\Gamma\gamma\Gamma$ can be written as a finite disjoint union of right cosets $L_i = \Gamma\gamma_i$ by virtue of the following lemma.

Lemma 3. *Let m be a natural number. The set $O_g(m) = \{\gamma \in \mathrm{Mat}(2g \times 2g, \mathbb{Z}) : \gamma^t J \gamma = mJ\}$ can be written as a finite disjoint union of right cosets. Every right coset has a representative of the form $(a, b; 0, d)$ with $a^t d = m \, 1_g$ and such that a has zeros below the diagonal*

So to each double coset $\Gamma\gamma\Gamma$ we can associate a finite formal sum of right cosets. Let $\mathcal{L}$ be the $\mathbb{Q}$-vector space of finite formal expressions $\sum_i c_i L_i$ with $L_i = \Gamma\gamma_i$ a right coset and $c_i \in \mathbb{Q}$. The map $H(\Gamma, G) \to \mathcal{L}$ is injective and induces an isomorphism $H(\Gamma, G) \cong \mathcal{L}^\Gamma$, where the action of Γ on $\mathcal{L}$ is $\Gamma\gamma_1 \mapsto \Gamma\gamma_1\gamma$.

We now make this into an algebra by specifying the product of $\Gamma\gamma\Gamma = \sum_i \Gamma\gamma_i$ and $\Gamma\delta\Gamma = \sum_j \Gamma\delta_j$ by

$$(\Gamma\gamma\Gamma) \cdot (\Gamma\delta\Gamma) = \sum_{i,j} \Gamma\gamma_i\delta_j.$$

To deal with these double cosets the following proposition is very helpful.

Proposition 4. *(Elementary divisors) Let $\gamma \in \mathrm{GSp}^+(2g, \mathbb{Q})$ be an element with integral entries. Then double coset $\Gamma\gamma\Gamma$ has a unique representative of the form*

$$\alpha = \mathrm{diag}(a_1, \ldots, a_g, d_1, \ldots, d_g)$$

with integers a_j, d_j satisfying $a_j > 0$, $a_j d_j = \eta(\gamma)$ for all j, and furthermore $a_g | d_g$, $a_j | a_{j+1}$ for $j = 1, \ldots, g - 1$.

On G we have the anti-involution

$$\gamma = \begin{pmatrix} a & b \\ c & d \end{pmatrix} \mapsto \gamma^\vee = \begin{pmatrix} d^t & -b^t \\ -c^t & a^t \end{pmatrix} = \eta(\gamma)\,\gamma^{-1}.$$

(Note that $\eta(\gamma^\vee) = \eta(\gamma)$.) Another involution is given by

$$\gamma \mapsto J\gamma J^{-1} = \begin{pmatrix} d & -c \\ -b & a \end{pmatrix} = \eta(\gamma)\gamma^{-t}.$$

Because of the proposition we have $\Gamma\gamma\Gamma = \Gamma\gamma^\vee\Gamma$ since we may choose γ diagonal and then $\gamma^\vee = J\gamma J^{-1}$ and $J \in \Gamma$. This implies that for a sum of right cosets $\Gamma\gamma\Gamma = \sum \Gamma\gamma_i$ we have $\Gamma\gamma\Gamma = \sum \gamma_i^\vee\Gamma$. And it is easy to see that $\gamma \mapsto \gamma^\vee$ defines an anti-involution of $H(\Gamma, G)$ which acts trivially so that the Hecke algebra is commutative.

We can decompose these diagonal matrices as a product of matrices so that in each of the factors only powers of one prime occur as non-zero entries. This leads to a decomposition

$$H(\Gamma, G) = \otimes_p H_p$$

as a product of local Hecke algebras

$$H_p = H(\Gamma, G \cap \mathrm{GL}(2g, \mathbb{Z}[1/p])),$$

where we allow in the denominators only powers of p. Now H_p has a subring H_p^0 generated by integral matrices. We have $H_p = H_p^0[1/T]$ with T the element defined by $T = \Gamma_g(p\,1_{2g})\Gamma_g$. By induction one proves the following theorem, cf., [4, 30].

Theorem 9. *The local Hecke algebra H_p^0 is generated by the element $T(p)$ given by $\Gamma_g \begin{pmatrix} 1_g & 0_g \\ 0_g & p\,1_g \end{pmatrix} \Gamma_g$ and the elements $T_i(p^2)$ for $i = 1, \ldots, g$ given by*

$$\Gamma_g \begin{pmatrix} 1_{g-i} & & & \\ & p1_i & & \\ & & p^2 1_{g-i} & \\ & & & p1_i \end{pmatrix} \Gamma_g$$

For completeness sake we also introduce the element $T_0(p^2)$ given by the double coset $\Gamma_p \begin{pmatrix} 1_g & 0 \\ 0_g & p^2 1_g \end{pmatrix} \Gamma_g$. Note that $T_g(p^2)$ equals the $T = \Gamma_g(p\,1_{2g})\Gamma_g$ given above.

Definition 7. *Let $T(m)$ be the element of $H(\Gamma, G)$ defined by the set $M = O_g(m)$ which is a finite disjoint union of double cosets.*

If $m = p$ is prime then $M = O_g(m)$ is one double coset and $T(m)$ coincides with $T(p)$, introduced above. For $m = p^2$ the set $O_g(p^2)$ is a union of $g + 1$ double cosets and the element $T(p^2)$ is a sum $\sum_{i=0}^{g} T_i(p^2)$.

The Hecke algebra can be made to act on the space of Siegel modular forms $M_\rho(\Gamma_g)$. We first define the 'slash operator'.

Definition 8. *Let $\rho : \mathrm{GL}(g, \mathbb{C}) \to \mathrm{End}(V)$ be a finite-dimensional irreducible complex representation corresponding to $(\lambda_1 \geq \ldots \geq \lambda_g)$. For a function $f : \mathcal{H}_g \to V$ and an element $\gamma \in \mathrm{GSp}^+(2g, \mathbb{Q})$ we set*

$$f|_{\gamma, \rho}(\tau) = \rho(c\tau + d)^{-1} f(\gamma(\tau)) \quad \gamma = \begin{pmatrix} * & * \\ c & d \end{pmatrix}.$$

(For a good action on integral cohomology one might wish to add a factor $\eta(\gamma)^{\sum \lambda_i - g(g+1)/2}$.) Note that $f|_{\gamma_1, \rho}|_{\gamma_2, \rho} = f|_{\gamma_1 \gamma_2, \rho}$. So if $g > 1$ then f is a modular form of weight ρ if and only if f is holomorphic and $f|_{\gamma, \rho} = f$ for every $\gamma \in \Gamma_g$.

Let now $M \subset \mathrm{GSp}(2g, \mathbb{Q})$ be a subset satisfying the two properties

1. $M = \sqcup_{i=1}^{h} \Gamma_g \gamma_i$ is a finite disjoint union of right cosets $\Gamma_g \gamma_i$;
2. $M \Gamma_g \subset M$.

The first condition implies that if for a modular form $f \in M_\rho$ we set

$$T_M f := \sum_{i=1}^{h} f|_{\gamma_i, \rho}$$

then this is independent of the choice of the representatives γ_i, while the second condition implies that $(T_M f)|_\gamma = T_M f$ for all $\gamma \in \Gamma_g$. Together these conditions imply that T_M is a linear operator on the space M_ρ.

Double cosets $\Gamma \gamma \Gamma$ satisfy condition 2) if $\Gamma = \Gamma_g$ and $\gamma \in \mathrm{Sp}(2g, \mathbb{Q})$ and also condition 1) by what was said above.

An important observation is that $\langle Tf, g \rangle = \langle f, T^\vee g \rangle$, where $\langle , \rangle$ gives the Petersson product and thus the Hecke operators define Hermitian operators on the space of cusp forms S_ρ.

Just as in the classical case $g = 1$ we can associate correspondences (i.e. divisors on $\mathcal{A}_g \times \mathcal{A}_g$) to Hecke operators. The correspondence associated to $T(p)$ sends a principally polarized abelian variety X to the sum $\sum X'$ of principally polarized X' which admit an isogeny $X \to X'$ with kernel an isotropic (for the Weil pairing) subgroup $H \subset X[p]$ of order p^g. Similarly, the correspondence associated to $T_i(p^2)$ sends X to the sum $\sum X'$ with the X' quotients X/H, where $H \subset X[p^2]$ is an isotropic subgroup of order p^{2g} with $H \cap X[p]$ of order p^{g+i}.

17 The Satake Isomorphism

We can identify the local Hecke algebra H_p with the $\mathbb{Q}$-algebra of $\mathbb{Q}$-valued locally constant functions on $\mathrm{GSp}(2g, \mathbb{Q}_p)$ with compact support and which are invariant under the (so-called hyperspecial maximal compact) subgroup $K = \mathrm{GSp}(2g, \mathbb{Z}_p)$ acting both from the left and right. The multiplication in this algebra is convolution $f_1 \cdot f_2 = \int_{\mathrm{GSp}(2g, \mathbb{Q}_p))} f_1(g) f_2(g^{-1}h) dg$, where dg denotes the unique Haar measure normalized such that the volume of K is 1. The correspondence is obtained by sending the double coset $K\gamma K$ to the characteristic function of $K\gamma K$. A compactly supported function in H_p is constant on double cosets and its support is a finite linear combination of characteristic functions of double cosets.

Note that Theorem 9 tells us that H_p is generated by the double cosets of diagonal matrices. In order to describe this algebra conveniently we compare it with the p-adic Hecke algebras of two subgroups, the diagonal torus and the Levi subgroup of the standard parabolic subgroup.

To be precise, recall the diagonal torus $\mathbb{T}$ of $\mathrm{GSp}(2g, \mathbb{Q})$ isomorphic to $\mathbb{G}_m^{g+1}$ and the Levi subgroup

$$M = \left\{ \begin{pmatrix} a & 0 \\ 0 & d \end{pmatrix} \in \mathrm{GSp}(2g, \mathbb{Q}) \right\}$$

of the standard parabolic $Q = \{(a, b; 0, d) \in \mathrm{GSp}(2g, \mathbb{Z})\}$ that stabilizes the first summand $\mathbb{Z}^g$ of $\mathbb{Z}^g \oplus \mathbb{Z}^g$. In particular for an element $(a, 0; 0, d) \in M$ we have $ad^t = \eta$ and the group M is isomorphic to $\mathrm{GL}(g) \times \mathbb{G}_m$. Let $Y \cong \mathbb{Z}^{g+1}$ be the co-character group of $\mathbb{T}_m$, i.e., $Y = \mathrm{Hom}(\mathbb{G}_m, \mathbb{T})$, cf. Section 12.

We can construct a local Hecke algebra $H_p(\mathbb{T}) = H_p(\mathbb{T}, \mathbb{T}_{\mathbb{Q}})$ for the group $\mathbb{T}$ too as the $\mathbb{Q}$-algebra of $\mathbb{Q}$-valued, bi-$\mathbb{T}(\mathbb{Z}_p)$-invariant, locally constant functions with compact support on $\mathbb{T}(\mathbb{Q}_p)$. This local Hecke-algebra is easy to describe: $H_p(\mathbb{T}) \cong \mathbb{Q}[Y]$, the group algebra over $\mathbb{Q}$ of Y where $\lambda \in Y$ corresponds to the characteristic function of the double coset $D_\lambda = K\lambda(p)K$. Concretely, $H_p(\mathbb{T})$ is isomorphic to the ring $\mathbb{Q}[(u_1/v_1)^{\pm}, \ldots, (u_g/v_g)^{\pm}, (v_1 \cdots v_g)^{\pm}]$ under a map that sends $(a_1, \ldots, a_g, c)$ to the element

$$(u_1/v_1)^{a_1} \cdots (u_g/v_g)^{a_g} (v_1 \cdots v_g)^c.$$

Similarly, we have a p-adic Hecke algebra $H_p(M) = H_p(M, M_{\mathbb{Q}})$ for M.

Recall that the Weyl group $W_G = N(\mathbb{T})/\mathbb{T}$, with $G = \mathrm{GSp}(2g, \mathbb{Q})$ and $N(\mathbb{T})$ the normalizer of $\mathbb{T}$ in G, acts. This group W_G is isomorphic to $S_g \ltimes (\mathbb{Z}/2\mathbb{Z})^g$, where the generator of the i-th factor $\mathbb{Z}/2\mathbb{Z}$ acts on a matrix of the form $\mathrm{diag}(\alpha_1, \ldots, \alpha_g, \delta_1, \ldots, \delta_g)$ by interchanging α_i and δ_i and the symmetric group S_g acts by permuting the α's and δ's. The Weyl group of M (normalizer this time in M) is isomorphic to the symmetric group S_g. The algebra of invariants $H_p(\mathbb{T})^{W_G}$ is of the form $\mathbb{Q}[y_0^{\pm}, y_1, \ldots, y_g]$, cf. [30].

We now give Satake's so-called spherical map of the Hecke algebra $H_p(\Gamma, G)$ to the Hecke algebras $H_p(M)$ and $H_p(\mathbb{T})$, cf., [17, 29, 40, 85]. The images will land in the W_M-invariant (resp. the W_G-invariant) part.

We first need the following characters. The Borel subgroup B of matrices $(a, b; 0, d)$ with a upper triangular and d lower triangular determines a set Φ^+

of positive roots in the set of all roots Φ (= characters that occur in the adjoint representation of G on $\mathrm{Lie}(B)$). We let $2\rho = \sum_{\Phi^+} \alpha$.

Define $e^{2\rho_n} : M \to \mathbb{G}_m$ by $\gamma = (a, 0; 0, d) \mapsto \det(a)^{g+1} \eta(\gamma)^{-g(g+1)/2}$, where the multiplier $\eta(\gamma)$ is defined by $a \cdot d^t = \eta(\gamma) 1_g$. (This corresponds to the adjoint action of $\mathbb{T}$ on the Lie algebra of the unipotent radical of P.) Secondly, we have the character $e^{2\rho_M} : \mathbb{T} \to \mathbb{G}_m$ given by

$$\mathrm{diag}(\alpha_1, \ldots, \alpha_g, \delta_1, \ldots, \delta_g) \mapsto \prod_{i=1}^{g} \alpha_i^{g+1-2i} = \prod_{i=1}^{g} \delta^{2i-(g+1)}.$$

and $2\rho_M$ is the sum of the positive roots in $\Phi_M^+ = \{a_i/a_j : 1 \le i < j \le g\}$. Together they give a character $e^{2\rho} : \mathbb{T} \to \mathbb{G}_m$ given by $e^{2\rho(t)} = e^{2\rho_n(t)} e^{2\rho_M(t)}$ for $t \in \mathbb{T}$; explicitly,

$$\mathrm{diag}(\alpha_1, \ldots, \alpha_g, \delta_1, \ldots, \delta_g) \mapsto \eta^{-g(g+1)/2} \prod_{i=1}^{g} \alpha_i^{2g+2-2i}.$$

Satake's spherical map $S_{G,M} : H_p(\Gamma, G) \to H_p(M)$ is defined by integrating

$$S_{G,M}(\phi)(m) = |e^{\rho_n(m)}| \int_{U(\mathbb{Q}_p)} \phi(mu) du,$$

where $|p| = 1/p$. Similarly, we have a map

$$S_{M,T} : H_p(M) \to H_p(\mathbb{T})$$

given by

$$S_T(\phi)(t) = |e^{\rho_M(t)}| \int_{M \cap N} \phi(tn) dn.$$

In [29] the authors define a 'twisted' version of these spherical maps where they put $|e^{2\rho_n(m)}|$ and $|e^{2\rho_M(t)}|$ instead of the multipliers above. In this way one avoids square roots of p. If one uses this twisted version one should also twist the action of the Weyl group on the co-character group Y of $\mathbb{T}$ by e^ρ too: in the usual action S_g permutes the a_i and d_i and the i-th generator τ_i of $(\mathbb{Z}/2\mathbb{Z})^g$ interchanges a_i and d_i. Under the twisted action τ_i sends (u_i, v_i) to $(p^{g+1-i} v_i, p^{i-g-1} u_i)$, while the permutation $(i\ i+1) \in S_g$ sends (u_i/v_i) to pu_{i+1}/v_{i+1}. The formula is $w \cdot \phi(t) = |e^{\rho(w^{-1}t) - \rho(t)}| \phi(w^{-1}t)$ for $w \in W$ and $t \in \mathbb{T}$, cf., [29].

The basic result is the following theorem.

Theorem 10. *Satake's spherical maps $S_{G,M}$ and $S_{M,T}$ define isomorphisms of $\mathbb{Q}$-algebras $H_p(G) \xrightarrow{\sim} H_p(\mathbb{T})^{W_G}$ and $H_p(M) \xrightarrow{\sim} H_p(\mathbb{T})^{W_M}$.*

For the untwisted version there is a similar result but one needs to tensor with $\mathbb{Q}(\sqrt{p})$. One can calculate these maps explicitly. A right coset $K\lambda(p)$ with

$\lambda \in Y$ is mapped under S_{GT} to $p^{\langle \lambda, \rho \rangle} \lambda$. Concretely, if $\gamma = \mathrm{diag}(p^{\alpha_1}, \ldots, p^{c - \alpha_g})$ then $S_{G,T}(K\gamma)$ equals

$$p^{cg(g+1)/4}(v_1 \cdots v_g)^c \prod_{i=1}^{g}(u_i / p^i v_i)^{\alpha_i}.$$

If we write a double coset $K\lambda(p)K$ as a finite sum of right cosets $K\gamma$ then we may take $\gamma = \lambda(p)$ as one of these coset representatives. Then the image of the double coset $K\lambda(p)K$ is a sum $p^{\langle \lambda, \rho \rangle} \lambda + \sum_{\mu} n_{\lambda, \mu} \mu$ where the μ satisfy $\mu < \lambda$ (i.e. $\lambda - \mu$ is positive on Φ^+) and the $n_{\lambda, \mu}$ are non-negative integers, cf., [17, 40].

18 Relations in the Hecke Algebra

We derive some relations in the Hecke algebras. We first define elements ϕ_i in the Hecke algebra $H_p(M)$ by

$$p^{i(i+1)/2}\phi_i = M(\mathbb{Z}_p) \begin{pmatrix} 1_{g-i} & & \\ & p1_g & \\ & & 1_i \end{pmatrix} M(\mathbb{Z}_p) \qquad i = 0, \ldots, g$$

From [4], p. 142–145 one can derive the following result.

Proposition 5. *We have* $S_{G,M}(T(p)) = \sum_{i=0}^{g} \phi_i$ *and for* $i = 1, \ldots, g$

$$S_{G,M}(T_i(p^2)) = \sum_{j,k \geq 0, j+i \leq k} m_{k-j}(i)p^{-\binom{k-j+1}{2}}\phi_j\phi_k,$$

where $m_h(i) = \#\{A \in \mathrm{Mat}(h \times h, \mathbb{F}_p) : A^t = A, \mathrm{corank}\,(A) = i\}$. *Moreover, for* $i = 0, \ldots, g$ *we have*

$$S_{M,T}(\phi_i) = (v_1 \cdots v_g)\sigma_i(u_1/v_1, \ldots, u_g/v_g),$$

where σ_i *denotes the elementary symmetric function of degree* i.

Example 1. $g = 1$. We have $T(p) \mapsto \phi_0 + \phi_1$, $T_0(p^2) \mapsto \phi_0^2 + ((p-1)/p)\phi_0\phi_1 + \phi_1^2$ and $T_1(p^2) \mapsto \phi_0\phi_1/p$. We derive that $T(p^2) = T_0(p^2) + T_1(p^2)$ satisfies the well-known relation $T(p^2) = T(p)^2 - pT_1(p^2)$.

$g = 2$. We find $T(p) \mapsto \phi_0 + \phi_1 + \phi_2$ and $T_1(p^2) \mapsto \frac{1}{p}\phi_0\phi_1 + \frac{p^2-1}{p^3}\phi_0\phi_2 + \frac{1}{p}\phi_1\phi_2$ and similarly $T_2(p^2) \mapsto \frac{1}{p^3}\phi_0\phi_2$.

We denote the element ϕ_0 corresponding to $(1_g, 0; 0, p1_g)$ by Frob. This element of $H_p(M)$ generates the fraction field of $H_p(M)$ over the fraction field of $H_p(\Gamma, G)$ as we can see from the calculation above. Indeed, we have that

$S_T(\phi_0) = v_1 \cdots v_g$ and this element of $H_p(\mathbb{T})$ is fixed by S_g, but not by any other element of W_G. In particular, it is a root of the polynomial

$$\prod_{w \in (\mathbb{Z}/2\mathbb{Z})^g} (X - w(\phi_0)) = \prod_{I \subset \{1,\ldots,g\}} \left(X - \prod_{i \in I} u_i \prod_{i \notin I} v_i \right).$$

For example, for $g = 1$ we find by elimination that ϕ_0 is a root of

$$X^2 - T(p)X + pT_1(p^2),$$

while for $g = 2$ we have that ϕ_0 is a root of

$$X^4 - T(p)X^3 + (pT_1(p^2) + (p^3 + p)T_2(p^2))X^2 - p^3\,T(p)T_2(p^2)X + p^6\,T_2(p^2)^2 .$$

Using the relation

$$T(p)^2 = T_0(p^2) + (p+1)T_1(p^2) + (p^3 + p^2 + p + 1)T_2(p^2)$$

this can be rewritten as a polynomial $F(X)$ given by

$$X^4 - T(p)X^3 + (T(p)^2 - T(p^2) - p^2 T_2(p^2))\,X^2 - p^3\,T(p)T_2(p^2)\,X + p^6\,T_2(p^2)^2 .$$

Moreover, in the power series ring over the Hecke ring of $\mathrm{Sp}(4,\mathbb{Q})$ one has the formal relation (cf., [93], [4], p. 152)

$$\sum_{i=0}^{\infty} T(p^i)\,z^i = \frac{1 - p^2\,T_2(p^2)\,z^2}{z^4 F(1/z)} .$$

For a slightly different approach we refer to a paper [67] by Krieg and a preprint by Ryan with an algorithm to calculate the images, cf., [82].

19 Satake Parameters

The usual argument that uses the Petersson product shows that the spaces S_ρ possess a basis of common eigenforms for the action of the Hecke algebra.

If F is a Siegel modular form in $M_\rho(\Gamma_g)$ for an irreducible representation $\rho = (\lambda_1, \ldots, \lambda_g)$ of $\mathrm{GL}(g, \mathbb{C})$ which is an eigenform of the Hecke algebra H then we get for each Hecke operator T an eigenvalue $\lambda_F(T) \in \mathbb{C}$, a real algebraic number. Now the determination of the local Hecke algebra $H_p \otimes \mathbb{C} \cong \mathbb{C}[Y]^{W_G}$ says that

$$\mathrm{Hom}_{\mathbb{C}}(H_p, \mathbb{C}) \cong (\mathbb{C}^*)^{g+1}/W_G .$$

In particular, for a fixed eigenform F the map $H_p \to \mathbb{C}$ given by $T \mapsto \lambda_F(T)$ is determined by (the W_G-orbit of) a $(g+1)$-tuple $(\alpha_0, \alpha_1, \ldots, \alpha_g)$ of non-zero complex numbers, the p-*Satake parameters* of F. So for $i = 1, \ldots, g$ the parameter α_i is the image of u_i/v_i and α_0 that of $v_1 \cdots v_g$ and $\tau_i \in W_G$ acts by $\tau_i(\alpha_0) = \alpha_0\alpha_i$, $\tau_i(\alpha_i) = 1/\alpha_i$ and $\tau_i(\alpha_j) = \alpha_j$ if $j \neq 0, i$. These Satake parameters satisfy the relation

$$\alpha_0^2\alpha_1 \cdots \alpha_g = p^{\sum_{i=1}^{g} \lambda_i - (g+1)g/2}.$$

This follows from the fact that $T_g(p^2)$, which corresponds to the double coset of $p \cdot 1_{2g}$, is mapped to $p^{-g(g+1)/2}(v_1 \cdots v_g)^2 \prod_{i=1}^{g}(u_i/v_i)$ as we saw above.

For example, if $f = \sum_n a(n)q^n \in S_k(\Gamma_1)$ is a normalized eigenform and if we write $a(p) = \beta + \bar{\beta}$ with $\beta\bar{\beta} = p^{k-1}$ then $(\alpha_0, \alpha_1) = (\beta, \bar{\beta}/\beta)$ or $(\alpha_0, \alpha_1) = (\bar{\beta}, \beta/\bar{\beta})$. Or if $f \in M_k(\Gamma_g)$ is the Siegel Eisenstein series of weight k then the Satake parameters at p are: $\alpha_0 = 1$, $\alpha_i = p^{k-i}$ for $i = 1, \ldots, g$.

The formulas from Proposition 5 give now formulas for the eigenvalues of the Hecke operators $T(p)$ and $T_i(p^2)$ in terms of these Satake parameters:

$$\lambda(p) = \alpha_0(1 + \sigma_1 + \ldots + \sigma_g)$$

and similarly

$$\lambda_i(p^2) = \sum_{j,k \geq 0, j+i \leq k}^{g} m_{k-j}(i)p^{-\binom{k-j+1}{2}}\alpha_0^2\sigma_i\sigma_j,$$

where σ_j is the jth elementary symmetric function in the α_i with $i = 1, \ldots, \alpha_g$ and the $m_h(i)$ are defined as in Proposition 5.

20 L-functions

It is customary to associate to an eigenform $f = \sum a(n)q^n \in M_k(\Gamma_1)$ of the Hecke algebra a Dirichlet series $\sum_{n \geq 1} a(n)n^{-s}$ with s a complex parameter whose real part is $> k/2 + 1$. It is well-known that for a cusp form this L-function admits a holomorphic continuation to the whole s-plane and satisfies a functional equation. The multiplicativity properties of the coefficients $a(n)$ ensure that we can write it formally as an Euler product

$$\sum_{n>0} a(n)n^{-s} = \prod_p (1 - a(p)p^{-s} + p^{k-1-2s})^{-1}.$$

In defining L-series for Siegel modular forms one uses Euler products.

Suppose now that $f \in M_\rho(\Gamma_g)$ is an eigenform of the Hecke algebra with eigenvalues $\lambda_f(T)$ for $T \in H_p^0$. Then the assignment $T \mapsto \lambda_f(T)$ defines an element of $\mathrm{Hom}_{\mathbb{C}}(H_p^0, \mathbb{C})$. We called the corresponding $(g+1)$-tuple of α's the p-Satake parameters of f. The fact that $\mathbb{Z}[Y]^{W_G}$ is also the representation ring of the complex dual group $\hat{G}$ of $G = \mathrm{GSP}(2g, \mathbb{Q})$ (determined by the dual 'root datum') is responsible for a connection with L-functions. In our case we can use the Satake parameters to define the following formal L-functions.

Firstly, there is the *spinor zeta function* $Z_f(s)$ with as Euler factor at p the expression $Z_{f,p}(p^{-s})^{-1}$ with $Z_{f,p}(t)$ given by

$$(1 - \alpha_0 t) \prod_{r=1}^{g} \prod_{1 \le i_1 < \cdots < i_r \le g} (1 - \alpha_0 \alpha_{i_1} \ldots \alpha_{i_r} t) = (1 - \alpha_0 t) \prod_I (1 - \alpha_0 \alpha_I t),$$

where the product has 2^g factors corresponding to the 2^g subsets $I \subseteq \{1, \ldots, g\}$. Secondly, there is the *standard zeta function* with as Euler factor $D_{f,p}(p^{-s})^{-1}$ at p the expression

$$D_{f,p}(t) = (1 - t) \prod_{i=1}^{g} (1 - \alpha_i t)(1 - \alpha_i^{-1} t).$$

For example, for $g = 1$ the spinor zeta function is $Z_f(s) = \sum a(n) n^{-s}$, the usual L-series and the standard zeta function $D_f(s - k + 1) = \prod(1 + p^{-s+k-1})^{-1} \sum a(n^2) n^{-s}$, that is related to the Rankin zeta function. For $g = 2$ and eigenform $f \in M_{j,k}(\Gamma_2)$ with $T(m)f = \lambda_f(m)f$ we have $Z_f(s) = \zeta(2s - j - 2k + 4) \sum_{m \in \mathbb{Z}_{>0}} \lambda_f(m) m^{-s}$.

We set

$$\Delta(f, s) = (2\pi)^{-gs} \pi^{-s/2} \Gamma \left(\frac{s + \epsilon}{2} \right) \prod_{j=1}^{g} \Gamma(s + k - j) D(f, s),$$

where $\epsilon = 0$ for g even and $\epsilon = 1$ for g odd. Then the function $\Delta(f, s)$ can be extended meromorphically to the whole s-plane and satisfies a functional equation $\Delta(f, s) = \Delta(f, 1 - s)$, cf. papers by Böcherer [13], Andrianov–Kalinin [3], Piatetski–Shapiro and Rallis [79]. If $f \in S_k(\Gamma_g)$ is a cusp form and $k \ge g$ then $\Delta(f, s)$ is holomorphic except for simple poles at $s = 0$ and $s = 1$. It is even holomorphic if the eigenform does not lie in the space generated by theta series coming from unimodular lattices of rank $2g$. Also for $k < g$ we have information about the poles, cf., [73]. Andrianov proved that for $g = 2$ the function $\Phi_f(s) = \Gamma(s)\Gamma(s - k + 2)(2\pi)^{-2s} Z_f(s)$ is meromorphic with only finitely many poles and satisfies a functional equation $\Phi_f(2k - 2 - s) = (-1)^k \Phi_f(s)$.

One instance where spinor zeta functions associated to Siegel classical modular forms of weight 2 occur is as L-functions associated to the 1-dimensional cohomology of simple abelian surfaces.

We end by giving two additional references: the lectures notes by Courtieu and Panchishkin [19] and a paper [104] by Yoshida on motives associated to Siegel modular forms.

21 Liftings

It is well-known that for a normalized cusp form which is an eigenform $f = \sum_{n \ge 1} a(n) q^n$ of weight k on Γ_1 we have the inequality $|a(p)| \le 2p^{(k-1)/2}$ for every prime p, or equivalently, the roots of the Euler factor $1-a(p)X+p^{k-1}X^2$ at p have absolute value $p^{-(k-1)/2}$. This was shown by Eichler for cusp forms of weight $k = 2$ on the congruence subgroups $\Gamma_0(N) \subset \mathrm{SL}(2,\mathbb{Z})$ and by Deligne for general k in two steps, by first reducing it to the Weil conjectures in 1968 ([20]) and then by proving the Weil conjectures in 1974.

For $g = 2$ the analogous Euler factor at p for an eigenform F of the Hecke algebra is the expression

$$\mathcal{F}_p = 1 - \lambda(p)X + (\lambda(p)^2 - \lambda(p^2) - p^{2k-4})X^2 - \lambda(p)p^{2k-3}X^3 + p^{4k-6}X^4 \, ,$$

with $\lambda(p)$ the eigenvalue of the cusp form $F \in S_k(\Gamma_2)$; cf., the polynomial at the end of Section 18. The tacit assumption of many mathematicians in the 1970's was that the absolute values of the roots of $\mathcal{F}_p$ were equal to $p^{-(2k-3)/2}$. For example, for $k = 3$ a classical cusp form F of weight 3 on a congruence subgroup $\Gamma_2(n)$ with $n \geq 3$ determines a holomorphic 3-form $F(\tau)\prod_{i \leq j} d\tau_{ij}$ on the complex 3-dimensional manifold $\Gamma_2(n)\backslash\mathcal{H}_2$ that can be extended to a compactification and we thus find an element of the cohomology group H^3, so we expect to find absolute value $p^{-3/2}$. But then in 1978 Kurokawa and independently H. Saito ([69]) found examples of Siegel modular forms of genus 2 contradicting this expectation. Their examples are the very first examples that one encounters, like the cusp form $\chi_{10} \in S_{10}(\Gamma_2)$. On the basis of explicit calculations Kurokawa guessed that

$$L(\chi_{10}, s) = \zeta(s - 9)\zeta(s - 8)L(f_{18}, s) \, ,$$

with $f_{18} = \Delta e_6 \in S_{18}(\Gamma_1)$ the normalized cusp form of weight 18 on $\mathrm{SL}(2,\mathbb{Z})$ and $L(\chi_{10}, s) = \prod_p \mathcal{F}(p^{-s})^{-1}$ the spinor L-function. For example, he found for $p = 2$

$$\mathcal{F}_2 = (1 - 2^8 X)(1 - 2^9 X)(1 + 528 \, X + 2^{17}X^2)$$

giving the absolute values p^8, p^9 and $p^{17/2}$ for the inverse roots. The examples he worked out suggested that in these cases $L(F_k, s) = \zeta(s - k + 1)\zeta(s - k + 2)L(f_{2k-2}, s)$ with $f_{2k-2} \in S_{2k-2}(\Gamma_1)$ a normalized cusp form and F_k a corresponding Siegel modular form of weight k which is an eigenform of the Hecke algebra. On the basis of this he conjectured the existence of a 'lift'

$$S_{2k-2}(\Gamma_1) \longrightarrow S_k(\Gamma_2), \quad f \mapsto F$$

with $L(F, s) = \zeta(s - k + 1)\zeta(s - k + 2)L(f, s)$. A little later, Maass identified in $M_k(\Gamma_2)$ a subspace ('Spezialschar', nowadays called the Maass subspace, cf., [71]) consisting of modular forms F with a Fourier development $F = \sum_{N \geq 0} a(N)e^{2\pi i \mathrm{Tr} N\tau}$ satisfying the property that $a(N)$ depends only on the discriminant $d(N)$ and the content $e(N)$, i.e., if we write

$$N = \begin{pmatrix} n & r/2 \\ r/2 & m \end{pmatrix}$$

then N corresponds to the positive definite quadratic form $[n, r, m] := nx^2 + rxy + my^2$ with discriminant $d = 4mn - r^2$ and content $e = \text{g.c.d.}(n, r, m)$. We shall write $a([n, r, m])$ for $a(N)$. The condition that F belongs to the Maass space can be formulated alternatively as

$$a([n, r, m]) = \sum_{d>0,\, d|(n,r,m)} d^{k-1} a([1, r/d, mn/d^2])$$

We shall write $M_k^*(\Gamma_2)$ or $S_k^*(\Gamma_2)$ for the Maass subspace of $M_k(\Gamma_2)$ or $S_k(\Gamma_2)$. It was then conjectured ('Saito–Kurokawa Conjecture') that there is a 1-1 correspondence between eigenforms in $S_{2k-2}(\Gamma_1)$ and eigenforms in the Maass space $S_k^*(\Gamma_2)$ given by an identity between their L-functions. More precisely, we now have the following theorem.

Theorem 11. *The Maass subspace $S_k^*(\Gamma_2)$ is invariant under the action of the Hecke algebra and there is a 1-1 correspondence between eigenspaces in $S_{2k-2}(\Gamma_1)$ and Hecke eigenspaces in $S_k^*(\Gamma_2)$ given by*

$$f \leftrightarrow F \quad \Longleftrightarrow \quad L(F, s) = \zeta(s - k + 1)\zeta(s - k + 2)L(f, s)$$

with $L(F, s)$ the spinor L-function of F.

The lion's share of the theorem is due to Maass, but it was completed by Andrianov and Zagier, see [2, 71, 109].

We can make an extended picture as follows. The map $F \mapsto \phi_{k,1}$ that sends a Siegel modular form to its first Fourier–Jacobi coefficient induces an isomorphism $M_k^*(\Gamma_2) \cong J_{k,1}$, the space of Jacobi forms, and the map $h = \sum c(n)q^n \mapsto \sum_{n \equiv -r^2 \,(\text{mod } 4)} c(n)q^{(n+r^2)/4}\zeta^r$ gives an isomorphism of the Kohnen plus space $M_{k-1/2}^+$ with $J_{k,1}$ fitting in a diagram

$$M_k^*(\Gamma_2) \xrightarrow{\sim} J_{k,1} \xleftarrow{\sim} M_{k-1/2}^+$$
$$\downarrow \cong$$
$$M_{2k-2}(\Gamma_1)$$

where the vertical map is the Kohnen isomorphism. Note that the vertical map is quite different from the horizontal two maps. The vertical isomorphism is not canonical at all, but depends on the choice of a discriminant D.

We now sketch a proof of Theorem 11. A classical Siegel modular form $F \in M_k(\Gamma_2)$ has a Fourier–Jacobi series $F(\tau, z, \tau') = \sum \phi_m(\tau, z)e^{2\pi i m \tau'}$ with $\phi_m(\tau, z) \in J_{k,m}$, the space of Jacobi forms of weight k and index m. The reader may check this by himself. We have on the Jacobi forms a sort of Hecke operators $V_m : J_{k,m} \to J_{k,ml}$ with $\phi|_{k,m}V_l(\tau, z)$ given explicitly by

$$l^{k-1} \sum_{\Gamma_1 \backslash O(l)} (c\tau + d)^{-k} e^{2\pi i m l(-cz^2/(c\tau+d))} \phi((a\tau + b)/(c\tau + d), lz/(c\tau + d)).$$

On coefficients, if $\phi = \sum_{n,r} c(n,r) q^n \zeta^r$ then

$$\phi|_{k,m} V_l = \sum_{n,r} \sum_{a|(n,r,l)} a^{k-1} c(nl/a^2, r/a) q^n \zeta^r .$$

One now checks using generators of Γ_2 that for $\phi \in J_{k,1}$ the expression

$$v(\phi) := \sum_{m \geq 0} (\phi|V_m)(\tau, z) e^{2\pi i m \tau'}$$

is a Siegel modular form in $M_k(\Gamma_2)$.

We then have a map $M_k(\Gamma_2) \to \oplus_{m=0}^\infty J_{k,m}$ by associating to a modular form its Fourier–Jacobi coefficients; we also have a map in the other direction $J_{k,1} \to M_k(\Gamma_2)$ given by $\phi \to v(\phi)$ and the composition

$$J_{k,1} \to M_k(\Gamma_2) \to \oplus_m J_{k,m} \xrightarrow{\text{pr}} J_{k,1}$$

is the identity. So $v : J_{k,1} \to M_k(\Gamma_2)$ is injective and the image consists of those modular forms F with the property that $\pi_m = \phi_1|V_m$. This implies the following relation for the Fourier coefficients for $[n, r, m] \neq [0, 0, 0]$

$$a([n, r, m]) = \sum_{d|(n,r,m)} d^{k-1} c((4mn - r^2)/d^2) ,$$

where $c(N)$ is given by

$$c(N) = \begin{cases} a([n, 0, 1]) & N = 4n \\ a([n, 1, 1]) & N = 4n - 1 . \end{cases}$$

In particular, we see that the image is the Maass subspace because

$$a([n, r, m]) = \sum_{d|(n,r,m)} d^{k-1} a([nm/d^2, r/d, 1]) .$$

On the other hand, it is known that $J_{k,1} \cong M_{k-1/2}^+$. Combination of the two isomorphisms yields what we want.

Duke and Imamoğlu conjectured in [23] a generalization of this and some evidence was given by Breulmann and Kuss [15]. Then Ikeda generalized the Saito–Kurokawa lift of modular forms from one variable to Siegel modular forms of degree 2 in [58] in 1999 under the condition that $g \equiv k \pmod 2$ to a lifting from an eigenform $f \in S_{2k}(\Gamma_1)$ to an eigenform $F \in S_{g+k}(\Gamma_{2g})$ such that the standard zeta function of F is given in terms of the usual L-function of f by

$$\zeta(s) \prod_{j=1}^{2g} L(f, s + k + g - j) .$$

The Satake parameters of F are $\beta_0, \beta_1, \ldots, \beta_{2g}$ with

$$\beta_0 = p^{gk-g(g+1)/2}, \; \beta_i = \alpha\, p^{i-1/2}, \; \beta_{g+i} = \alpha^{-1} p^{i-1/2} \quad \text{for} \quad i = 1, \ldots, g$$

with $f = \sum a(n) q^n$ and

$$(1 - \alpha p^{k-1/2} X)(1 - \alpha^{-1} p^{k-1/2} X) = 1 - a(p) X + p^{2k-1} X^2,$$

cf., [74]. (In particular, such lifts do not satisfy the Ramanujan inequality.) Kohnen ([62]) has interpreted it as an explicit linear map $S^+_{k+1/2} \longrightarrow S_{k+g}(\Gamma_{2g})$ given by

$$f = \sum_{(-1)^k n \equiv 0, 1 \,(\bmod\, 4)} c(n) q^n F \mapsto \sum_N a(N) e^{2\pi \operatorname{Tri} N\tau},$$

with $a(N)$ given by an expression $\sum_{a|f_N} a^{k-1} \phi(a, N) c(|D_N|/a^2)$ and $\phi(a, N)$ an explicitly given integer-valued numbertheoretic function.

One defines also a Maass space with $M_k^*(\Gamma_g)$ consisting of F such that $a(N) = a(N')$ if the discriminants of N and N' are the same and in addition $\phi(a, N) = \phi(a, N')$ for all divisors a of $f_N = f_{N'}$. Under the additional assumption that $g \equiv 0, 1 \,(\bmod\, 4)$ Kohnen and Kojima prove in [64] that the image of the lifting is the Maass space.

Example 2. Let $k = 6$ and $g = 2$. Then the Ikeda lift is a map from $S_{12}(\Gamma_1) \to S_8(\Gamma_4)$ and the image of Δ is a cusp form that vanishes on the closure of the Jacobian locus (i.e., the abelian 4-folds that are Jacobians of curves of genus 4), [15]. Or take $k = g = 6$ and get a lift $S_{12}(\Gamma_1) \to S_{12}(\Gamma_{12})$. This lifted form occurs in the paper [14].

Miyawaki observed in [72] that the standard L-function of a non-zero cusp form F of weight 12 on Γ_3 is a product $D_\Delta(F, s) L(\phi_{20}, s + 10) L(\phi_{20}, s + 9)$, with $\Delta \in S_{12}(\Gamma_1)$ and $\phi_{20} \in S_{20}(\Gamma_1)$ the normalized Hecke eigenforms of weight 12 and 20. He conjectured a lifting and his idea was refined by Ikeda to the following conjecture.

Conjecture 1. *(Miyawaki–Ikeda) Let k and n be natural numbers with $k - n$ even. Furthermore, let $f \in S_{2k}(\Gamma_1)$ be a normalized Hecke eigenform. Then there exists for every eigenform $g \in S_{k+n+r}(\Gamma_r)$ with $n, r \geq 1$ a Siegel modular eigenform $\mathcal{F}_{f,g} \in S_{k+n+r}(\Gamma_{2n+r})$ such that*

$$D_{\mathcal{F}_{f,g}}(s) = Z_g(s) \prod_{j=1}^{2n} L_f(s + k + n - j),$$

with $L_f = Z_f$ the usual L-function.

In [59] Ikeda constructs a lifting from Siegel modular cusp forms of degree r to Siegel cusp forms of degree $r + 2n$. This is a partial confirmation of this conjecture.

Finally, I would like to mention a conjectured lifting from vector-valued Siegel modular forms of half-integral weight to vector-valued Siegel modular forms of integral weight due to Ibukiyama. He predicts in the case of genus $g = 2$ for even $j \geq 0$ and $k \geq 3$ an isomorphism

$$S^+_{j,k-1/2}(\Gamma_0(4), \psi) \xrightarrow{\sim} S_{2k-6,j+3}(\Gamma_2)$$

which should generalize the Shimura–Kohnen lifting $S^+_{k-1/2}(\Gamma_0(4)) \cong S_{2k-2}(\Gamma_1)$, see [51]. Here $\psi(\gamma) = \left(\frac{-4}{\det(d)}\right)$.

22 The Moduli Space of Principally Polarized Abelian Varieties

It is a fundamental fact, due to Mumford, that the moduli space of principally polarized abelian varieties exists as an algebraic stack $\mathcal{A}_g$ over the integers. The orbifold $\Gamma_g \backslash \mathcal{H}_g$ is the complex fibre $\mathcal{A}_g(\mathbb{C})$ of this algebraic stack. This fact has very deep consequences for the arithmetic theory of Siegel modular forms, but an exposition of this exceeds the framework of these lectures. Also the various compactifications, the Baily–Borel or Satake compactification and the toroidal compactifications constructed by Igusa and Mumford et. al. exist over $\mathbb{Z}$ as was shown by Faltings. We refer to an extensive, but very condensed survey of this theory in [29]. In particular, Faltings constructed the Satake compactification over $\mathbb{Z}$ as the image of a toroidal compactification $\tilde{\mathcal{A}}_g$ by the sections of a sufficiently big power of $\det(\mathbb{E})$, the determinant of the Hodge bundle. A corollary of Faltings' results is that the ring of classical Siegel modular forms with integral Fourier coefficients is finitely generated over $\mathbb{Z}$.

In the following sections we shall sketch how one can use some of these facts to extract information on the Hecke eigenvalues of Siegel modular forms.

The action of the Galois group of $\mathbb{Q}$ on the points of $\mathcal{A}_g(\bar{\mathbb{Q}})$ that correspond to abelian varieties with complex multiplication is described in Shimura's theory of canonical models. This theory can also explain the integrality of the eigenvalues of Hecke operators. For this we refer to two papers by Deligne, see [21, 22].

23 Elliptic Curves over Finite Fields

Suppose we did not have the elementary approach to $g = 1$ modular forms using holomorphic functions on the upper half plane like the Eisenstein series and Δ. How would we get the arithmetic information hidden in the Fourier coefficients of Hecke eigenforms? Would we encounter Δ?

We claim that one would by playing with elliptic curves over finite fields. Let $\mathbb{F}_q$ with $q = p^m$ be a finite field of characteristic p and cardinality q. An elliptic curve E defined over $\mathbb{F}_q$ can be given as an affine curve by an equation

$$y^2 + a_1 xy + a_3 y = x^3 + a_2 x^2 + a_4 x + a_6 ,$$

with $a_i \in \mathbb{F}_q$ and with non-zero discriminant (a polynomial in the coefficients). We can then count the number $\#E(\mathbb{F}_q)$ of $\mathbb{F}_q$-rational points of E. A result of Hasse tells us that $\#E(\mathbb{F}_q)$ is of the form $q + 1 - \alpha - \bar{\alpha}$ for some algebraic integer α with $|\alpha| = \sqrt{q}$. We can do this for all elliptic curves E defined over $\mathbb{F}_q$ up to $\mathbb{F}_q$-isomorphism and we could ask (as Birch did in [10]) for the average of $\#E(\mathbb{F}_q)$, or better for

$$\sum_E \frac{q + 1 - \#E(\mathbb{F}_q)}{\#\mathrm{Aut}_{\mathbb{F}_q}(E)} ,$$

where $\mathrm{Aut}_{\mathbb{F}_q}(E)$ is the group of $\mathbb{F}_q$-automorphisms of E, or more generally we could ask for the average of the expression

$$h(k, E) := \alpha^k + \alpha^{k-1} \bar{\alpha} + \ldots + \alpha \bar{\alpha}^{k-1} + \bar{\alpha}^k ,$$

i.e. we sum

$$\sigma_k(q) = -\sum_E \frac{h(k, E)}{\#\mathrm{Aut}_{\mathbb{F}_q}(E)}$$

where the sum is over all elliptic curves E defined over $\mathbb{F}_q$ up to $\mathbb{F}_q$-isomorphism. (As a rule of thumb, whenever one counts mathematical objects one should count them with weight $1/\#\mathrm{Aut}$ with Aut the group of automorphisms of the object.) If we do this for $\mathbb{F}_3$ we get the following table, where we also give the j-invariant of the curve $y^2 = f$

f	$\#E(k)$	$1/\#\mathrm{Aut}_k(E)$	j
$x^3 + x^2 + 1$	6	$1/2$	-1
$x^3 + x^2 - 1$	3	$1/2$	1
$x^3 - x^2 + 1$	5	$1/2$	1
$x^3 - x^2 - 1$	2	$1/2$	-1
$x^3 + x$	4	$1/2$	0
$x^3 - x$	4	$1/6$	0
$x^3 - x + 1$	7	$1/6$	0
$x^3 - x - 1$	1	$1/6$	0

and obtain the following frequencies for the number of $\mathbb{F}_3$-rational points:

n	1	2	3	4	5	6	7
freq	$1/6$	$1/2$	$1/2$	$2/3$	$1/2$	$1/2$	$1/6$

Note that $\sum 1/\mathrm{Aut}_{\mathbb{F}_q}(E) = q$ and $\sum_{E:\ j(E)=j} 1/\mathrm{Aut}_{\mathbb{F}_q}(E) = 1$ (see [36] for a proof); so a 'physical point' of the moduli space contributes 1.

If we work this out not only for $p = 3$, but for several primes ($p = 2, 3, 5, 7$ and 11) we get the following values:

p	2	3	5	7	11
σ_{10}	-23	253	4831	-16743	534613

Anyone who remembers the cusp form of weight 12

$$\Delta = \sum_{n>0} \tau(n)q^n = q - 24\,q^2 + 252\,q^3 - 3520\,q^4 + 4830\,q^5 + \ldots$$

will not fail to notice that $\sigma_{10}(p) = \tau(p) + 1$ for the primes listed in this example. And in fact, the relation $\sigma_{10}(p) = \tau(p) + 1$ holds for all primes p. The reason behind this is that the cohomology of the nth power of the universal elliptic curve $\mathcal{E} \to \mathcal{A}_1$ is expressed in terms of cusp forms on $\mathrm{SL}(2,\mathbb{Z})$. To describe this we recall the local system $\mathbb{W}$ on $\mathcal{A}_1$ associated to η^{-1} times the standard representation of $\mathrm{GSp}(2,\mathbb{Q})$ in Section 12. The fibre of this local system over a point $[E]$ given by the elliptic curve E can be identified with the cohomology group $H^1(E,\mathbb{Q})$. Or consider the universal elliptic curve (in the orbifold sense) $\pi : \mathcal{E} \to \mathcal{A}_1$ obtained as the quotient $\mathrm{SL}(2,\mathbb{Z}) \times \mathbb{Z}^2 \backslash \mathcal{H}_1 \times \mathbb{C}$, where the action of $(a,b;c,d) \in \mathrm{SL}(2,\mathbb{C})$ on $(\tau,z) \in \mathcal{H}_1 \times \mathbb{C}$ is $((a\tau + b)/(c\tau + d), (c\tau + d)^{-1}z)$. Associating to an elliptic curve its homology $H_1(E,\mathbb{Q})$ defines a local system that can be obtained as a quotient $\mathrm{SL}(2,\mathbb{Z}) \backslash \mathcal{H}_1 \times \mathbb{Q}^2$. Then the dual of this local system is $\mathbb{W} := R^1\pi_*\mathbb{Q}$. We now put

$$\mathbb{W}^k := \mathrm{Sym}^k(\mathbb{W}),$$

a local system with a $k + 1$-dimensional fibre for $k \geq 0$. We now have the following cohomological interpretation of cusp forms on $\mathrm{SL}(2,\mathbb{Z})$, cf. [20].

Theorem 12. *(Eichler–Shimura) For even $k \in \mathbb{Z}_{\geq 2}$ we have an isomorphism of the compactly supported cohomology of $\mathbb{W}^k$*

$$H^1_c(\mathcal{A}_1, \mathbb{W}^k \otimes \mathbb{C}) \cong S_{k+2} \oplus \bar{S}_{k+2} \oplus \mathbb{C}$$

with S_{k+2} the space of cusp forms of weight $k + 2$ on $\mathrm{SL}(2,\mathbb{Z})$ and $\bar{S}_{k+2}$ the complex conjugate of this space.

Replacing $\mathbb{W}$ by $\mathbb{W}_{\mathbb{R}}$ we have the exact sequence

$$0 \to \mathbb{E} \to \mathbb{W} \otimes_{\mathbb{R}} O \to \mathbb{E}^\vee \to 0$$

with O the structure sheaf and an induced map $\mathbb{E}^{\otimes k} \to \mathbb{W}^k \otimes_{\mathbb{R}} O$. Now the de Rham resolution

$$0 \to \mathbb{W}^k \otimes_{\mathbb{R}} \mathbb{C} \to \mathbb{W}^k \otimes_{\mathbb{R}} O \xrightarrow{d} \mathbb{W}^k \otimes \Omega^1 \to 0$$

defines a connecting homomorphism

$$H^0(\mathcal{A}_1, \Omega^1(\mathbb{W}^k)) \to H^1(\mathcal{A}_1, \mathbb{W}^k \otimes \mathbb{C}).$$

The right hand space has a natural complex conjugation and we thus find also a complex conjugate map

$$\overline{H^0(\mathcal{A}_1, \Omega^1(\mathbb{W}^k))} \to H^1(\mathcal{A}_1, \mathbb{W}^k \otimes \mathbb{C}).$$

A cusp form $f \in S_{k+2}$ defines a section of $H^0(\mathcal{A}_1, \Omega^1(\mathbb{W}^k))$ by putting $f(\tau) \mapsto f(\tau)d\tau dz^k$. We thus have a cohomological interpretation of the space of cusp forms.

As observed above the moduli space $\mathcal{A}_1$ is defined over the integers $\mathbb{Z}$. This means that we also have the moduli space $\mathcal{A}_1 \otimes \mathbb{F}_p$ of elliptic curves in characteristic $p > 0$. It is well-known that one can obtain a lot of information about cohomology by counting points over finite fields. (Here we work with ℓ-adic étale cohomology for $\ell \neq p$.) And, indeed, there exists an analogue of the Eichler–Shimura isomorphism in characteristic p and the relation $\sigma_{10}(p) = \tau(p) + 1$ is a manifestation of this. In fact a good notation for writing this relation is

$$H_c^1(\mathcal{A}_1, \mathbb{W}^{10}) = S[12] + 1,$$

where the formula

$$H_c^1(\mathcal{A}_1, \mathbb{W}^{2k}) \cong S[2k+2] + 1 \quad \text{for } k \geq 1$$

may be interpreted complex-analytically as the Eichler–Shimura isomorphism and in characteristic p as the relation

$$\sigma_{2k}(p) = 1 + \text{Trace of } T(p) \text{ on } S_{2k+2}.$$

(A better interpretation is as a relation in a suitable K-group and with $S[2k+2]$ as the motive associated to S_{2k+2}. This motive can be constructed in the kth power of $\mathcal{E}$ as done by Scholl [87] or using moduli space of n-pointed elliptic curves as done by Consani and Faber, [18].)

This 1 in the formula $H_c^1(\mathcal{A}_1, \mathbb{W}^{2k}) \cong S[2k+2] + 1$ is really a nuisance. To get rid of it in a conceptual way we consider the natural map

$$H_c^1(\mathcal{A}_1, \mathbb{W}^k) \to H^1(\mathcal{A}_1, \mathbb{W}^k)$$

the image of which is called the *interior cohomology* and denoted by $H_!^1(\mathcal{A}_1, \mathbb{W}^k)$. We thus have an elegant and sophisticated form of the Eichler–Shimura isomorphism

$$H_c^1(\mathcal{A}_1, \mathbb{W}^k) = S[k+2] + 1, \quad H_!^1(\mathcal{A}_1, \mathbb{W}^k) = S[k+2].$$

The 1 is the 1 in $1 + p^{k+1}$, the eigenvalue of the action of $T(p)$ on the Eisenstein series E_{k+2} of weight $k+2$ on $\mathrm{SL}(2, \mathbb{Z})$.

The moral of this is that we can obtain information on the traces of Hecke operators on the space S_{k+2} by calculating $\sigma_k(p)$, i.e., by counting points on elliptic curves over $\mathbb{F}_p$. Even from a purely computational point of view this is not a bad approach to calculating the traces of Hecke operators.

24 Counting Points on Curves of Genus 2

With the example of $g = 1$ in mind it is natural to ask whether also for $g = 2$ we could obtain information on modular forms using curves of genus 2 over finite fields. In joint work with Carel Faber ([27]) we showed that we can.

For $g = 2$ the quotient space $\Gamma_2 \backslash \mathcal{H}_2$ is the analytic space of the moduli space $\mathcal{A}_2$ of principally polarized abelian surfaces. A principally polarized abelian surface is the Jacobian of a smooth projective irreducible algebraic curve or it is a product of two elliptic curves. If the characteristic is not 2 a curve of genus 2 can be given as an affine curve with equation $y^2 = f(x)$ with f a polynomial of degree 5 or 6 without multiple zeros.

The moduli space $\mathcal{A}_2$ exists over $\mathbb{Z}$ and provides us with a moduli space $\mathcal{A}_2 \otimes \mathbb{F}_p$ for every characteristic $p > 0$. Also here we have a local system which is the analogue of the local system $\mathbb{W}$ that we saw for $g = 1$:

$$\mathbb{V} := \mathrm{GSp}(4, \mathbb{Z}) \backslash \mathcal{H}_2 \times \mathbb{Q}^4 \,,$$

where the action of $\gamma = (a, b; c, d) \in \mathrm{GSp}(4, \mathbb{Z})$ is given by η^{-1} times the standard representation. Or in more functorial terms, we consider the universal family $\pi : \mathcal{X}_2 \to \mathcal{A}_2$ and then $\mathbb{V}$ is the direct image $R^1 \pi_*(\mathbb{Q})$. The fibre of this local system over the point $[X]$ corresponding to the polarized abelian surface X is $H^1(X, \mathbb{Q})$. The local system $\mathbb{V}$ comes equipped with a symplectic pairing $\mathbb{V} \times \mathbb{V} \to \mathbb{Q}(-1)$. Just as for $g = 1$ where we made the local systems $\mathbb{W}^k$ out of the basic one $\mathbb{W}$ we can construct more local systems out of $\mathbb{V}$ but now parametrized by two indices l and m with $l \geq m \geq 0$. Namely, the irreducible representations of $\mathrm{Sp}(4, \mathbb{Q})$ are parametrized by such pairs (l, m) and we thus have local systems $\mathbb{V}_{l,m}$ with $l \geq m \geq 0$ such that $\mathbb{V}_{l,0} = \mathrm{Sym}^l(\mathbb{V})$ and $\mathbb{V}_{1,1}$ is the 'primitive part' of $\wedge^2 \mathbb{V}$. A local system $\mathbb{V}_{l,m}$ is called *regular* if $l > m > 0$.

Just as in the case $g = 1$ we are now interested in the cohomology of the local systems $\mathbb{V}_{l,m}$. We put

$$e_c(\mathcal{A}_2, \mathbb{V}_{l,m}) = \sum_i (-1)^i [H_c^i(\mathcal{A}_2, \mathbb{V}_{l,m})] \,.$$

Here we consider the alternating sum of the cohomology groups with compact support in the Grothendieck group of mixed Hodge structures.

We also have an ℓ-adic analogue of this that can be used in positive characteristic. It is obtained from $R^1 \pi_*(\mathbb{Q}_\ell)$ and lives over $\mathcal{A}_2 \otimes \mathbb{Z}[1/\ell]$; we consider the étale cohomology of this sheaf. We simply use the same name $\mathbb{V}_{l,m}$ and assume that ℓ is different from the characteristic p.

Using a theorem of Getzler [37] (on $\mathcal{M}_2$) tells us what the Euler characteristic $\sum_i (-1)^i \dim H_c^i(\mathcal{A}_2, \mathbb{V}_{l,m})$ over $\mathbb{C}$ is. This Euler characteristic equals the Euler characteristic of the ℓ-adic variant over a finite field, cf., [9].

The first observation is that because of the action of the hyperelliptic involution these cohomology groups are zero for $l + m$ odd.

Our strategy is now to make a list of all $\mathbb{F}_q$-isomorphism classes of curves of genus 2 over $\mathbb{F}_q$ and to determine for each of them $\#\mathrm{Aut}_{\mathbb{F}_q}(C)$ and the characteristic polynomial of Frobenius. So for each curve C we determine algebraic integers $\alpha_1, \bar{\alpha}_1, \alpha_2, \bar{\alpha}_2$ of absolute value $\sqrt{q}$ such that

$$\#C(\mathbb{F}_{q^i}) = q^i + 1 - \alpha_1^i - \bar{\alpha}_1^i - \alpha_2^i - \bar{\alpha}_2^i$$

for all $i \geq 1$. These α's can be calculated using this identity for $i = 1$ and $i = 2$. We also must calculate the contribution from the degenerate curves of genus 2, i.e., the contribution from the principally polarized abelian surfaces that are products of elliptic curves.

Having done that we are able to calculate the trace of Frobenius on the alternating sum of $H_c^i(\mathcal{A}_2 \otimes \mathbb{F}_q, \mathbb{V}_{l,m})$, where by $\mathbb{V}_{l,m}$ we mean the ℓ-adic variant, a smooth ℓ-adic sheaf on $\mathcal{A}_2 \otimes \mathbb{F}_q$. In practice, it means that we sum a certain symmetric expression in the α's divided by $\#\mathrm{Aut}_{\mathbb{F}_q}(C)$, analoguous to the $\sigma_k(q)$ for genus 1.

What does this tell us about Siegel modular forms of degree $g = 2$? To get the connection with modular forms we have to replace the compactly supported cohomology by the interior cohomology, i.e., by the image of $H_c^i(\mathcal{A}_2, \mathbb{V}_{l,m}) \to H^i(\mathcal{A}_2, \mathbb{V}_{l,m})$ which is denoted by $H_!^i(\mathcal{A}_2, \mathbb{V}_{l,m})$. So let us define

$$e_{\mathrm{Eis}}(\mathcal{A}_2, \mathbb{V}_{l,m}) = e_c(\mathcal{A}_2, \mathbb{V}_{l,m}) - e_!(\mathcal{A}_2, \mathbb{V}_{l,m}).$$

If we do the same thing for $g = 1$ we find $e_{\mathrm{Eis}}(\mathcal{A}_1, \mathbb{W}^k) = -1$ for even $k > 0$.

Let $\mathbb{L}$ be the 1-dimensional Tate Hodge structure of weight 2. It corresponds to the second cohomology of $\mathbb{P}^1$. In terms of counting points one reads q for $\mathbb{L}$. Our first result is (cf., [27])

Theorem 13. *Let (l, m) be regular. Then $e_{\mathrm{Eis}}(\mathcal{A}_2, \mathbb{V}_{l,m})$ is given by*

$$-S[l+3] - s_{l+m+4}\mathbb{L}^{m+1} + S[m+2] + s_{l-m+2} \cdot 1 + \begin{cases} 1 & l \text{ even} \\ 0 & l \text{ odd,} \end{cases}$$

where $s_n = \dim S_n(\Gamma_1)$.

Faltings has shown (see [29]) that $H_!^3(\mathcal{A}_2, \mathbb{V}_{l,m})$ possesses a Hodge filtration

$$0 \subset F^{l+m+3} \subset F^{l+2} \subset F^{m+1} \subset F^0 = H_!^3(\mathcal{A}_2, \mathbb{V}_{l,m}).$$

Moreover, if (l, m) is regular then $H_!^i(\mathcal{A}_2, \mathbb{V}_{l,m}) = (0)$ for $i \neq 3$. Furthermore, Faltings shows that

$$F^{l+m+3} \cong S_{l-m,m+3}(\Gamma_2).$$

Here $S_{j,k}(\Gamma_2)$ is the space of Siegel modular forms for the representation $\mathrm{Sym}^j \otimes \det^k$ of $\mathrm{GL}(2, \mathbb{C})$. This is the sought-for connection with vector valued Siegel modular forms and the analogue of $H_!^1(\mathcal{A}_1, \mathbb{W}^k) = F^0 \supset F^{k+1} \cong S_{k+2}(\Gamma_1)$ for $g = 1$. Faltings gives an interpretation of all the steps in the Hodge filtration in terms of the cohomology of the bundles $\mathcal{W}(\lambda)$.

However, although for $g = 1$ the Eichler–Shimura isomorphism tells us that we know $H^1_!(\mathcal{A}_1, \mathbb{W}^k)$ once we know $S_{k+2}(\Gamma_1)$, for $g = 2$ there might be pieces of cohomology hiding in $F^{l+2} \subset F^{m+1}$ that are not detectable in F^{l+m+3} or in F^0/F^{m+1} and indeed there is such cohomology. The contribution to this part of the cohomology is called the contribution from *endoscopic lifting from* $N = \mathrm{GL}(2) \times \mathrm{GL}(2)/\mathbb{G}_m$.

We conjecture on the basis of our numerical calculations that this endoscopic contribution is as follows.

Conjecture 2. *Let* (l, m) *be regular. Then the endoscopic contribution is given by*

$$e_{\mathrm{endo}}(\mathcal{A}_2, \mathbb{V}_{l,m}) = -s_{l+m+4}S[l - m + 2]\,\mathbb{L}^{m+1}\,.$$

There is a very extensive literature on endoscopic lifting (cf. [68]), but a precise result on the image in our case seems to be absent. Experts on endoscopic lifting should be able to prove this conjecture. Actually, since we know the Euler characteristics of the interior cohomology and have Tsushima's dimension formula it suffices to construct a subspace of dimension $2s_{l+m+4}s_{l-m+2}$ in the endoscopic part via endoscopic lifting for regular (l, m).

In terms of Galois representations a Siegel modular form (with rational Fourier coefficients) should correspond to a rank 4 part of the cohomology or a 4-dimensional irreducible Galois representation. A modular form in the endoscopic part corresponds to a rank 2 part and a 2-dimensional Galois representation. Modular forms coming from the Saito–Kurokawa lift give 4-dimensional representations that split off two 1-dimensional pieces.

In analogy with the case of $g = 1$ we now set

$$S[l - m, m + 3] := H^3_!(\mathcal{A}_2, \mathbb{V}_{l,m}) - H^3_{\mathrm{endo}}(\mathcal{A}_2, \mathbb{V}_{l,m})\,.$$

This should be a motive analogous to the motive $S[k]$ we encountered for $g = 1$ and lives in a power of the universal abelian surface over $\mathcal{A}_2$. The trace of Frobenius on étale ℓ-adic $H^3_!(\mathcal{A}_2, \mathbb{V}_{l,m}) - H^3_{\mathrm{endo}}(\mathcal{A}_2, \mathbb{V}_{l,m})$ should be the trace of the Hecke operator $T(p)$ on the space of modular forms $S_{l-m,m+3}$.

25 The Ring of Vector-Valued Siegel Modular Forms for Genus 2

The quest for vector-valued Siegel modular forms starts with genus 2. We can consider the direct sum $M = \oplus_\rho M_\rho(\Gamma_2)$ (see Section 3), where ρ runs through the set of irreducible polynomial representations of $\mathrm{GL}(2, \mathbb{C})$. Each such ρ is given by a pair (j, k) such that $\rho = \mathrm{Sym}^j(W) \otimes \det(W)^k$, with W the standard representation of $\mathrm{GL}(2, \mathbb{C})$. (Note that in the earlier notation we have $(\lambda_1 - \lambda_2, \lambda_2) = (j, k)$.) So we may write $M = \oplus_{j,k \geq 0} M_{j,k}(\Gamma_2)$ and we know that $M_{j,k}(\Gamma_2) = (0)$ if j is odd. If F and F' are Siegel modular forms of weights (j, k) and (j', k') then the product is a modular forms of weight of weight $(j+j', k+k')$. The multiplication is obtained from the canonical map $\mathrm{Sym}^{j_1}(W)\otimes\det(W)^{k_1}\otimes\mathrm{Sym}^{j_2}(W)\otimes\det(W)^{k_2} \to \mathrm{Sym}^{j_1+j_2}(W)\otimes\det(W)^{k_1+k_2}$ obtained from multiplying polynomials in two variables.

There is the Siegel operator that goes from $M_{j,k}(\Gamma_2)$ to $M_{j+k}(\Gamma_1)$. For $j > 0$ the Siegel operator gives a map to $S_{j+k}(\Gamma_1)$ and for $j > 0, k > 4$ the map $\Phi : M_{j,k}(\Gamma_2) \to S_{j+k}(\Gamma_1)$ is surjective. For these facts on the Siegel operator we refer to Arakawa's paper [6]. The Siegel operator is multiplicative: $\Phi(F \cdot F') = \Phi(F)\Phi(F')$.

There is a dimension formula for $\dim M_{j,k}(\Gamma_2)$, due to Tsushima, [101]. But apart from this not much is known about vector-valued Siegel modular forms. The direct sum $\oplus_k M_{j,k}(\Gamma_2)$ for fixed j is a module over the ring $M^{\mathrm{cl}} = \oplus M_{0,k}(\Gamma_2)$ of classical Siegel modular forms and we know generators of this module for $j = 2$ and $j = 4$ and even $j = 6$ due to Satoh and Ibukiyama, cf. [48, 49, 86].

One way to construct vector-valued Siegel modular forms from classical Siegel modular forms is differentiation, the simplest example being given by a pair $f \in M_a(\Gamma_2)$, $g \in M_b(\Gamma_2)$ for which one sets

$$[f, g] := \frac{1}{b} f \nabla g - \frac{1}{a} g \nabla f$$

with ∇f defined by

$$2\pi i \nabla f = a \, (2iy)^{-1} f + \begin{pmatrix} \partial/\partial\tau_{11} & \partial/\partial\tau_{12} \\ \partial/\partial\tau_{12} & \partial/\partial\tau_{22} \end{pmatrix} f.$$

The point is that $[f, g]$ is then a modular form in $M_{2,a+b}(\Gamma_2)$. Using this operation (an instance of Cohen–Rankin operators) Satoh showed in [86] that $\oplus_{k\equiv 0(2)} M_{2,k}$ is generated over the ring $\oplus_k M_k(\Gamma_2)$ of classical Siegel modular forms by such $[f, g]$ with f and g classical Siegel modular forms.

We give a little table with dimensions for $\dim S_{j,k}(\Gamma_2)$ for $4 \le k \le 20$, $0 \le j \le 18$ with j even:

j\k	4	5	6	7	8	9	10	11	12	13	14	15	16	17	18	19	20
0	0	0	0	0	0	0	1	0	1	0	1	0	2	0	2	0	3
2	0	0	0	0	0	0	0	0	0	0	1	0	2	0	2	0	3
4	0	0	0	0	0	0	1	0	1	0	2	1	3	1	4	2	6
6	0	0	0	0	1	0	1	1	2	1	3	2	5	3	7	4	9
8	0	0	0	0	1	1	2	1	3	2	5	4	7	5	9	7	13
10	0	0	0	0	0	1	2	1	3	2	5	5	8	6	11	9	15
12	0	0	1	1	2	2	4	4	6	5	9	8	13	11	17	15	22
14	0	0	0	1	2	2	4	4	6	6	10	10	15	13	19	18	26
16	0	0	1	1	3	3	6	5	9	8	13	13	19	17	25	23	33
18	0	1	1	2	4	5	7	8	11	11	17	17	23	23	31	30	40

The ring $\oplus_{j,k} M_{j,k}(\Gamma_2)$ is not finitely generated as was explained to me by Christian Grundh. Here is his argument.

Lemma 4. *The ring $\oplus_{j,k} M_{j,k}(\Gamma_2)$ is not finitely generated.*

Proof. Suppose that g_n for $n = 1, \ldots, r$ are the generators with weights (j_n, k_n). If we have a modular form g of weight (j, k) with $j > \max(j_n, n = 1, \ldots, r)$ then g is a sum of products of g_n, two of which at least have $j_n > 0$, hence by the properties of Φ we see that then $\Phi(g)$ is a sum of products of cusp forms, hence lies in the ideal generated by Δ^2 of the ring of elliptic modular forms. But for $j > 0, k > 4$ the map $\Phi : M_{j,k}(\Gamma_2) \to S_{j+k}(\Gamma_1)$ is surjective, so we have forms g in $M_{j,k}(\Gamma_2)$ that land in the ideal generated by Δ, but not in the ideal generated by Δ^2. Thus the ring cannot be generated by g_n for $n = 1, \ldots, r$.

Just as Δ is the first cusp form for $g = 1$ that one encounters the first vector-valued cusp form that one encounters for $g = 2$ is the generator of $S_{6,8}(\Gamma_2)$. The adjective 'first' refers to the fact that the weight of the local system $\mathbb{V}_{j+k-3,k-3}$ is $j + 2k - 6$. Our calculations (modulo the endoscopic conjecture given in Section 24) allow the determination of the eigenvalues $\lambda(p)$ and $\lambda(p^2)$ for $p = 2, 3, 5, 7$. We then can calculate the characteristic polynomial of Frobenius and even the slopes of it on $S_{6,8}(\Gamma_2)$.

p	$\lambda(p)$	$\lambda(p^2)$	slopes
2	0	-57344	$13/2, 25/2$
3	-27000	143765361	$3, 7, 12, 16$
5	2843100	-7734928874375	$2, 7, 12, 17$
7	-107822000	4057621173384801	$0, 6, 13, 19$

At our request Ibukiyama ([48]) has constructed a vector-valued Siegel modular form $0 \neq F \in S_{6,8}$, using a theta series for the lattice

$$\Gamma = \{x \in \mathbb{Q}^{16} : 2x_i \in \mathbb{Z}, x_i - x_j \in \mathbb{Z}, \sum_{i=1}^{16} x_i \in 2\mathbb{Z}\}.$$

One puts $a = (2, i, i, i, i, 0, \ldots, 0) \in \mathbb{C}^{16}$ and one denotes by $(\, ,)$ the usual scalar product. If $F = (F_0, \ldots, F_6)$ is the vector of functions on $\mathcal{H}_2$ defined by

$$F_\nu = \sum_{x,y \in \Gamma} (x, a)^{6-\nu} (y, a)^\nu e^{\pi i ((x,x)\tau_{11} + 2(x,y)\tau_{12} + (y,y)\tau_{22})} \qquad (\nu = 0, \ldots, 6)$$

with $\tau = (\tau_{11}, \tau_{12}; \tau_{12}, \tau_{22}) \in \mathcal{H}_2$, then Ibukiyama's result is that $F \neq 0$ and $F \in S_{6,8}$. The vanishing of $\lambda(2)$ in the table above agrees with this.

Here are two more examples of 1-dimensional spaces, the space $S_{18,5}$ and the last one, $S_{28,4}$. In these examples and the other ones we assume the validity of our conjecture on the endoscopic contribution. The eigenvalues $\lambda(p)$ grow approximately like $p^{(j+2k-3)/2}$, i.e. $p^{25/2}$ and $p^{33/2}$.

p	$\lambda(p)$ on $S_{18,5}$	$\lambda(p)$ on $S_{28,4}$
2	-2880	35040
3	-538920	30776760
5	118939500	522308049900
7	1043249200	18814963644400
11	-9077287359096	132158356344353064
13	-133873858788740	-1710588414695522180
17	667196591802660	-17044541241181641180
19	2075242468196920	8882130949720048073220
23	-8558834216776560	$-4334264380661701885752 0$
29	64653981488634780	$-17266319209397250361482 0$
31	-5977672283905752896	1826186223285615270299584
37	56922208975445092780	$-2974751686265520483949154 0$

In principle our database allows for the calculation of the traces of the Hecke operators $T(p)$ with $p \leq 37$ on the spaces $S_{j,k}$ for all values j, k. In the cases at hand these numbers tend to be 'smooth', i.e., they are highly composite numbers as we illustrate with the two 1-dimensional spaces $S_{j,k}$ for $(j, k) = (8, 8)$ and $(12, 6)$ (where the trace equals the eigenvalue of $T(p)$).

p	$\lambda(p)$ on $S_{8,8}$	$\lambda(p)$ on $S_{12,6}$
2	$2^6 \cdot 3 \cdot 7$	$-2^4 \cdot 3 \cdot 5$
3	$-2^3 \cdot 3^2 \cdot 89$	$2^3 \cdot 3^5 \cdot 5 \cdot 7$
5	$-2^2 \cdot 3 \cdot 5^2 \cdot 13^2 \cdot 607$	$2^2 \cdot 3 \cdot 5^2 \cdot 7 \cdot 79 \cdot 89$
7	$2^4 \cdot 7 \cdot 109 \cdot 36973$	$-2^4 \cdot 5^2 \cdot 7 \cdot 119633$
11	$2^3 \cdot 3 \cdot 4759 \cdot 114089$	$2^3 \cdot 3 \cdot 23 \cdot 2267 \cdot 2861$
13	$-2^2 \cdot 13 \cdot 17 \cdot 109 \cdot 3404113$	$2^2 \cdot 5 \cdot 7 \cdot 13 \cdot 50083049$
17	$2^2 \cdot 3^2 \cdot 17 \cdot 41 \cdot 1307 \cdot 168331$	$-2^2 \cdot 3^2 \cdot 5 \cdot 7 \cdot 13 \cdot 47 \cdot 14320807$
19	$-2^3 \cdot 5 \cdot 74707 \cdot 9443867$	$-2^3 \cdot 5 \cdot 7^3 \cdot 19 \cdot 2377 \cdot 35603$

Satoh had calculated a few eigenvalues of Hecke operators $T(m)$ acting on $S_{14,2}(\Gamma_2)$, (for $m = 2, 3, 4, 5, 9$ and 25) cf. [86], and our values agree with his.

26 Harder's Conjecture

In his study of the contribution of the boundary of the moduli space to the cohomology of local systems on the symplectic group, more precisely of the Eisenstein cohomology, Harder arrived at a conjectural congruence between modular forms for $g = 1$ and Siegel modular forms for $g = 2$, cf., [44, 45]. The second reference is his colloquium talk in Bonn (February 2003) which can be found in this volume and where this conjectural relationship was formulated in precise terms. One can view his conjectured congruences as a generalization of the famous congruence for the Fourier coefficients of the $g = 1$ cusp form $\Delta = \sum \tau(n) q^n$ of weight 12

$$\tau(p) \equiv p^{11} + 1 \,(\mathrm{mod}\,691).$$

To formulate it we start with a $g = 1$ cusp form $f \in S_r(\Gamma_1)$ of weight r that is a normalized eigenform of the Hecke operators. We write $f = \sum_{n \geq 1} a(n) q^n$ with $a(n) = 1$. To f we can associate the L-series $L(f, s)$ defined by $L(f, s) = \sum_{n \geq 1} a(n)/n^s$ for complex s with real part $> k/2 + 1$. If we define $\Lambda(f, s)$ by

$$\Lambda(f, s) = \frac{\Gamma(s)}{(2\pi)^s} L(f, s) = \int_0^\infty f(iy) y^{s-1} dy$$

then $\Lambda(f, s)$ admits a holomorphic continuation to the whole s-plane and satisfies a functional equation $\Lambda(f, s) = i^k \Lambda(f, k - s)$. It is customary to call the values $\Lambda(f, t)$ for $t = k - 1, k - 2, \ldots, 0$ the *critical values*. In view of the functional equation we may restrict to the values $t = k - 1, \ldots, k/2$.

A basic result due to Manin and Vishik is the following.

Theorem 14. *There exist two real numbers ('periods') ω_+, ω_- such that the ratios*
$$\Lambda(f, k - 1)/\omega_-, \ \Lambda(f, k - 2)/\omega_+, \ldots, \Lambda(f, k/2)/\omega_{(-1)^{k/2}}$$
are in the field of Fourier coeffients $\mathbb{Q}_f = \mathbb{Q}(a(n) : n \in \mathbb{Z}_{\geq 1})$.

If the Fourier coefficients are rational integers we may normalize these ratios so that we get integers in a minimal way. In practice one observes that one usually finds many small primes dividing these coordinates. By small we mean here less than k (or something close to this). Occasionally, there is a larger prime dividing these critical values of $\Lambda(f, s)$.

Instead of calculating the integrals one may use a slightly different approach by employing the so-called period polynomials, [65], which are defined for $f \in S_k(\Gamma_1)$ by $r = i r^+ + r^-$ with

$$r^+(f) = \sum_{0 \leq n \leq k-2, n \text{ even}} (-1)^{n/2} \left(\binom{k-2}{n} \right) r_n(f) X^{k-2-n}.$$

and

$$r^-(f) = \sum_{0 < n < k-2, n \text{ odd}} (-1)^{(n-1)/2} \left(\binom{k-2}{n} \right) r_n(f) X^{k-2-n}$$

with $r_n(f) = \int_0^\infty f(it) t^n dt$ for $n = 0, \ldots, k - 2$. Then the coefficients of these period polynomials give up to 'small' primes the critical L-values. These can be calculated purely algebraically and these are the ones that I used. By slight abuse of notation I denote these ratios again by the same symbols $(\Lambda(f, k - 1) : \Lambda(f, k - 3) : \ldots)$. See also [25] for more on the critical values.

For example, if we do this for $f = \Delta \in S_{12}$ then we get

$$(\Lambda(f,10) : \Lambda(f,8) : \Lambda(f,6)) = (48 : 25 : 20)$$

and see only 'small' primes. The first example where we see larger primes is the normalized eigenform $f = \Delta e_4 e_6 \in S_{22}$. We find for the even critical values

$$(\Lambda(f,20) : \ldots : \Lambda(f,12)) = (2^5 \cdot 3^3 \cdot 5 \cdot 19 : 2^3 \cdot 7 \cdot 13^2 : 3 \cdot 5 \cdot 7 \cdot 13 : 2 \cdot 3 \cdot 41 : 2 \cdot 3 \cdot 7)$$

where obviously 41 is the exception. We shall write for short $41 | \Lambda(f,14)$. What is the meaning of these exceptional primes dividing the critical values?

Harder made the following conjecture.

Conjecture 3. *(Harder's Conjecture) Let $f \in S_r(\Gamma_1)$ be a normalized eigenform with field of Fourier coefficients $\mathbb{Q}_f$. If a 'large' prime ℓ of $\mathbb{Q}_f$ divides a critical value $\Lambda(f,t)$ then there exists a Siegel modular form $F \in S_{j,k}(\Gamma_2)$ of genus 2 and weight (j,k) with $j = 2t - r - 2$ and $k = r - t + 2$ that is an eigenform for the Hecke algebra with eigenvalue $\lambda(p)$ for $T(p)$ with field $\mathbb{Q}_F$ of eigenvalues $\lambda(p)$ and such that for a suitable prime ℓ' of the compositum L of $\mathbb{Q}_f$ and $\mathbb{Q}_F$ dividing ℓ one has*

$$\lambda(p) \equiv p^{k-2} + a(p) + p^{j+k-1} \pmod{\ell'}$$

for all primes p.

(Here the $\lambda(p)$ are algebraic integers lying in a totally real field $\mathbb{Q}_F$. Harder formulated the conjecture for the case $L = \mathbb{Q}$.)

For example, if $f = \Delta e_4 e_6 \in S_{22}(\Gamma_1)$ is the unique normalized cusp form of weight 22 then $41 | \Lambda(f,14)$, so Harder predicts that the space $S_{4,10}$ should contain a non-zero eigenform F with eigenvalues $\lambda(p)$ satisfying $\lambda(p) \equiv p^8 + a(p) + p^{13} \pmod{41}$ for all p. A mimimum consistency is that at least $\dim S_{4,10}(\Gamma_2) \neq 0$; as it turns out this dimension is 1.

27 Evidence for Harder's Conjecture

Since we can calculate the trace of the Hecke operators $T(p)$ on the spaces $S_{j,k}(\Gamma_2)$ for all primes $p \leq 37$ (modulo the conjecture on the endoscopic contribution) we can try to check the conjecture by Harder (and gain evidence for the conjecture on the endoscopic contribution at the same time). As we just saw, the first case where we have a 'large' prime dividing a critical L-value is the eigenform $f = \Delta e_{10} \in S_{22}(\Gamma_1)$ of weight 22. Here the prime 41 divides the critical values $L(f,14)/\Omega^+$. The conjecture predicts a congruence between the Fourier coefficients of $f = \sum_{n=1}^{\infty} a(n)q^n$ and the eigenvalues $\lambda(p)$ of a form F in the 1-dimensional space $S_{4,10}(\Gamma_2)$. We give the tables with the eigenvalues $a(p)$ of f and $\lambda(p)$ of $F \in S_{4,10}(\Gamma_2)$ for the primes $p \leq 37$.

p	$a(p)$	$\lambda(p)$
2	-288	-1680
3	-128844	55080
5	21640950	-7338900
7	-768078808	609422800
11	-94724929188	25358200824
13	-80621789794	-263384451140
17	3052282930002	-2146704955740
19	-7920788351740	43021727413960
23	-73845437470344	-233610984201360
29	-4253031736469010	-545371828324260
31	1900541176310432	830680103136064
37	22191429912035222	11555498201265580

Proposition 6. *The congruence* $\lambda(p) \equiv p^8 + a(p) + p^{13} \pmod{41}$ *for the eigenvalues* $\lambda(p)$ *and* $a(p)$ *on* $S_{4,10}(\Gamma_2)$ *and* $S_{22}(\Gamma_1)$ *holds for all primes* $p \le 37$.

In this way we can check Harder's conjecture for many cases given in the tables below in the following sense. If both $\dim S_r(\Gamma_1) = 1$ and $\dim S_{j,k}(\Gamma_2) = 1$ and if ℓ is a prime $> r$ dividing the critical L-value then we checked the congruence $\lambda(p) - a(p) - p^{j+k-1} - p^{k-2} \equiv 0 \pmod{\ell}$ for all primes $p \le 37$. In case $\dim S_r(\Gamma_1) = 2$ and $\dim S_{j,k}(\Gamma_2) = 1$ I checked that in the quadratic field $\mathbb{Q}(a(p))$ the expression $\lambda(p) - a(p) - p^{j+k-1} - p^{k-2}$ has a norm divisible by ℓ for all primes $p \le 37$. With a bit of additional effort one can check the congruence in the real quadratic field. For example, take $r = 24$ and let

$$f = \sum a(n)q^n = q - (54 - 12\sqrt{144169})\, q^2 + \cdots$$

be a normalized eigenform in $S_{24}(\Gamma_1)$. In the quadratic field $\mathbb{Q}(\sqrt{144169})$ the prime 73 splits as $\pi \cdot \pi'$ with $\pi = (73, 53 + 36\sqrt{144169})$. Let $\lambda(p)$ be the eigenvalue under $T(p)$ of the generator of $S_{12,7}(\Gamma_2)$. Then we can check the congruence

$$\lambda(p) \equiv p^5 + a(p) + p^{18} \pmod{\pi}$$

for all $p \le 37$.

In case $\dim S_{j,k}(\Gamma_2) = 2$ I can calculate the characteristic polynomial g of $T(2)$. In general this is an irreducible polynomial g of degree 8 over $\mathbb{Q}$. The corresponding number field L possesses just one subfield L of degree 2 over $\mathbb{Q}$ and g decomposes in two polynomials of degree 4 that are irreducible over K. I then checked that the expression $\lambda(p) - a(p) - p^{j+k-1} - p^{k-2}$ has a norm in the composite field $(\mathbb{Q}(a(2)), K)$ which is divisible by our congruence prime ℓ.

For example, we treat the case of the local system $V_{18,6}$ with $(\ell, m) = (18, 6)$. The characteristic polynomial g of Frobenius at the prime 2 is:

$$1 + t_1 X + t_2 X^2 + t_3 X^3 + t_4 X^4 + 2^{27} t_3 X^5 + 2^{54} t_2 X^6 + 2^{81} t_1 X^7 + 2^{108} X^8.$$

with the coefficients $t_1 = 12432$, $t_2 = 193574912$, $t_3 = 3043199287296$ and $t_4 = 31380514975776768$. The corresponding degree 8 field extension K of $\mathbb{Q}$ has one quadratic subfield $\mathbb{Q}(\sqrt{7 \cdot 3607})$. Our polynomial g splits into the product of a quartic polynomial h

$$18014398509481984\, X^4 + (834297397248 - 9663676416\sqrt{25249})\, X^3 +$$
$$(142913536 - 110592\sqrt{25249})\, X^2 + (6216 - 72\sqrt{25249})\, X + 1$$

and its conjugate over this quadratic subfield $\mathbb{Q}(\sqrt{25249})$ and we get $\lambda(2) = -6216 \pm 72\sqrt{25249}$. The normalized eigenform in S_{28} has Fourier coefficient $a(2) = -4140 \pm 108\sqrt{18209}$ and one checks that the norm of

$$6216 + 72\sqrt{25249} + 2^7 + 2^{20} - (4140 + 108\sqrt{18209})$$

in the field $\mathbb{Q}(\sqrt{25249}, \sqrt{18209})$ is divisible by 4057 as predicted by Harder.

But there are cases where the characteristic polynomial g decomposes. These are the cases $(j, k) = (18, 7)$ where we have two factors of degree 4 and $(j, k) = (8, 13)$ where g is a product of four quadratic factors. In the cases $(j, k) = (18, 7)$ there is a congruence modulo 3779. In fact, g decomposes as the product of

$$288230376151711744X^4 - 4252017623040X^3 + 45752320X^2 - 7920X + 1$$

and

$$288230376151711744X^4 + 17575006175232X^3 + 857571328X^2 + 32736X + 1$$

and one calculates

$$\mathrm{Norm}(4320 + 96\sqrt{51349} + 2^{24} + 2^5 + 32736) = 282720345772032$$

and this is divisible by 3779. In the cases $(j, k, r) = (32, 4, 38)$ there are two congruence primes and one finds indeed a congruence for both of them.

The following table lists the congruence primes in question. All of these are checked in the sense explained above.

Let me finish by expressing the hope that these explicit examples will convince the reader that Siegel modular forms are not less fascinating than elliptic modular forms and moreover that in this corner of nature there are many exciting secrets that await discovery.

r	$\dim(S_r)$	(j,k)	$\dim(S_{j,k})$	L-value	primes
20	1	$(6,8)$	1	$2^2 \cdot 3 \cdot 11^2$	
22	1	$(4,10)$	1	$-2 \cdot 3 \cdot 17 \cdot 41$	41
22	1	$(8,8)$	1	$3 \cdot 7 \cdot 13 \cdot 17$	
22	1	$(12,6)$	1	$-2 \cdot 7 \cdot 13^2$	
24	2	$(12,7)$	1	$2^4 \cdot 5 \cdot 7^2 \cdot 11 \cdot 73$	73
24	2	$(6,10)$	1	$3 \cdot 11^2 \cdot 13^2 \cdot 17$	
24	2	$(8,9)$	1	$3 \cdot 7^2 \cdot 11 \cdot 19 \cdot 179$	179
26	1	$(4,12)$	1	$2 \cdot 11 \cdot 17 \cdot 19$	
26	1	$(6,11)$	1	$3 \cdot 5 \cdot 11 \cdot 19$	
26	1	$(10,9)$	1	$-2 \cdot 7 \cdot 11 \cdot 29$	29
26	1	$(14,7)$	1	$5 \cdot 7 \cdot 97$	97
26	1	$(16,6)$	1	$-2 \cdot 11 \cdot 17 \cdot 19$	
26	1	$(8,10)$	2	$-3^2 \cdot 7 \cdot 11 \cdot 19$	
26	1	$(12,8)$	2	$3 \cdot 5^2 \cdot 11 \cdot 17$	
28	2	$(2,14)$	1	$2^3 \cdot 5^2 \cdot 13^2 \cdot 17^2 \cdot 19 \cdot 23$	
28	2	$(16,7)$	1	$2^5 \cdot 3^4 \cdot 5 \cdot 7 \cdot 13 \cdot 367$	367
28	1	$(14,8)$	2	$2^4 \cdot 11 \cdot 13^2 \cdot 17 \cdot 19 \cdot 23 \cdot 647$	647
28	2	$(12,9)$	2	$2^3 \cdot 7 \cdot 11 \cdot 13 \cdot 23 \cdot 4057$	4057
28	2	$(8,11)$	1	$5 \cdot 11^2 \cdot 13 \cdot 23 \cdot 2027$	2027
28	2	$(18,6)$	1	$2^4 \cdot 3^2 \cdot 5^2 \cdot 11 \cdot 13^2 \cdot 17^2 \cdot 19$	
28	1	$(10,10)$	2	$2^2 \cdot 5^2 \cdot 11^2 \cdot 13^2 \cdot 17 \cdot 23 \cdot 157$	157
28	2	$(6,12)$	2	$5 \cdot 11^2 \cdot 13^2 \cdot 19 \cdot 23 \cdot 823$	823
28	2	$(20,5)$	1	$2^9 \cdot 3^4 \cdot 5 \cdot 193$	193
30	2	$(14,9)$	2	$2^8 \cdot 3 \cdot 5 \cdot 13 \cdot 1039$	1039
30	2	$(6,13)$	1	$2^4 \cdot 5 \cdot 11 \cdot 13 \cdot 19 \cdot 23$	
30	2	$(10,11)$	1	$3^4 \cdot 11 \cdot 13 \cdot 23 \cdot 97$	97
30	2	$(24,4)$	1	$2^{10} \cdot 3^4 \cdot 5^5 \cdot 7 \cdot 97$	97
30	2	$(20,6)$	2	$2^6 \cdot 3^3 \cdot 7 \cdot 11 \cdot 13 \cdot 17 \cdot 19 \cdot 23 \cdot 593$	593
30	2	$(4,14)$	2	$3^2 \cdot 5 \cdot 7^2 \cdot 13 \cdot 19^2 \cdot 23 \cdot 4289$	4289
30	2	$(18,7)$	2	$2^4 \cdot 3^2 \cdot 5 \cdot 11 \cdot 3779$	3779
32	2	$(4,15)$	1	$2^2 \cdot 5 \cdot 7^2 \cdot 13 \cdot 19 \cdot 23 \cdot 61$	61
32	2	$(2,16)$	2	$3^3 \cdot 5^2 \cdot 7^2 \cdot 19^2 \cdot 23 \cdot 211$	211
32	2	$(22,6)$	2	$2^3 \cdot 3^3 \cdot 5 \cdot 7 \cdot 13 \cdot 17 \cdot 19 \cdot 23 \cdot 7687$	7687
32	2	$(24,5)$	2	$2^9 \cdot 3^5 \cdot 5 \cdot 3119$	3119
32	2	$(8,13)$	2	$2 \cdot 7^3 \cdot 11^3 \cdot 13^2 \cdot 23$	
34	2	$(10,13)$	2	$2^3 \cdot 3^2 \cdot 5 \cdot 7 \cdot 13^2 \cdot 23^2 \cdot 29^2$	
34	2	$(28,4)$	1	$2^{10} \cdot 3^8 \cdot 5^5 \cdot 7 \cdot 103$	103
34	2	$(26,5)$	2	$2^{11} \cdot 3^3 \cdot 5^3 \cdot 15511$	15511
34	2	$(6,15)$	2	$2 \cdot 5^2 \cdot 7 \cdot 13 \cdot 23^2 \cdot 29 \cdot 233$	233
38	2	$(32,4)$	2	$2^8 \cdot 3^8 \cdot 5^4 \cdot 7^2 \cdot 67 \cdot 83$	67, 83

References

1. A.N. Andrianov: Quadratic forms and Hecke operators. Grundlehren der Mathematik 289, Springer Verlag, 1987.
2. A.N. Andrianov: Modular descent and the Saito–Kurokawa conjecture. *Invent. Math.* **53** (1979), p. 267–280.
3. A.N. Andrianov, V.L. Kalinin: On the analytic properties of standard zeta functions of Siegel modular forms. *Math. USSR Sb.* **35** (1979), p. 1–17.
4. A.N. Andrianov, V.G. Zhuravlev: Modular forms and Hecke operators. Translated from the 1990 Russian original by Neal Koblitz. Translations of Mathematical Monographs, 145. AMS, Providence, RI, 1995.
5. H. Aoki: Estimating Siegel modular forms of genus 2 using Jacobi forms. *J. Math. Kyoto Univ.* **40** (2000), p. 581–588.
6. T. Arakawa: Vector valued Siegel's modular forms of degree 2 and the associated Andrianov *L*-functions, *Manuscr. Math.* **44** (1983) p. 155–185.
7. A. Ash, D. Mumford, M. Rapoport, Y. Tai: Smooth compactification of locally symmetric varieties. *Lie Groups*: History, Frontiers and Applications, Vol. IV. Math. Sci. Press, Brookline, Mass., 1975.
8. W. Baily, A. Borel: Compactification of arithmetic quotients of bounded symmetric domains. *Ann. of Math,* **84** (1966), p. 442–528.
9. J. Bergström, G. van der Geer: The Euler characteristic of local systems on the moduli of curves and abelian varieties of genus 3. `arXiv:0705.0293`
10. B. Birch: How the number of points of an elliptic curve over a fixed prime field varies. *J. London Math. Soc.* **43** (1968), p. 57–60.
11. D. Blasius, J.D. Rogawski: Zeta functions of Shimura varieties. In: *Motives (2)*, U. Jannsen, S. Kleiman, J.-P. Serre, Eds., Proc. Symp. Pure Math. **55** (1994), p. 447–524.
12. S. Böcherer: Siegel modular forms and theta series. *Proc. Symp. Pure Math.* **49**, Part 2, (1989), p. 3–17.
13. S. Böcherer: Über die Funktionalgleichung automorpher *L*-Funktionen zur Siegelschen Modulgruppe. *J. Reine Angew. Math.* **362** (1985), p. 146–168.
14. R.E. Borcherds, E. Freitag, R. Weissauer: A Siegel cusp form of degree 12 and weight 12. *J. Reine Angew. Math.* **494** (1998), p. 141–153.
15. S. Breulmann, M. Kuss: On a conjecture of Duke-Imamoğlu. *Proc. A.M.S.* **128** (2000), p. 1595–1604.
16. H. Braun: Eine Frau und die Mathematik 1933–1940. Der Beginn einer wissenschaftlichen Laufbahn. Herausgegeben von Max Koecher. Berlin etc.: Springer-Verlag, 1990.
17. P. Cartier: Representations of *p*-adic groups: A survey. Proc. Symp. Pure Math. **33**,1, p. 111–155.
18. C. Consani, C. Faber: On the cusp form motives in genus 1 and level 1. In: Moduli and arithmetic geometry (Kyoto, 2004). Advanced Studies in Pure Mathematics 45, Math. Soc. of Japan, 2006, p. 297–314.
19. M. Courtieu, A. Panchishkin: *Non-Archimedean L-functions and arithmetical Siegel modular forms.* Second edition. Lecture Notes in Mathematics, 1471. Springer-Verlag, Berlin, 2004.
20. P. Deligne: Formes modulaires et représentations ℓ-adiques. Sém. Bourbaki 1968/9, no. 355. Lecture Notes in Math. 179 (1971), p. 139–172.

21. P. Deligne: Travaux de Shimura. Séminaire Bourbaki 1971. 23ème année (1970/71), Exp. No. 389, pp. 123–165. Lecture Notes in Math., Vol. 244, Springer, Berlin, 1971.

22. P. Deligne: Variétés de Shimura: Interpretation modulaire et techniques de construction de modèles canoniques. Automorphic forms, representations and L-functions, Part 2, pp. 247–289, Proc. Sympos. Pure Math., XXXIII, Amer. Math. Soc., Providence, R.I., 1979.

23. W. Duke, Ö. Imamoğlu: A converse theorem and the Saito–Kurokawa lift. Int. Math. Res. Notices 7 (1996), p. 347–355.

24. W. Duke, Ö. Imamoğlu: Siegel modular forms of small weight. Math. Annalen 310 (1998), p. 73–82.

25. N. Dummigan: Period ratios of modular forms. Math. Ann. 318 (2000), p. 621–636.

26. M. Eichler, D. Zagier: The theory of Jacobi forms. Progress in Mathematics, 55. Birkhäuser Boston, Inc., Boston, MA, 1985.

27. C. Faber, G. van der Geer: Sur la cohomologie des systèmes locaux sur les espaces de modules des courbes de genre 2 et des surfaces abéliennes. I, II C. R. Math. Acad. Sci. Paris 338, (2004) No.5, p. 381–384 and No.6, 467–470.

28. G. Faltings: On the cohomology of locally symmetric Hermitian spaces. Paul Dubreil and Marie-Paule Malliavin algebra seminar, Paris, 1982, 55–98, Lecture Notes in Math., 1029, Springer, Berlin, 1983.

29. G. Faltings, C-L. Chai: Degeneration of abelian varieties. Ergebnisse der Math. 22. Springer Verlag 1990.

30. E. Freitag: Siegelsche Modulfunktionen. Grundlehren der Mathematischen Wissenschaften 254. Springer-Verlag, Berlin

31. E. Freitag: Singular modular forms and theta relations. Lecture Notes in Math. 1487. Springer Verlag

32. E. Freitag: Eine Verschwindungssatz für automorphe Formen zur Siegelschen Modulgruppe. Math. Zeitschrift 165, (1979), p. 11–18.

33. E. Freitag: Zur theorie der Modulformen zweiten Grades. Nachr. Akad. Wiss. Göttingen Math.-Phys. Kl. II (1965) p. 151–157.

34. W. Fulton, J. Harris: Representation theory. A first course. Graduate Texts in Mathematics, 129. Readings in Mathematics. Springer-Verlag, New York, 1991.

35. G. van der Geer: Hilbert Modular Surfaces. Springer Verlag 1987.

36. G. van der Geer, M. van der Vlugt: Supersingular curves of genus 2 over finite fields of characteristic 2. Math. Nachr. 159 (1992), p. 73–81.

37. E. Getzler: Euler characteristics of local systems on $\mathcal{M}_2$. Compositio Math. 132 (2002), 121–135.

38. R. Godement: Fonctions automorphes, vol. 1. Seminaire H. Cartan, 1957/8 Paris.

39. E. Gottschling: Explizite Bestimmung der Randflächen des Fundamentalbereiches der Modulgruppe zweiten Grades. Math. Ann. 138, 1959, p. 103–124.

40. B.H. Gross: On the Satake isomorphism. Galois representations in arithmetic algebraic geometry (Durham, 1996), p. 223–237, London Math. Soc. Lecture Note Ser., 254, Cambridge Univ. Press, Cambridge, 1998.

41. C. Grundh: Master Thesis. Stockholm.

42. S. Grushevsky: Geometry of $\mathcal{A}_g$ and its compactifications. To appear in Proc. Symp. Pure Math.

43. W.F. Hammond: On the graded ring of Siegel modular forms of genus two. *Amer. J. Math.* **87** (1965), p. 502–506.

44. G. Harder: Eisensteinkohomologie und die Konstruktion gemischter Motive. Lecture Notes in Mathematics, 1562. Springer-Verlag, Berlin, 1993.

45. G. Harder: A congruence between a Siegel and an elliptic modular form. Manuscript, February 2003. In this volume.

46. M. Harris: *Arithmetic vector bundles on Shimura varieties*. Automorphic forms of several variables (Katata, 1983), p. 138–159, Progr. Math., 46, Birkhäuser Boston, Boston, MA, 1984.

47. K. Hulek, G.K. Sankaran: The geometry of Siegel modular varieties. Higher dimensional birational geometry (Kyoto, 1997), p. 89–156, Adv. Stud. Pure Math., 35, Math. Soc. Japan, Tokyo, 2002.

48. T. Ibukiyama: Vector valued Siegel modular forms of symmetric tensor representations of degree 2. Unpublished preprint.

49. T. Ibukiyama: Vector valued Siegel modular forms of $\det^k \mathrm{Sym}(4)$ and $\det^k \mathrm{Sym}(6)$. Unpublished preprint.

50. T. Ibukiyama: Letter to G. van der Geer, July 2001.

51. T. Ibukiyama: Construction of vector valued Siegel modular forms and conjecture on Shimura correspondence. Preprint 2004.

52. J. Igusa: On Siegel modular forms of genus 2. *Am. J. Math.* **84** (1962), p. 612–649.

53. J. Igusa: Modular forms and projective invariants. *Am. J. Math.* **89** (1967), p. 817–855.

54. J. Igusa: On the ring of modular forms of degree two over $\mathbb{Z}$. *Am. J. Math.* **101** (1979), p. 149–183.

55. J. Igusa: Schottky's invariant and quadratic forms. E. B. Christoffel (Aachen/Monschau, 1979), p. 352–362, Birkhäuser, Basel-Boston, Mass., 1981.

56. J. Igusa: Theta Functions. Springer Verlag. Die Grundlehren der mathematischen Wissenschaften, Band 194. Springer-Verlag, New York-Heidelberg, 1972.

57. J. Igusa: On Jacobi's derivative formula and its generalizations. *Amer. J. Math.* **102** (1980), no. 2, p. 409–446.

58. T. Ikeda: On the lifting of elliptic cusp forms to Siegel cusp forms of degree $2n$. *Ann. of Math.* (2) **154** (2001), p. 641–681.

59. T. Ikeda: Pullback of the lifting of elliptic cusp forms and Miyawaki's conjecture. *Duke Math. Journal* **131**, (2006), p. 469–497.

60. R. de Jong: Falting's Delta invariant of a hyperelliptic Riemann surface. In: Number Fields and Function Fields, two parallel worlds. (Eds. G. van der Geer, B. Moonen, R. Schoof). Progress in Math. 239, Birkhäuser 2005.

61. H. Klingen: Introductory lectures on Siegel modular forms. Cambridge Studies in advanced mathematics 20. Cambridge University Press 1990.

62. W. Kohnen: Lifting modular forms of half-integral weight to Siegel modular forms of even genus. *Math. Ann.* **322**, (2003), p. 787–809.

63. M. Koecher: Zur Theorie der Modulformen n-ten Grades. I. *Math. Zeitschrift* **59** (1954), p. 399–416.

64. W. Kohnen, H. Kojima: A Maass space in higher genus. *Compositio Math.* **141** (2005), 313–322.

65. W. Kohnen, D. Zagier: Modular forms with rational periods. In: Modular forms (Durham, 1983), p. 197–249. Ellis Horwood Ser. Math. Appl.: Statist. Oper. Res., Horwood, Chichester, 1984.

244   G. van der Geer

66. B. Kostant: Lie algebra cohomology and the generalized Borel-Weil theorem. *Ann. of Math.* **74** (1961), p. 329–387.
67. A. Krieg: Das Vertauschungsgesetz zwischen Hecke-Operatoren und dem Siegelschen ϕ-Operator. *Arch. Math.* **46** (1986), p. 323–329.
68. S.S. Kudla, S. Rallis: A regularized Siegel-Weil formula: the first term identity. *Annals of Math.* **140** (1994), 1–80.
69. N. Kurokawa: Examples of eigenvalues of Hecke operators on Siegel cusp forms of degree two. *Invent. Math.* **49** (1978), p. 149–165.
70. H. Maass: Siegel's modular forms and Dirichlet series. Lecture Notes in Math. 216, Springer Verlag, 1971.
71. H. Maass: Über eine Spezialschar von Modulformen zweiten Grades. *Invent. Math.* **52** (1979), p. 95–104. II *Invent. Math.* **53** (1979), p. 249–253. III *Invent. Math.* **53** (1979), p. 255–265.
72. I. Miyawaki: Numerical examples of Siegel cusp forms of degree 3 and their zeta-functions. *Mem. Fac. Sci. Kyushu Univ. Ser. A* **46** (1992), p. 307–339.
73. S. Mizumoto: Poles and residues of standard L-functions attached to Siegel modular forms. *Math. Annalen* **289** (1991), p. 589–612.
74. K. Murokawa: Relations between symmetric power L-functions and spinor L-functions attached to Ikeda lifts. *Kodai Math. J.* **25** (2002), p. 61–71.
75. D. Mumford: On the Kodaira dimension of the Siegel modular variety. *Algebraic geometry - open problems* (Ravello, 1982), p. 348–375, Lecture Notes in Math., 997, Springer, Berlin, 1983.
76. D. Mumford: Abelian varieties. Tata Institute of Fundamental Research Studies in Mathematics, No. 5 Published for the Tata Institute of Fundamental Research, Bombay; Oxford University Press, London 1970.
77. Y. Namikawa: Toroidal compactification of Siegel spaces. Lecture Notes in Mathematics, 812. Springer, Berlin, 1980.
78. G. Nebe, B. Venkov: On Siegel modular forms of weight 12. *J. reine angew. Math.* **351** (2001), p. 49–60.
79. I. Piatetski–Shapiro, S. Rallis: L-functions of automorphic forms on simple classical groups. In: *Modular forms* (Durham, 1983), p. 251–261, Ellis Horwood, Horwood, Chichester, 1984.
80. C. Poor: Schottky's form and the hyperelliptic locus. *Proc. Amer. Math. Soc.* **124** (1996), 1987–1991.
81. B. Riemann: Theorie der Abel'schen Funktionen. *J. für die reine und angew. Math.* **54** (1857), p. 101–155.
82. N.C. Ryan: Computing the Satake p-parameters of Siegel modular forms. math.NT/0411393.
83. R. Salvati Manni: On the holomorphic differential forms of the Siegel modular variety. *Arch. Math. (Basel)* **53** (1989), no. 4, 363–372.
84. R. Salvati–Manni: On the holomorphic differential forms of the Siegel modular variety. II. *Math. Z.* **204** (1990), no. 4, 475–484.
85. I. Satake: On the compactification of the Siegel space. *J. Indian Math. Soc.* **20** (1956), p. 259–281.
86. T. Satoh: On certain vector valued Siegel modular forms of degree 2. *Math. Ann.* **274** (1986) p. 335–352.
87. A.J. Scholl: Motives for modular forms. *Invent. Math.* **100** (1990), p. 419–430.
88. J. Schwermer: On Euler products and residual Eisenstein cohomology classes for Siegel modular varieties. *Forum Math.* **7** (1995), p. 1–28.

89. J.-P. Serre: Rigidité du foncteur de Jacobi d'échelon $n \geq 3$. Appendice d'exposé 17, Séminaire Henri Cartan 13e année, 1960/61.

90. C.L. Siegel: Einführung in die Theorie der Modulfunktionen n-ten Grades. *Math. Annalen* **116** (1939), p. 617–657 (= Gesammelte Abhandlungen, II, p. 97–137).

91. C.L. Siegel: Symplectic geometry. *Am. J. of Math.* **65** (1943), p. 1–86 (= Gesammelte Abhandlungen, II, p. 274–359. Springer Verlag.)

92. G. Shimura: Introduction to the arithmetic theory of automorphic functions. Reprint of the 1971 original. Publications of the Math. So. of Japan, 11. Kanô Memorial Lectures, 1. Princeton University Press, Princeton, NJ, 1994.

93. G. Shimura: On modular correspondences for $\mathrm{Sp}(n, \mathbb{Z})$ and their congruence relations. *Proc. Ac. Sci. USA* **49**, (1963), p. 824–828.

94. G. Shimura: Arithmeticity in the theory of automorphic forms. Mathematical Surveys and Monographs, 82. American Mathematical Society, Providence, RI, 2000.

95. G. Shimura: Arithmetic and analytic theories of quadratic forms and Clifford groups. Mathematical Surveys and Monographs, 109. American Mathematical Society, Providence, RI, 2004.

96. C.L. Siegel: Über die analytische Theorie der quadratischen Formen. *Annals of Math.* **36** (1935), 527–606.

97. C.L. Siegel: Symplectic geometry. *Amer. J. Math.* **65** (1943), p. 1–86.

98. C.L. Siegel: Zur Theorie der Modulfunktionen n-ten Grades. *Comm. Pure Appl. Math.* **8** (1955), p. 677–681.

99. R. Taylor: On the ℓ-adic cohomology of Siegel threefolds. *Invent. Math.* **114** (1993), 289–310.

100. R. Tsushima: A formula for the dimension of spaces of Siegel cusp forms of degree three. *Am. J. Math.* **102** (1980), p. 937–977.

101. R. Tsushima: An explicit dimension formula for the spaces of generalized automorphic forms with respect to $\mathrm{Sp}(2, \mathbb{Z})$. *Proc. Jap. Acad.* **59A** (1983), 139–142.

102. S. Tsuyumine: On Siegel modular forms of degree three. *Amer. J. Math.* **108** (1986), p. 755–862. Addendum. *Amer. J. Math.* **108** (1986), p. 1001–1003.

103. T. Veenstra: Siegel modular forms, L-functions and Satake parameters. *J. Number Theory* **87** (2001), p. 15–30.

104. H. Yoshida: Motives and Siegel modular forms. *American Journal of Math.* **123** (2001), p. 1171–1197.

105. M. Weissman: Multiplying modular forms. Preprint, 2007.

106. R. Weissauer: Vektorwertige Modulformen kleinen Gewichts. *Journal für die reine und angewandte Math.* **343** (1983), p. 184–202.

107. R. Weissauer: Stabile Modulformen und Eisensteinreihen. Lecture Notes in Mathematics, 1219. Springer-Verlag, Berlin, 1986.

108. E. Witt: Eine Identität zwischen Modulformen zweiten Grades. *Math. Sem. Hamburg* **14** (1941), p. 323–337.

109. D. Zagier: Sur la conjecture de Saito–Kurokawa (d'après H. Maass). Seminaire Delange-Pisot-Poitou, Paris 1979–80, pp. 371–394, Progr. Math., 12, Birkhäuser, Boston, Mass., 1981.

A Congruence Between a Siegel and an Elliptic Modular Form

Günter Harder

Mathematisches Institut, Universität Bonn, Beringstraße 1, 53115 Bonn, Germany
E-mail: harder@math.uni-bonn.de

Preface

The winter semester 2002/2003 was the last semester before my retirement from the university. It also happened that I was the chairman of the Colloquium and the speaker foreseen for February 7 had to cancel his visit.

At about the same time I found some numerical support for a very general conjecture relating divisibilities of certain special values of L-functions to congruences between modular forms. I have been thinking about this kind of relationship for many years, but I never had any idea how one could find experimental evidence. But in the early 2003 C. Faber and G. van der Geer had written a program that produced lists of eigenvalues of Hecke operators on some special Siegel modular forms. After a few days of suspense we could compare their list with my list of eigenvalues of elliptic modular forms and verify the congruence in our examples.

I was very exited about this and spontaneously invited myself to give the Colloquium lecture, which is documented in the text below. (Bonn Spring 2007)

1 Elliptic and Siegel Modular Forms

I have to recall some well known facts from the classical theory of modular forms. We have the upper half plane

$$\mathbb{H} = \{z \mid x + iy \text{ with } y > 0\} \,.$$

On this upper half plane we have an action of $Sl_2(\mathbb{R})$, which is given by

$$Sl_2(\mathbb{R}) \times \mathbb{H} \longrightarrow \mathbb{H}$$
$$\left(\begin{pmatrix} a & b \\ c & d \end{pmatrix}, z \right) \mapsto \frac{az+b}{cz+d} \,.$$

The stabilizer of $i \in \mathbb{H}$ is the maximal compact subgroup $SO(2)$ and we can identify $\mathbb{H} = Sl_2(\mathbb{R})/SO(2)$. Let k be a positive (even) integer. A *holomorphic modular form of weight k with respect to $Sl_2(\mathbb{Z})$* is a holomorphic function $f : \mathbb{H} \to \mathbb{C}$, which satisfies

$$f\left(\frac{az+b}{cz+d}\right) = (cz+d)^k f(z)$$

for all matrices

$$\begin{pmatrix} a & b \\ c & d \end{pmatrix} \in Sl_2(\mathbb{Z}) \,,$$

and which satisfies a growth condition. To formulate this growth condition we restrict f to a "neighborhood of infinity" $\mathbb{H}(c) = \{z | \Im(z) > c\}$. On this neighborhood the group $\Gamma_\infty = \begin{pmatrix} 1 & n \\ 0 & 1 \end{pmatrix}$ with $n \in \mathbb{Z}$ acts and the map $z \mapsto e^{2\pi i z}$ identifies $\Gamma_\infty \backslash \mathbb{H}(c)$ to a punctured disk. Since f satisfies $f(z) = f(z+1)$ we can view its restriction to $\Gamma_\infty \backslash \mathbb{H}(c)$ as a function in the variable q. The growth condition requires that f has a (Fourier or Laurent) expansion

$$f(q) = a_0 + a_1 q + a_2 q^2 \dots \,,$$

i.e. it extends to a holomorphic function on the disk. If $a_0 = 0$, then f is called a *cusp form*.

Remark: The quotient $Sl_2(\mathbb{Z})\backslash\mathbb{H}$ has the structure of a Riemann surface, which can be compactified to a compact Riemann surface $\overline{Sl_2(\mathbb{Z})\backslash\mathbb{H}}$ by adding one point at ∞. We write the maximal compact subgroup

$$SO(2) = U(1) = K = \left\{ e(\phi) \mid e(\phi) = \begin{pmatrix} \cos(\phi) & \sin(\phi) \\ -\sin(\phi) & \cos(\phi) \end{pmatrix} \right\} \,.$$

Since $\mathbb{H} = Sl_2(\mathbb{R})/SO(2)$, the representation $\rho_k : SO(2) \to \mathbb{C}^\times$, which is given by $e(\phi) \mapsto e(\phi)^k$ defines a $Sl_2(\mathbb{R})$-invariant holomorphic line bundle $\mathcal{L}_k$ on $\mathbb{H}$, this gives us a line bundle, also called $\mathcal{L}_k$, on $Sl_2(\mathbb{Z})\backslash\mathbb{H}$. This line bundle can be extended in a specific way to a line bundle on the compactification. Then the space of modular forms of weight k can be canonically identified with the space of sections $H^0(\overline{Sl_2(\mathbb{Z})\backslash\mathbb{H}}, \mathcal{L}_k)$.

We have the two modular forms of weight 4 and 6

$$E_4(z) = \frac{1}{2} \sum_{(c,d)=1} \frac{1}{(cz+d)^4} \,,$$

$$E_6(z) = \frac{1}{2} \sum_{(c,d)=1} \frac{1}{(cz+d)^6} \,,$$

and then we have the q-expansions

$$E_4(q) = 1 + 240q + 2160q^2 + 6720q^3 + 17520q^4 + 30240q^5 \dots$$
$$E_6(q) = 1 - 504q - 16632q^2 - 122976q^3 - 532728q^4 - 1575504q^5 + \dots \,.$$

The space of cusp forms has dimension 1 for the values $k = 12, 16, 18, 20, 22, 26$. The modular form

$$\Delta(z) = \frac{E_4(q)^3 - E_6(q)^2}{12^3} = q - 24q^2 + 252q^3 - 1472q^4 + 4830q^5 + \ldots$$

is the generator of the space of cusp forms of weight 12.

The space of cusp forms of weight 22 is generated by

$$f(q) = \frac{E_6(q)E_4(q)^4 - E_6(q)^3 \cdot E_4(q)}{12^3} = q - 288q^2 - 128844q^3 - 2014208q^4$$
$$+ 21640950q^5 + 37107072q^6 - 768078808q^7 + 1184071680q^8$$
$$+ 6140423133q^9 - 6232593600q^{10} - 94724929188q^{11} \pm \ldots .$$

Now I have to say a few words on Siegel modular forms. We start from a lattice

$$L = \mathbb{Z}^4 = \mathbb{Z}e_1 \oplus \mathbb{Z}e_2 \oplus \mathbb{Z}f_2 \oplus \mathbb{Z}f_1$$

on which we have an alternating pairing which on the basis vectors is given by

$$\langle e_1, f_1 \rangle = \langle e_2, f_2 \rangle = -\langle f_1, e_1 \rangle = -\langle f_2, e_2 \rangle = 1 \; ,$$

and all other values of the pairing are zero. The group of automorphisms of this symplectic form is a semi-simple group scheme $Sp_2/\operatorname{Spec}(\mathbb{Z})$. This is the symplectic group of genus 2.

Its group of real points

$$Sp_2(\mathbb{R}) = \{g \in GL_4(\mathbb{R}) \mid \langle gx, gy \rangle = \langle x, y \rangle\}$$

contains $U(2)$ as a maximal compact subgroup and we can form the quotient space

$$\mathbb{H}_2 = Sp_2(\mathbb{R})/U(2) \; .$$

This is the space of symmetric 2×2 matrices

$$Z = X + iY$$

with complex entries whose imaginary part Y is positive definite. Hence we have a complex structure on this space. This complex structure can also be seen in the following way: let $P(\mathbb{C})$ in $Sp_2(\mathbb{C})$ be the stabilizer of the isotropic plane $\{e_1 - if_1, e_2 - if_2\} \subset \mathbb{C}^4$, then we have an open embedding

$$\mathbb{H}_2 = Sp_2(\mathbb{R})/U(2) \hookrightarrow Sp_2(\mathbb{C})/P(\mathbb{C}) \; ,$$

the group $SU(2)$ is the group of real points of $P(\mathbb{C})$ intersected with its complex conjugate $\bar{P}(\mathbb{C})$. The object on the right is the Grassmann variety of isotropic complex planes in $(\mathbb{C}^4, \langle \; , \; \rangle)$. It is projective and of dimension 3. The group $\Gamma = Sp_2(\mathbb{Z})$ acts upon $\mathbb{H}_2$ and the quotient $\Gamma \backslash \mathbb{H}_2$ is a quasiprojective algebraic variety over $\mathbb{C}$. We have a homomorphism

$P(\mathbb{C}) \to GL_2(\mathbb{C})$. For any pair of integers $i \geq 0, j$ the holomorphic representation

$$\rho : GL_2(\mathbb{C}) \longrightarrow \mathrm{Sym}^i(\mathbb{C}^2) \otimes \det^j$$

defines a holomorphic vector bundle $\mathcal{E}_{ij}$ on the flag variety $Sp_2(\mathbb{C})/P(\mathbb{C})$ which is $Sp_2(\mathbb{C})$-equivariant. Hence its restriction – also called $\mathcal{E}_{ij}$ – to $\mathbb{H}_2$ is a $Sp_2(\mathbb{R})$ equivariant holomorphic bundle on $\mathbb{H}_2$ and hence descends to a holomorphic bundle on $\Gamma\backslash\mathbb{H}_2$. We can consider the space of holomorphic sections

$$H^0(\Gamma\backslash\mathbb{H}_2, \mathcal{E}_{ij}) ,$$

and define the subspace of modular forms M_{ij} (which satisfy some growth condition) and the subspace S_{ij} of cusp forms; these are rapidly decreasing at infinity. These spaces are called the spaces of modular forms (cusp forms) of weight i, j. (See remark above.)

There are formulas by R. Tsushima for the dimensions of these spaces S_{ij} (Riemann–Roch–Hirzebruch or the trace formula), and for small values i, j the dimensions are zero. We say that i, j is a regular pair if $i > 0, j > 3$. We have 29 cases of regular pairs i, j where S_{ij} is of dimension one.

2 The Hecke Algebra and a Congruence

Whenever we have such a space of modular forms we have an action of the algebra of Hecke operators on it. This is an algebra generated by operators T_p (for $Sl_2(\mathbb{Z})$) and $T_p^{(\nu)}$, $\nu = 1, 2$ (for $Sp_2(\mathbb{Z})$), which are attached to a prime p and which induce endomorphisms $T_p^{(\nu)} : S_{ij} \to S_{ij}$, and which commute with each other. If we pick a prime, then we can consider the matrix

$$\begin{pmatrix} p & & & 0 \\ & p & & \\ & & 1 & \\ 0 & & & 1 \end{pmatrix}$$

which is in $GSp_2(\mathbb{Q})$, and if $f(Z) \in H^0(\Gamma\backslash\mathbb{H}_2, \mathcal{E}_{ij})$, then $f\left(\begin{pmatrix} p & & & 0 \\ & p & & \\ & & 1 & \\ 0 & & & 1 \end{pmatrix} Z \right)$

is not invariant under Γ; it is only a section in

$$H^0(\Gamma_0(p)\backslash\mathbb{H}_2, \mathcal{E}_{ij}) ,$$

where $\Gamma_0(p) \subset \Gamma$ is a subgroup of finite index. We can form a trace by summing over $\Gamma_0(p)\backslash\Gamma$ and up to a normalizing factor this will be our operator

$$T_p^{(1)} : S_{ij} \longrightarrow S_{ij} .$$

If now $\dim S_{ij} = 1$, then the operator $T_p^{(1)} : S_{ij} \to S_{ij}$ induces the multiplication by a number $\lambda(p)$ on S_{ij} and if $j \geq 3$, then we get a sequence of integers

$$\{\lambda(p)\}_{p \in \mathrm{Primes}}$$

which, of course, depends on i, j.

We also have the Hecke operators for classical modular forms, and our cusp form f of weight 22 is also an eigenform for the operators T_p. In this case the situation is simple. Because f is normalized, i.e. $a_1 = 1$, we have

$$T_p f = a_p f$$

where a_p is the p-th Fourier coefficient. We have $\dim S_{4,10} = 1$ and formulate the conjecture

Conjecture: *For $S_{4,10}$ we have a congruence*

$$\lambda(p) \equiv p^8 + a_p + p^{13} \quad \mathrm{mod}\, 41 \quad \textit{for all primes } p .$$

Prop. *The conjecture holds for $2 \leq p \leq 11$.*

One might say that this is really not so much evidence for the conjecture. But here are certain numbers, namely, $4, 10, 22, 8, 13$ and 41, which seem to be somewhat arbitrary. I did not play with these numbers until I found a congruence. I picked all these numbers in advance and only then I checked the congruence, which I expected to be true for this specific choice.

The congruence is a generalization of a classical congruence. If we write the Δ-function

$$\Delta(z) = q - 24q^2 + 252q^3 - 1472q^4 + 4830q^5 \pm \ldots = \sum_{n=1}^{\infty} \tau(n)q^n ,$$

then we have the famous Ramanujan congruence

$$\tau(p) \equiv p^{11} + 1 \quad \mathrm{mod}\, 691 \quad \text{for all primes } p .$$

But there is a difference: Usually people interpret this last congruence as a congruence between the q-expansions of two modular forms, namely the Δ-function and the Eisenstein series $E_{12}(z)$. Since the Fourier coefficients are the same as the eigenvalues of the Hecke operators we also get the congruence between the eigenvalues. For the congruence between the Siegel modular form and the elliptic modular form we only have a congruence between Hecke eigenvalues. I do not see a congruence between Fourier coefficients.

I want to say something about the numbers, how I get them and I want to say a few words about the meaning of this congruence.

3 The Special Values of the *L*-function

We start from our modular cusp form of weight 22

$$f(q) = q - 288q^2 - 128844q^3 - 2014208q^4 + 21640950q^5 + \ldots = \sum_{n=1}^{\infty} a_n q^n ,$$

we have its associated *L*-function $L(f,s) = \sum_{n=1}^{\infty} \frac{a_n}{n^s}$, and because f is an eigenform for the Hecke algebra this *L*-function has an Euler product expansion

$$L(f,s) = \sum_{n=1}^{\infty} \frac{a_n}{n^s} = \prod_{p} \frac{1}{1 - a_p p^{-s} + p^{21-2s}} .$$

Actually it is better to consider the Mellin transform

$$\int_0^{\infty} f(iy) y^s \frac{dy}{y} = \frac{\Gamma(s)}{(2\pi)^s} \cdot L(f,s) = \Lambda(f,s) .$$

From this integral representation we easily get the functional equation

$$\Lambda(f, 22-s) = -\Lambda(f,s) .$$

Now we consider the "special" values $\Lambda(f,21), \Lambda(f,20), \ldots, \Lambda(f,11)$. It follows from the theory of modular symbols (Manin–Vishik) that there exist two real numbers $\Omega_-, \Omega_+ \neq 0$ (the *periods*) such that

$$\frac{\Lambda(f,21)}{\Omega_-}, \quad \frac{\Lambda(f,20)}{\Omega_+}, \quad \frac{\Lambda(f,19)}{\Omega_-}, \quad \ldots \in \mathbb{Q}$$

These periods are only defined up to elements in $\mathbb{Q}^\times$, but a closer look allows us to pin them down up to a factor in $\mathbb{Z}^\times = \{\pm 1\}$. In this case we can simply try to normalize them such that

$$\left\{ \frac{\Lambda(f,21)}{\Omega_-}, \quad \frac{\Lambda(f,19)}{\Omega_-}, \quad \ldots, \quad \frac{\Lambda(f,11)}{\Omega_-} \right\}$$

and

$$\left\{ \frac{\Lambda(f,20)}{\Omega_+}, \quad \ldots, \quad \frac{\Lambda(f,14)}{\Omega_+}, \quad \frac{\Lambda(f,12)}{\Omega_+} \right\}$$

are sets of co-prime integers. Of course, it is not so difficult to produce these lists of integers. (From this list we conclude that the normalization of Ω_- was not the right one. This is related to the fact that

$$131 \cdot 593 \mid \zeta(-21) ,$$

and this produces a congruence between f and an Eisenstein series

$$a_p \equiv p^{21} + 1 \mod 131 \cdot 593 .$$

This forces us to replace Ω_- by $131 \cdot 593 \cdot \Omega_-$.) With this modification the list for the odd case is

$$\{2^5 \cdot 3^3 \cdot 5^6 \cdot 7 \cdot 13 \cdot 17 \cdot 19/(131 \cdot 593),\ 2^5 \cdot 3 \cdot 5^2 \cdot 13 \cdot 17,\ 2 \cdot 3 \cdot 5^3 \cdot 7 \cdot 13,\ 2 \cdot 5^2 \cdot 13 \cdot 17,\ 5^3 7,\ 0\}$$

and for the even case

$$\{2^5 \cdot 3^3 \cdot 5 \cdot 19,\ 2^3 \cdot 7 \cdot 13^2,\ 3 \cdot 5 \cdot 7 \cdot 13,\ 2 \cdot 3 \cdot 41,\ 2 \cdot 3 \cdot 7\}\ .$$

We have exactly one "large" prime dividing a value. This is

$$41\ \Big|\ \frac{\Lambda(f,14)}{\Omega_+}\ ,$$

and this divisibility is the source for the congruence above.

4 Cohomology with Coefficients

To explain this connection I have to recall some other facts from the theory of Siegel modular varieties. The space

$$\Gamma \backslash \mathbb{H}_2$$

can be interpreted as the parameter space of principally polarized abelian surface over $\mathbb{C}$. Roughly, we can attach to a point in $\mathbb{H}_2$ a triple

$$\langle L, \langle\,,\rangle\,, I \rangle = \mathcal{A}_I$$

where I is a complex structure on $L \otimes \mathbb{R}$, which is an isometry for the pairing and s.t. the associated hermitian form is positive definite. (I personally prefer to view $\mathbb{H}_2$ as the space of such complex structures on $L \otimes \mathbb{R}$.) This $\mathcal{A}_I$ is an abelian surface and $\mathcal{A}_I \cong \mathcal{A}_{I'}$ if there is a $\gamma \in \Gamma$ such that $\gamma I = I'$, and this γ provides an isomorphism $\gamma^* : \mathcal{A}_I \cong \mathcal{A}_{I'}$. Here we encounter a minor difficulty, because γ is not unique, and γ^* depends on the choice of γ. Therefore we can not attach an abelian variety to a point $\tilde{I} \in \Gamma \backslash \mathbb{H}_2$. But if we pass to a suitably small normal congruence subgroup $\Gamma' \subset \Gamma$ then it is clear that we have a family $\pi : \mathcal{A} \to \Gamma' \backslash \mathbb{H}_2$ of principally polarized abelian varieties over $\Gamma' \backslash \mathbb{H}_2$. Then the family of cohomology groups $H^1(\mathcal{A}_{\tilde{I}}, \mathbb{Z})$ defines a local system of free $\mathbb{Z}$-modules of rank 4 over $\Gamma' \backslash \mathbb{H}_2$. This local system descends to a sheaf on $\Gamma \backslash \mathbb{H}_2$.

This sheaf is also obtained from the standard representation

$$\rho_{10} : \Gamma \longrightarrow Gl(L) = Gl(\mathcal{M}_{1,0})\ .$$

We define a representation $\rho_{01} : \Gamma \to Gl(\mathcal{M}_{0,1})$ where the module $\mathcal{M}_{0,1}$ is defined by

$$\Lambda^2 \mathcal{M}_{1,0} = \mathcal{M}_{0,1} \oplus \mathbb{Z}\ .$$

We can form the modules $\mathrm{Sym}^m(\mathcal{M}_{1,0}) \otimes \mathrm{Sym}^n(\mathcal{M}_{0,1})$ and these modules have a unique submodule (or quotient)

$$\rho_{m,n} : \Gamma \to Gl(\mathcal{M}_{m,n}) ,$$

which is defined be the requirement that it has the largest dominant weight amoung all highest weights of submodules.

(The representations ρ_{10} (resp. ρ_{01}) have highest weight γ_β (resp. γ_α), which are the two fundamental dominant weights. Then $\mathcal{M}_{m,n}$ is the unique irreducible submodule with highest weight $\lambda = m\gamma_\beta + n\gamma_\alpha$. At this point is an ambiguity: Instead of taking the submodule we could take the unique irreducible quotient having this fundamental weight (actually this ambiguity already occurs when we form the symmetric products). Then the submodule will map injectively into the quotient and the image is a submodule of finite index. This index will be only divisible by "small" primes $\leq m, n$, they do not play a role in our considerations, in other words it does not matter whether we take the submodule or the quotient.)

These representations yield sheaves $\tilde{\mathcal{M}}_{m,n}$ of $\mathbb{Z}$-modules. For an open set $U \subset \Gamma\backslash\mathbb{H}_2$ and its inverse image $\tilde{U} \subset \mathbb{H}_2$ we have

$$\tilde{\mathcal{M}}_{m,n}(U) = \{f : \tilde{U} \to \mathcal{M}_{m,n} | f \quad \text{locally constant and}$$
$$f(\gamma u) = \rho_{n,m}(\gamma)f(u) \quad \text{for all} \quad \gamma \in \Gamma\} .$$

For m even, these modules give us sheaves $\tilde{\mathcal{M}}_{m,n}$ on the space $\Gamma\backslash\mathbb{H}_2$ which are almost local systems. We can consider the cohomology groups

$$H_c^i(\Gamma\backslash\mathbb{H}_2, \tilde{\mathcal{M}}_{m,n}), H^i(\Gamma\backslash\mathbb{H}_2, \tilde{\mathcal{M}}_{m,n})$$

where $H_c^\bullet$ denotes the cohomology with compact support. These cohomology groups sit in an exact sequence

$$\to H^{i-1}(\partial\overline{(\Gamma\backslash\mathbb{H}_2)}, \tilde{\mathcal{M}}_{m,n}) \to H_c^i(\Gamma\backslash\mathbb{H}_2, \tilde{\mathcal{M}}_{m,n}) \to H^i(\Gamma\backslash\mathbb{H}_2, \tilde{\mathcal{M}}_{m,n})$$
$$\to H^i(\partial\overline{(\Gamma\backslash\mathbb{H}_2)}, \tilde{\mathcal{M}}_{m,n}) \to ,$$

where $\partial\overline{(\Gamma\backslash\mathbb{H}_2)}$ is the boundary of the Borel–Serre compactification.

We denote by $H_!^i(\Gamma\backslash\mathbb{H}_2, \tilde{\mathcal{M}}_{m,n})$ the image of the cohomology with compact supports in the cohomology. The coefficient system $\tilde{\mathcal{M}}_{m,n}$ is called regular if $n, m > 0$. In this case all cohomology groups $H_!^i(\Gamma\backslash\mathbb{H}_2, \tilde{\mathcal{M}}_{m,n} \otimes \mathbb{Q})$ vanish for $i \neq 3$. We have a Hodge filtration on $H_!^3(\Gamma\backslash\mathbb{H}_2, \tilde{\mathcal{M}}_{m,n} \otimes \mathbb{C})$, and the lowest step of this filtration is given by (Faltings)

$$S_{m,n+3} \hookrightarrow H_!^3(\Gamma\backslash\mathbb{H}_2, \tilde{\mathcal{M}}_{m,n} \otimes \mathbb{C}) .$$

To get the connection to the conjecture we choose $m = 4$, $n = 7$.

Now I formulate a second conjecture. We invert some small primes (say ≤ 22), and we denote the resulting ring by $R = \mathbb{Z}[\frac{1}{2}, \ldots, \frac{1}{19}]$.

Then we get an exact sequence (*assumption*)

$$0 \to H^3(\Gamma \backslash \mathbb{H}_2, \tilde{\mathcal{M}}_{4,7} \otimes R) \to H^3(\Gamma \backslash \mathbb{H}_2, \tilde{\mathcal{M}}_{4,7} \otimes R)$$
$$\to H^3\left(\partial\overline{(\Gamma \backslash \mathbb{H}_2)}, \tilde{\mathcal{M}}_{4,7} \otimes R\right) \to 0 \,,$$

I know that this is true if I replace R by $\mathbb{Q}$.

Now we can show that we have an action of the Hecke operators on these modules and we have

$$H^3(\partial\overline{(\Gamma \backslash \mathbb{H}_2)}, \tilde{\mathcal{M}}_{4,7} \otimes R) = R$$

where $T^{(1)}_{(p)}$ acts on R with the eigenvalue

$$p^8 + a_p + p^{13} \,.$$

(For this assertion I refer to my lecture notes volume or [Modsym].)

Now we formulate another *assumption*

$$H^3_!(\Gamma \backslash \mathbb{H}_2, \tilde{\mathcal{M}}_{4,7} \otimes R) \cong R^4 \,.$$

Then we know that $T^{(1)}_p$ acts as a scalar by multiplication by $\lambda(p)$ on this cohomology group. We have $\lambda(2) \neq 2^8 + a_2 + 2^{13}$ and hence we can decompose

$$H^3(\Gamma \backslash \mathbb{H}_2, \tilde{\mathcal{M}}_{4,7} \otimes \mathbb{Q}) = H^3_!(\Gamma \backslash \mathbb{H}_2, \tilde{\mathcal{M}}_{4,7} \otimes \mathbb{Q}) \oplus H^3_{\text{Eis}}(\Gamma \backslash \mathbb{H}_2, \tilde{\mathcal{M}}_{4,7} \otimes \mathbb{Q}) \,.$$

Now the main assertion of the second conjecture is:

If we intersect this decomposition with the integral cohomology, then

$$H^3(\Gamma \backslash \mathbb{H}_2, \tilde{\mathcal{M}}_{4,7} \otimes R) \supset H^3_!(\Gamma \backslash \mathbb{H}_2, \tilde{\mathcal{M}}_{4,7} \otimes R) \oplus H^3_{\text{Eis}}(\Gamma \backslash \mathbb{H}_2, \tilde{\mathcal{M}}_{4,7} \otimes R) \,.$$

and the index of the direct sum in $H^3(\Gamma \backslash \mathbb{H}_2, \tilde{\mathcal{M}}_{4,7} \otimes R)$ *is divisible by* 41. *(The denominator of the Eisenstein class is divisible by* 41.)

The point with this second conjecture is that it implies the first conjecture and it can be verified on a computer. To do this we have to find a way to compute the cohomology groups. This can be done by using a suitable acyclic covering of $\Gamma \backslash \mathbb{H}_2$, and then the cohomology is computed from the Čech complex of this covering. We could also try to use a cell decomposition. This method will allow us to check the two assumptions. I think the problem will be that the number of cells will not be so big, but we have nontrivial coefficient systems, its dimension in a general point is 1820. It will be still more difficult to implement the action of the Hecke operator, because one has to pass to a finer cell decomposition, which also computes the cohomology and where the Hecke operators can be implemented as a homomorphism between the two complexes.

Of course, we could also compute mod 41, then we find $\lambda(2) \equiv 2^8 + a_2 + 2^{13}$ mod 41, and our conjecture would say that $T^{(1)}_2$ mod 41 is not diagonalizable.

Why didn't I do this earlier? In my lecture notes volume (Chap III, 3.1) I discuss the above conjecture in greater generality and I raise the question whether computer experiments should be made. There I say that these computations would "...einen beträchtlichen Aufwand erfordern, aber die Frage entscheiden, ob es sich lohnt, das Problem zu behandeln".

Recently I got some kind of unexpected help from C. Faber und G. van der Geer. They produced some tables of eigenvalues $\lambda(p)$ for certain local systems $\mathcal{M}_{m,n}$ with some small values n, m.

They make use of the fact that $\Gamma \backslash \mathbb{H}_2$ is actually the set of the complex points of a quasiprojective scheme $\mathcal{A}_2 / \operatorname{Spec}(\mathbb{Z})$, and that our local systems $\tilde{\mathcal{M}}_{m,n}$ have an algebraic-geometric meaning. They are "motivic" sheaves, and it is not quite clear what that means. But in any case we can pick a prime ℓ and then $\tilde{\mathcal{M}}_{m,n} \otimes \mathbb{Z}_\ell$ will be an ℓ-adic sheaf on $\mathcal{A}_2$. Then we have the Grothendieck fixed point formula

$$\operatorname{tr}\left(\Phi_p \mid H_c^\bullet(\mathcal{A}_2 \times_\mathbb{Z} \bar{\mathbb{F}}_p, \tilde{\mathcal{M}}_{m,n} \otimes \mathbb{Z}_\ell)\right) = \sum_{x \in \mathcal{A}_2(\mathbb{F}_p)} \operatorname{tr}\left(\Phi_p \mid \tilde{\mathcal{M}}_{m,n,x}\right),$$

where Φ_p is the Frobenius at p. The right hand side can be computed because we have the modular interpretation.

The left hand side consists of several pieces (Eisenstein cohomology, endoscopic contributions, if $m = 0$ there may be some Saito–Kurokawa lifts), and the trace of Φ_p on these pieces can be computed explicitly (for small n, m) and can be expressed in terms of modular forms for $Sl_2(\mathbb{Z})$ and in terms of algebraic Hecke characters. This can be brought to the right hand side, and the resulting expression can be computed explicitly for small values n, m.

Then we are left with the "genuine" part in $H_c^3(\mathcal{A}_2 \times_\mathbb{Z} \bar{\mathbb{F}}_p, \tilde{\mathcal{M}}_{m,n} \otimes \mathbb{Z}_\ell)$, and this part will be of rank $4 \cdot \dim S_{m,n+3}$. If now $\dim S_{m,n+3} = 1$, then this "genuine" part will be of rank 4 and we have

$$\operatorname{tr}\left(\Phi_p \mid H_{\text{genuine}}^3(\mathcal{A}_2 \times_\mathbb{Z} \bar{\mathbb{F}}_p, \tilde{\mathcal{M}}_{m,n} \otimes \mathbb{Z}_\ell)\right) = \lambda(p).$$

But now the $\lambda(p)$ can be computed from the right hand side, if we take the effort to compute the sum over $\mathcal{A}_2(\mathbb{F}_p)$ and the non "genuine" traces.

After I saw the preprint by C. Faber and G. van der Geer I realized that I might be able to check the first conjecture in a special case. I had to go through the values $\frac{L(f,k)}{\Omega_\varepsilon(k)}$ for the modular cusp form f of weight ≤ 22. (For higher weights except 26 the dimension of these space are ≥ 2. The eigenvalues of the Hecke operators are algebraic integers and also the normalized L-values will be algebraic integers, and the computations will be much more complicated.) I had to find a "large" prime dividing one of the values, and I found for our form of weight 22

$$41 \mid \frac{L(f, 14)}{\Omega_+}.$$

I computed the numbers $4, 7$ and $7 + 3 = 10$ from these data and wrote an e-mail to G. van der Geer inquiring the dimension of $S_{4,10}$. Several answers were possible. The dimension could be zero. This would be devastating. The dimension could be >1, this would mean a horrible additional computational effort. But the answer was

Re: Kohomologie lokaler Systemen
Lieber Guenter,
die Dimension ist dann 1. Die ersten Eigenwerte sind wie folgt:

$$-2^4 \cdot 3 \cdot 5 \cdot 7$$
$$2^3 \cdot 3^4 \cdot 5 \cdot 17$$
$$-2^2 \cdot 3 \cdot 5^2 \cdot 17 \cdot 1439$$
$$2^4 \cdot 5^2 \cdot 7^2 \cdot 17 \cdot 31 \cdot 59$$
$$2^3 \cdot 3 \cdot 11 \cdot 17 \cdot 5650223$$

d.h. fuer die Primzahlen 2, 3, 5, 7, 11.
Mit bestem Gruss, Gerard

I read this message in my office in the Beringstrasse and I had the values of the a_p at home on my laptop. After two oral examinations of computer science students I went home and checked the numbers. I was extremely pleased when I found that the congruences hold.

(Actually van der Geer was also pleased because he considered it as confirmation of his computations with Faber. (I have multiplied the values in his table by -1, probably this has to be done because the trace occurs in odd degree))

5 Why the Denominator?

We stick to the case $\mathcal{M}_{4,7}$, and f is still our modular cusp form of weight 22. If we had a splitting under the Hecke-algebra

$$H^3(\Gamma \backslash \mathbb{H}_2, \tilde{\mathcal{M}}_{4,7} \otimes R) = H^3_!(\Gamma \backslash \mathbb{H}_2, \tilde{\mathcal{M}}_{4,7} \otimes R) \oplus H^3(\partial(\overline{\Gamma \backslash \mathbb{H}_2}), \tilde{\mathcal{M}}_{4,7} \otimes R) \,,$$

then we could construct a mixed Tate motive $\mathcal{X}(f)$ which sits in an exact sequence

$$0 \to R(-8) \to \mathcal{X}(f) \to R(-13) \to 0$$

and hence defines an element in the extension group

$$[\mathcal{X}(f)] \in \mathrm{Ext}^1_{\mathcal{MM}}(R(-13), R(-8)) = \mathrm{Ext}^1_{\mathcal{MM}}(R(-5), R(0)) \,.$$

(For more details see [Mixmot].) This Ext^1 group is some kind of undefined object, but we can attach to our object $\mathcal{X}(f)$ elements in two other extension groups, namely:

(i) an extension class in the category of mixed Hodge structures

$$[\mathcal{X}(f)]_{BdRh} \in \text{Ext}^1_{BdRh}(R(-13), R(-8)) = \text{Ext}^1_{BdRh}(R(-5), R(0)) = \mathbb{R}$$

(See MixMot 1.5.2). It is some kind of general belief that those elements in the extension group of mixed Hodge structures, which come from mixed motives $\mathcal{X}$ over $\mathbb{Z}$, are in fact elements of the form

$$[\mathcal{X}_{BdRh}] = a(\mathcal{X})\zeta'(-4) \text{ with } a(\mathcal{X}) \in \mathbb{Q}.$$

This last conjecture can be verified in our particular case we have the formula

$$[\mathcal{X}_{BdRh}(f)] = c \cdot \frac{\frac{\Lambda(f,13)}{\Omega_-}}{\frac{\Lambda(f,14)}{\Omega_+}} \zeta'(-4)$$

where c is a rational number containing only small primes.

(ii) For any prime ℓ we can attach an ℓ-adic extension class

$$[\mathcal{X}(f)]_\ell \in \text{Ext}^1_{Gal}(R_\ell(-13), R_\ell(-8)) = H^1(\text{Gal}(\bar{\mathbb{Q}}/\mathbb{Q},)R_\ell(5))$$

and this cohomology group contains certain specific elements $c_\ell(5)$, these are the Soulé elements. These elements should also be generators of the image of

$$\text{Ext}^1_{\mathcal{MM}}(R_\ell(-13), R_\ell(-8)) \to H^1(\text{Gal}(\bar{\mathbb{Q}}/\mathbb{Q}), R_\ell(5))$$

if we tensor by $\mathbb{Q}$. We write the Galois cohomology group multiplicatively and now it is general belief that we must have

$$[\mathcal{X}(f)]_\ell = c_\ell(5)^{c \cdot \frac{\frac{\Lambda(f,13)}{\Omega_-}}{\frac{\Lambda(f,14)}{\Omega_+}}}.$$

From now on we choose $\ell = 41$ (of course we could replace ℓ by 41 in the following considerations but this causes some confusion), then the value of the ℓ-adic ζ-function $\zeta_\ell(5) \neq 0 \mod \ell$ and this implies that $c_\ell(5)$ is a primitive element in the Galois cohomology group. But the rational exponent has ℓ in its denominator, this contradicts the existence of our mixed motive and this motive has been constructed under the assumption that ℓ does not divide the denominator of the Eisenstein class.

6 Arithmetic Implications

We get a diagram (still $\ell = 41$) of ℓ-adic Galois modules

$$0 \to H^3_!(\mathcal{A}_2 \times_\mathbb{Z} \bar{\mathbb{Q}}, \tilde{\mathcal{M}}_{4,7} \otimes R_\ell) \to H^3(\mathcal{A}_2 \times_\mathbb{Z} \bar{\mathbb{Q}}, \tilde{\mathcal{M}}_{4,7} \otimes R_\ell) \to R_\ell(-13) \to 0$$

$$\cup \qquad\qquad r \nearrow$$

$$H^3_!(\mathcal{A}_2 \times_\mathbb{Z} \bar{\mathbb{Q}}, \tilde{\mathcal{M}}_{4,7} \otimes R_\ell) \oplus R_\ell(-13)$$

where the image of the homomorphism r is contained in $\ell R_\ell(-13)$. This gives us an injective homomorphism

$$\psi : \mathbb{Z}/(\ell)(-13) \longleftrightarrow H^3_!(\mathcal{A}_2 \times_\mathbb{Z} \bar{\mathbb{Q}}, \tilde{\mathcal{M}}_{4,7} \otimes \mathbb{Z}/(\ell)) .$$

The module $H^3_!(\mathcal{A}_2 \times_\mathbb{Z} \bar{\mathbb{Q}}, \tilde{\mathcal{M}}_{4,7} \otimes \mathbb{Z}/(\ell))$ is of dimension 4 over $\mathbb{F}_\ell$ and the cup product provides a non -degenerated pairing of this module with itself into $\mathbb{Z}/(\ell)(-21)$. The orthogonal complement Y of the image $\psi(\mathbb{Z}/(\ell)(-13))$ is of dimension 3 over $\mathbb{F}_\ell$ and we get two exact sequences

$$0 \to \mathbb{Z}/(\ell)(-13) \to Y \to X \to 0$$

and

$$0 \to X \to H^3_!(\mathcal{A}_2 \times_\mathbb{Z} \bar{\mathbb{Q}}, \tilde{\mathcal{M}}_{4,7} \otimes \mathbb{Z}/(\ell))/\psi(\mathbb{Z}/(\ell)(-13)) \to \mathbb{Z}/(\ell)(-8) \to 0 .$$

The module X is actually the reduction of the ℓ-adic representation attached to f mod ℓ. It also has a non degenerate pairing with itself with values in $\mathbb{Z}/(\ell)(-21)$ and the two sequences are dual to each other. The sequences give us two extension classes, the first one a class

$$[Y] \in \text{Ext}^1_{Gal}(X, \mathbb{Z}/(\ell)(-13)) = H^1(\text{Gal}(\bar{\mathbb{Q}}/\mathbb{Q}), \text{Hom}(X, \mathbb{Z}/(\ell)(-13))) \xrightarrow{\sim}$$
$$H^1(\text{Gal}(\bar{\mathbb{Q}}/\mathbb{Q}), X \otimes \mathbb{Z}/(\ell)(8))$$

and under the isomorphism $[Y]$ is mapped to the extension class of the second sequence.

Now we can hope that this extension class is actually an element in the Selmer group of the Scholl-Deligne motive $M(f)$ attached to f, and that it is in fact an element of order ℓ. If this turns out to be the case, then we have produced an element in the Selmer group whose existence is predicted by the general philosophy of the Bloch–Kato–Birch–Swinnerton Dyer conjecture.

References

Fa-vdG   Faber, Carel; van der Geer, Gerard: Sur la cohomologie des systèmes locaux sur les espaces de modules des courbes de genre 2 et des surfaces abéliennes.I, C. R. Math. Acad. Sci. Paris 338 (2004), no. 5, 381–384. II, no. 6, 467–470

Ha-Eis   G. Harder, Eisensteinkohomologie und die Konstruktion gemischter Motive, SLN 1562
The following two manuscripts on mixed motives and modular symbols can be found on my home page
        www.math.uni-bonn.de/people/harder/
in my ftp-directory folder Eisenstein.

Mixmot   Modular construction of mixed motives II

Modsym   Modular Symbols and Special Values of Automorphic L-Functions

Appendix

In the meanwhile C. Faber, G. van der Geer and I did some further computations. We have still another one dimensional space of modular cusp forms, this is spanned by the modular form

$$g(q) = \Delta(q)E_6(q)E_4(q)^2 =$$

$$q - 48q^2 - 195804q^3 - 33552128q^4 - 741989850q^5 + 9398592q^6 + \ldots$$

of weight 26. We have the following divisibilities by "large" primes

$$43 \Big| \frac{L(g, 23)}{\Omega_-}, \quad 97 \Big| \frac{L(g, 21)}{\Omega_-}, \quad 29 \Big| \frac{L(g, 19)}{\Omega_-}.$$

The corresponding spaces of modular forms $S_{18,5}, S_{14,7}, S_{10,9}$ have dimension 1.

Now let ℓ be one of the primes 41, 43, 97, 29. Let

$$f(q) = q + a_2 q^2 + a_3 q^3 \ldots$$

be the corresponding modular form of weight 22 or 26. Let $S_{i,j}$ be the corresponding one dimensional space of Siegel modular forms and let $\tilde{\mathcal{M}}_{i,j-3} = \tilde{\mathcal{M}}_{m,n}$ be the corresponding local system. The Hecke algebra acts on the cohomology

$$H^3_!(\mathcal{A}_2 \times_{\mathbb{Z}} \bar{\mathbb{Q}}, \tilde{\mathcal{M}}_{m,n} \otimes R_\ell)$$

and we should find an isotypical submodule of rank 4 on which the Hecke operators act by the scalar by which they acts on $S_{i,j}$. The local Hecke algebra at a prime p is generated by two Hecke operators $T_{p,\alpha}, T_{p,\beta}$ which correspond to the double classes

$$Sp_2(\mathbb{Z}_p) \begin{pmatrix} p & & & 0 \\ & p & & \\ & & 1 & \\ 0 & & & 1 \end{pmatrix} Sp_2(\mathbb{Z}_p) \text{ and } Sp_2(\mathbb{Z}_p) \begin{pmatrix} p^2 & & & 0 \\ & p & & \\ & & p & \\ 0 & & & 1 \end{pmatrix} Sp_2(\mathbb{Z}_p).$$

(See [Ha-Eis] 3.1.2.1) So we get sequences of eigenvalues

$$\{\lambda_\alpha(p), \lambda_\beta(p)\}_{p \in \text{Primes}}$$

Now we state the conjecture that in all four cases we have congruences

$$\lambda_\alpha(p) \equiv p^{n+1} + a_p + p^{n+m+2} \mod \ell$$

and

$$\lambda_\beta(p) \equiv a_p(1 + p^{m+1}) + (p^2 - 1)p^{n_\alpha + n_\beta} \mod \ell$$

for all primes p.

For our four primes ℓ above and the corresponding modular forms the conjecture for $\lambda_\alpha(p)$ has been checked for all $p \leq 37$.

The general rule is: If k is even and f an eigenform of weight k. Let $K = \mathbb{Q}(f)$ be its field of definition. Let us assume that a large prime $\mathfrak{l}$ divides $L(f, \nu)/\Omega_{\epsilon(\nu)}$. Then we solve the equations

$$k = 2n + m + 4, \nu = n + m + 3 \, .$$

Then we can construct an Eisenstein class in $H_!^3(\Gamma\backslash\mathbb{H}_2, \tilde{\mathcal{M}}_{m,n} \otimes K)$ whose denominator is divisible by $\mathfrak{l}$.

Added on April 3, 2003 (the day when the first Abel-Prize was given to J.-P. Serre):
I also checked congruences for the modular cusp forms of weight 24. In this case we have two eigenforms

$$f(q) = \sum_n^\infty a_n q^n = q - (540 - 12\sqrt{144169})q^2 + (169740 + 576\sqrt{144169})q^3 \ldots$$

where we take the positive root-, and we have the conjugate eigenform

$$f'(q) = \sum_n^\infty a_n' q^n \, .$$

We put $\omega = \frac{1+\sqrt{144169}}{2}$. In this case we find periods $\Omega_\pm, \Omega_\pm'$ such that

$$\frac{L(f,k)}{\Omega_{\epsilon(k)}} \in \mathbb{Z}[\omega], \frac{L(f',k)}{\Omega'_{\epsilon(k)}} \in \mathbb{Z}[\omega] \, .$$

We normalize the periods such that these numbers for a fixed choice of the sign $\epsilon(k)$ are coprime and such that $\frac{L(f,k)}{\Omega_{\epsilon(k)}}$ and $\frac{L(f',k)}{\Omega'_{\epsilon(k)}}$ are conjugate.

The primes 73 and 179 split in $\mathbb{Q}(\sqrt{144169})$ and for 73 the decomposition is

$$\mathfrak{l} = (73, 53 + 36\sqrt{144169}), \mathfrak{l}' = (73, 53 - 36\sqrt{144169})$$

$$(73) = \mathfrak{l}\mathfrak{l}' \, .$$

We find $\frac{L(f,19)}{\Omega_-} \in \mathfrak{l}$. The corresponding space $S_{12,7}$ has dimension 1, if $\lambda(p)$ is the sequence of eigenvalues the congruence

$$\lambda(p) \equiv p^5 + a_p + p^{18} \quad \mathrm{mod}\ \mathfrak{l}$$

has been checked for all primes $p \leq 19$. Of course we get a second congruence if we conjugate it.

For 179 we have a splitting

$$(179) = \mathfrak{l}\mathfrak{l}' \, ,$$

with $\mathfrak{l} = (179, 54+61\sqrt{144169})$, $\mathfrak{l}' = (179, 54-61\sqrt{144169})$. We find $\frac{L(f,17)}{\Omega_-} \in \mathfrak{l}$, again the corresponding space $S_{8,9}$ has dimension 1. If $\lambda(p)$ is the sequence of eigenvalues the congruences

$$\lambda(p) \equiv p^7 + a_p + p^{16} \mod \mathfrak{l}$$

and of course its conjugates have been checked for the same set of primes p. (There is a slight risk that I mixed up the two primes $\mathfrak{l}, \mathfrak{l}'$.)

Added on March 25, 2005:
When lecturing on this subject, I had sometimes difficulties to get the numbers right. Therefore I formulate the rules:

We start from an elliptic modular form f for $Sl_2(\mathbb{Z})$ which is of (even) weight k, it should be an eigenform for the Hecke-algebra. Then its eigenvalues generate a field $\mathbb{Q}(f)$.

Then we look at the values

$$\left\{ \frac{\Lambda(f, k-1)}{\Omega_+}, \frac{\Lambda(f, k-3)}{\Omega_+}, \ldots, \frac{\Lambda(f, k-\nu)}{\Omega_+}, \ldots, \frac{\Lambda(f, k-\mu(k))}{\Omega_+} \right\}$$

and

$$\left\{ \frac{\Lambda(f, k-2)}{\Omega_+}, \frac{\Lambda(f, k-4)}{\Omega_+}, \ldots, \frac{\Lambda(f, k-\nu)}{\Omega_+}, \ldots, \frac{\Lambda(f, k-\mu'(k))}{\Omega_+} \right\},$$

where in the first row the ν are odd and in the second row they are even. The last value is the one which is nearest to the central point $\frac{k}{2}$ from above.

Then we look for large primes

$$\ell \Big| \frac{\Lambda(f, k-\nu)}{\Omega_{\epsilon(\nu)}}.$$

Now we choose the highest weight $\lambda = m\gamma_\beta + n\gamma_\alpha$. The numbers m, n must satisfy

$$2n + m + 3 = n + 1 + n + m + 2 = k - 1 \quad \text{and} \quad n + m + 3 = k - \nu$$

hence we get

$$n = \nu - 1 \quad \text{and} \quad m = k - 2\nu - 2$$

Index

abelian variety 202
anisotropic vector 128
automorphic Green function 164, 168

Baily–Borel compactification 109, 205, 226
Baily–Borel topology 109
Betti number 117
Birch–Swinnerton-Dyer conjecture 80, 95, 97
Borcherds product 83, 103, 156
  CM values 173
  local 153
boundary component 139
  rational 139
boundary point 109
Braun 194

canonical automorphism 131
canonical involution 131
canonical model 226
Cayley transform 185
Chow group 159
Chowla–Selberg formula 84, 85, 103
class field theory 75
class numbers 106
  Hurwitz 73–75, 81
  of binary quadratic forms 8, 41, 74, 81, 100
  of imaginary quadratic fields 8, 41
class polynomial 72, 73, 83
classical Siegel modular form 189
Clifford algebra 130
  center 132

even 131
Clifford group 133
Clifford norm 131
CM cycle 169, 173
CM extension 171
co-rank of a modular form 195
Cohen–Kuznetsov series 53–55, 102
compact dual 187
compactifications 204
complex multiplication (CM) 67–99, 102, 103, 169
  CM modular forms 93–99, 103
  elliptic curves with CM 67, 90–92
converse theorem
  strong 166
  weak 165
critical value 236
cusp 107
cusp form 23, 115, 192

Dedekind zeta function 120
  special value 123
desingularization 110
diagonal 122
different 106
Dirichlet character 17, 32, 78, 84
discriminant
  of a quadratic field 106
  of a quadratic space 129, 132
discriminant function $\Delta(z)$ 11, 20–22, 36, 212, 249
divisor sum 121
Doi–Naganuma lift 149